METHODS IN MOLECULAR BIOLOGY

Series Editor
John M. Walker
School of Life and Medical Sciences
University of Hertfordshire
Hatfield, Hertfordshire, UK

For further volumes:
http://www.springer.com/series/7651

For over 35 years, biological scientists have come to rely on the research protocols and methodologies in the critically acclaimed *Methods in Molecular Biology* series. The series was the first to introduce the step-by-step protocols approach that has become the standard in all biomedical protocol publishing. Each protocol is provided in readily-reproducible step-by-step fashion, opening with an introductory overview, a list of the materials and reagents needed to complete the experiment, and followed by a detailed procedure that is supported with a helpful notes section offering tips and tricks of the trade as well as troubleshooting advice. These hallmark features were introduced by series editor Dr. John Walker and constitute the key ingredient in each and every volume of the *Methods in Molecular Biology* series. Tested and trusted, comprehensive and reliable, all protocols from the series are indexed in PubMed.

Transcriptome Data Analysis

Edited by

Rajeev K. Azad

Department of Biological Sciences and BioDiscovery Institute, University of North Texas, Denton, TX, USA

Humana Press

Editor
Rajeev K. Azad
Department of Biological Sciences
and BioDiscovery Institute
University of North Texas
Denton, TX, USA

ISSN 1064-3745 ISSN 1940-6029 (electronic)
Methods in Molecular Biology
ISBN 978-1-0716-3885-9 ISBN 978-1-0716-3886-6 (eBook)
https://doi.org/10.1007/978-1-0716-3886-6

Cover Illustration Caption: For more information, see Chapter 12, Figure 2.

This Humana imprint is published by the registered company Springer Science+Business Media, LLC, part of Springer Nature.
The registered company address is: 1 New York Plaza, New York, NY 10004, U.S.A.

Preface

Advances in transcriptomics have revolutionized the fields of biology and medicine. Use of high-throughput sequencing technologies has resulted in mountains of transcriptomic data, which has necessitated the development of efficient methods for their analysis and interpretation. This volume presents a comprehensive description of the advances in transcriptomics, with a focus on methods and pipelines for transcriptome data analysis. In addition to the step-by-step description of well-established RNA sequencing (RNA-Seq) data analysis protocols, specialized pipelines have also been described, such as multi-omics data integration and analysis, gene interaction network construction, single-cell trajectory inference, detection of structural variants, and application of machine learning, to mention a few. This Methods in Molecular Biology series provides a useful resource for educators and researchers to help familiarize them with the new developments in the field, learn usage of the protocols for transcriptome data analysis, and implement the tools or pipelines to address relevant problems of their interest.

Denton, TX, USA *Rajeev K. Azad*

Contents

Contributors

SHANDAR AHMAD • *School of Computational and Integrative Sciences, Jawaharlal Nehru University, New Delhi, India*

ANA PAULA ALONSO • *Department of Biological Sciences & BioDiscovery Institute, University of North Texas, Denton, TX, USA*

CINTIA L. ARIAS • *Department of Biological Sciences & BioDiscovery Institute, University of North Texas, Denton, TX, USA*

SIMRAN ARORA • *Bioinformatics Laboratory, National Institute of Plant Genome Research (NIPGR), New Delhi, India*

BRIAN G. AYRE • *Department of Biological Sciences and BioDiscovery Institute, University of North Texas, Denton, TX, USA*

RAJEEV K. AZAD • *Department of Biological Sciences and BioDiscovery Institute, University of North Texas, Denton, TX, USA*

DAVID J. BURKS • *Department of Biological Sciences, University of North Texas, Denton, TX, USA*

MOOKYUNG CHEON • *Dementia Research Group, Korea Brain Research Institute (KBRI), Daegu, Republic of Korea*

PRAGYA CHITKARA • *Bioinformatics Laboratory, National Institute of Plant Genome Research (NIPGR), New Delhi, India*

JAMIE DIXSON • *Department of Biological Sciences and BioDiscovery Institute, University of North Texas, Denton, TX, USA*

LAURA L. ELO • *Turku Bioscience Centre, University of Turku and Åbo Akademi University, Turku, Finland; Institute of Biomedicine, University of Turku, Turku, Finland*

WEAM FALLATAH • *Department of Biological Sciences, University of North Texas, Denton, TX, USA*

HAFSA FAROOQ • *Department of Computer Science, Georgia State University, Atlanta, GA, USA*

MARK GRINSHPON • *Department of Mathematics and Statistics, Georgia State University, Atlanta, GA, USA*

SWATI GUPTA • *Department of Computational and Data Sciences, Indian Institute of Science, Bengaluru, Karnataka, India*

KOHEI HAGIWARA • *Department of Computational Biology, St. Jude Children's Research Hospital, Memphis, TN, USA*

FIZA HAMID • *Bioinformatics Laboratory, National Institute of Plant Genome Research (NIPGR), New Delhi, India*

RAYMOND J. HENDRICKS • *Department of Biological Sciences, University of North Texas, Denton, TX, USA*

ROYA HOSSEINI • *Department of Computer Science, Georgia State University, Atlanta, GA, USA*

ALMAS JABEEN • *School of Computational and Integrative Sciences, Jawaharlal Nehru University, New Delhi, India*

PUDUR JAGADEESWARAN • *Department of Biological Sciences, University of North Texas, Denton, TX, USA*

CHRISTOPHER R. JOHNSTON • *Department of Biological Sciences & BioDiscovery Institute, University of North Texas, Denton, TX, USA*
SINI JUNTTILA • *Turku Bioscience Centre, University of Turku and Åbo Akademi University, Turku, Finland*
MANISHA KALSAN • *School of Computational and Integrative Sciences, Jawaharlal Nehru University, New Delhi, India*
YONGJIN KIM • *Dementia Research Group, Korea Brain Research Institute (KBRI), Daegu, Republic of Korea*
AMBARISH KUMAR • *Department of Biological Sciences and BioDiscovery Institute, University of North Texas, Denton, TX, USA*
SHAILESH KUMAR • *Bioinformatics Laboratory, National Institute of Plant Genome Research (NIPGR), New Delhi, India*
TAEGUN KWON • *Department of Biological Sciences, University of North Texas, Denton, TX, USA*
MARY L. LADAGE • *Department of Biological Sciences, University of North Texas, Denton, TX, USA*
ZIYI LI • *Department of Biostatistics, The University of Texas MD Anderson Cancer Center, Houston, TX, USA*
JABILA MARY • *Department of Biological Sciences, University of North Texas, Denton, TX, USA*
RÓISÍN C. MCGARRY • *Department of Biological Sciences and BioDiscovery Institute, University of North Texas, Denton, TX, USA*
SAIFUN RIPA NAHAR • *Department of Biological Sciences, University of North Texas, Denton, TX, USA*
PAMELA A. PADILLA • *Department of Biological Sciences, University of North Texas, Denton, TX, USA*
STEFANO MARIA PAGNOTTA • *Department of Science and Technology, Università degli Studi del Sannio, Benevento, Italy*
SRI LAKSHMI BHAVANI PAGOLU • *Centre for Computational Natural Sciences and Bioinformatics, International Institute of Information Technology, Hyderabad, Telangana, India*
DEBNATH PAL • *Department of Computational and Data Sciences, Indian Institute of Science, Bengaluru, Karnataka, India*
NITA PAREKH • *Centre for Computational Natural Sciences and Bioinformatics, International Institute of Information Technology, Hyderabad, Telangana, India*
XIAOLAN RAO • *State Key Laboratory of Biocatalysis and Enzyme Engineering, School of Life Sciences, Hubei University, Wuhan, China*
JOSE ROBLEDO • *Department of Biological Sciences, University of North Texas, Denton, TX, USA*
FILIPP RONDEL • *Department of Computer Science, Georgia State University, Atlanta, GA, USA*
MANUEL A. RUIZ • *Department of Biological Sciences, University of North Texas, Denton, TX, USA*
SOHAM SENGUPTA • *Center for Pediatric Neurological Disease Research, St. Jude Children's Research Hospital, Memphis, TN, USA*
ANKITA SINGH • *Department of Surgical Disciplines, All India Institute of Medical Science (AIIMS), New Delhi, India*

VIKRAM SINGH • *Centre for Computational Biology and Bioinformatics, Central University of Himachal Pradesh, Dharamshala, Himachal Pradesh, India*
JOHANNES SMOLANDER • *Turku Bioscience Centre, University of Turku and Åbo Akademi University, Turku, Finland*
ANTÓNIO G. G. SOUSA • *Turku Bioscience Centre, University of Turku and Åbo Akademi University, Turku, Finland*
VIVEK THAKUR • *Department of Systems & Computational Biology, School of Life Sciences, University of Hyderabad, Hyderabad, Telangana, India*
A. T. VIVEK • *Bioinformatics Laboratory, National Institute of Plant Genome Research (NIPGR), New Delhi, India*
PENG YANG • *Department of Statistics, Rice University, Houston, TX, USA*
SHAFAQUE ZAHRA • *Bioinformatics Laboratory, National Institute of Plant Genome Research (NIPGR), New Delhi, India*
ALEX ZELIKOVSKY • *Department of Computer Science, Georgia State University, Atlanta, GA, USA*
JINGHUI ZHANG • *Department of Computational Biology, St. Jude Children's Research Hospital, Memphis, TN, USA*

Chapter 1

An RNA-Seq Data Analysis Pipeline

Soham Sengupta and Rajeev K. Azad

Abstract

In this chapter, we present an established pipeline for analyzing RNA-Seq data, which involves a step-by-step flow starting from raw data obtained from a sequencer and culminating in the identification of differentially expressed genes with their functional characterization. The pipeline is divided into three sections, each addressing crucial stages of the analysis process. The first section covers the initial steps of the pipeline, including downloading of the data of interest and performing quality control assessment. Assessment ensures that the data used for analysis is reliable and suitable for downstream analyses. In the second section, gene-level quantification is performed, which entails quantification of expression levels of genes in the samples. The third and final section is focused on differential expression analysis, which involves comparing gene expression levels between two or more conditions. This step helps identify genes that show significant differences in expression levels under different experimental conditions. To facilitate accessibility and reproducibility, we have provided an online repository containing all scripts and files. Additionally, custom scripts are available, enabling users to modify the pipeline's output for various downstream analyses. By following this pipeline, researchers can effectively analyze RNA-Seq data and gain valuable insights into gene expression patterns and, furthermore, the understanding of biological processes.

Key words RNA-Seq, Alignment, Expression quantification, Differential gene expression

1 Introduction to RNA-Seq and Transcriptomics

RNA sequencing (RNA-Seq) has revolutionized the field of transcriptomics by enabling comprehensive and unbiased profiling of gene expression. RNA-Seq technology has facilitated high-throughput sequencing of RNA molecules. This chapter details a workflow for processing the RNA-Seq data that is received in the form of FASTQ files. The pipeline commences with the handling of raw sequence reads, which are then subjected to thorough quality assessment and preprocessing. These crucial steps are undertaken to ensure reliable estimates of transcript abundance and robust quantification of differential regulation, which are the main objectives of RNA-Seq studies. An online repository for this protocol has been created at GitHub, which provides access to all scripts and

Rajeev K. Azad (ed.), *Transcriptome Data Analysis*, Methods in Molecular Biology, vol. 2812,
https://doi.org/10.1007/978-1-0716-3886-6_1,

data files: https://github.com/sohamsg90/Protocol-for-RNAseq-analysis. Furthermore, custom scripts have been developed to facilitate manipulation of pipeline output for manifold downstream analyses. In this tutorial for RNA-Seq analysis, we use paired-end reads from the model organism *Arabidopsis thaliana* and demonstrate processing of the data and outcomes for each step in the pipeline.

2 Materials and Methods

A suite of tools is required at each step of data processing (Table 1). Most of these tools are compatible with GNU/Linux command line environment and macOS. However, with the help of Windows Subsystem for Linux, these tools can also be run in the Windows operating systems.

2.1 Read Download, Preparation, and Quality Assessment

For this tutorial, we consider the FASTQ read files from a published article [5]. The sequence read files are hosted at https://www.ncbi.nlm.nih.gov/geo/query/acc.cgi?acc=GSE134391. The FASTQ files can be downloaded using fastq-dump. fastq-dump is a tool for downloading sequencing read data from NCBI's Sequence Read Archive (SRA) as FASTQ files. An alternative approach is using the https://sra-explorer.info/# website to generate bash scripts making the download more streamlined and organized. This dataset consists of samples for one control and three mutants. For each sample, there are three biological replicates.

The initial step of the procedure is to assess the quality of the reads acquired from the sequencer. The presence of adapter contamination can lead to substantial issues during transcriptome assembly, such as the production of chimeric reads. Additionally, low-quality read ends can considerably reduce the alignment mapping rate. Raw reads are typically stored in the FASTQ format, which includes both the sequence and the quality information for

Table 1
List of software tools required to perform RNA-Seq data analysis

Software	Link
DESeq2 [1]	https://bioconductor.org/packages/release/bioc/html/DESeq2.html
FastQC [2]	https://www.bioinformatics.babraham.ac.uk/projects/fastqc/
SRA Explorer	https://sra-explorer.info/#
STAR [3]	https://github.com/alexdobin/STAR
Trimmomatic [4]	http://www.usadellab.org/cms/?page¼trimmomatic

each base of each read. In this tutorial, the files provided are in the compressed FASTQ format (**.fastq.gz*). To conduct the quality assessment, we utilize here the FastQC tool. The command line for analyzing FASTQ files is as follows:

```
$ mkdir FastQC
$ fastqc *.fastq.gz -o FastQC/
```

Upon execution, this process generates an HTML report, which can be easily viewed using any standard Internet browser. The report categorizes the results into "pass," "caution," or "fail," denoted by green, yellow, or red, respectively, for each tested category. The FastQC manual offers valuable insights for each category, along with guidance on addressing any flagged issues in the yellow or red categories.

Among all the tested categories, two are of paramount importance: "Per Base Sequence Quality" and "Adapter Content." The "Per Base Sequence Quality" evaluation flags a warning if the lower quartile score for any base position across all reads falls below 10 or if the median for a base position is less than 25. If the lower quartile score drops below 5 or the median goes below 20, FastQC reports a failure. It's worth noting that the quality of data tends to degrade over the course of runs, and longer read lengths experience a more pronounced drop in quality across each read. Additionally, issues encountered during sequencing, such as flow cell interference, affect base quality earlier in the read.

On the other hand, the "Adapter Content" category is valuable for identifying adapter sequence contamination and appropriately removing it from each read where present.

2.2 Adapter Trimming and Adapter Removal

After reviewing the FastQC report for the presence or absence of adapter sequences, the Trimmomatic tool is used to trim low-quality ends from the raw reads. Trimmomatic also has the capability to identify and eliminate any remaining adapter sequences from the reads when provided with a FASTA file containing the adapter sequence. It is important to note that different commands are required for single-end and paired-end reads. In this tutorial, we use the paired-end command for the input paired-end read files *CT1_1.fastq.gz* and *CT1_2.fastq.gz*.

```
$ java -jar trimmomatic-0.38.jar PE CT1_1.fastq.gz CT1_2.
fastq.gz
CT1_1_Trimmed.fastq.gz CT1_1_Orphan.fastq.gz CT1_2_Trimmed.
fastq.gz
CT1_2_Orphan.fastq.gz ILLUMINACLIP :adapter.fa:2:30:10
LEADING:3 TRAILING:3
SLIDINGWINDOW:4:15 MINLEN:36 HEADCROP: 14
```

Upon execution, this command will produce four output files. The files named *_Trimmed.fastq.gz (*CT1_1_Trimmed.fastq.gz*,

CT1_2_Trimmed.fastq.gz) will be utilized for downstream analysis. On the other hand, the files named *_Orphan.fastq.gz will contain mate pairs where one of the two associated reads did not meet the specified cutoffs and was consequently removed. While these orphaned reads may have limited utility for our current purposes, in situations where data is scarce, they might still offer valuable information to researchers despite lacking their paired mate.

2.3 Alignment to the Reference Genome

In response to the growing demand for RNA-Seq tools for more reliable quantification of transcript abundance, several RNA-Seq alignment tools have emerged in recent years. For this tutorial, we have selected STAR as our gene-level aligner based on numerous comparative studies that have consistently identified it as the most accurate freely available program for RNA-Seq read alignment, even at its default parameter setting.

Furthermore, as of the time of writing, STAR offers the added benefit of providing raw counts directly without requiring an intermediary program to analyze the BAM files. This feature makes STAR an appealing choice, especially for users who may be less experienced in this field. The inclusion of the GeneCounts parameter, which was not available in earlier versions of STAR, is crucial for the successful completion of this tutorial.

Before aligning the FASTQ reads to the reference genome, the first step is to create a genome index for STAR. This process involves using the whole genomic sequence (genome FASTA file) and the associated annotation (GTF) files specific to the organism of interest. The genome FASTA file contains the DNA sequences for all the chromosomes, while the GTF file contains tab-delimited texts providing information about the gene structure, including the coordinates/indices of genes, exons, and introns in the chromosomes, as organized in the genome FASTA file.

For commonly studied organisms, these files can be obtained by downloading them from their respective websites or accessing public databases like Ensembl, NCBI, and others.

In this tutorial, the reference genome of *Arabidopsis thaliana* is used, downloaded from Ensembl Plants using the command line (genome file: Arabidopsis_thaliana.TAIR10.dna.toplevel.fa.gz):

```
$ wget ftp://ftp.ensemblgenomes.org/pub/plants/release-57/
fasta/arabidopsis_thaliana/dna/Arabidopsis_thaliana.TAIR10.
dna.toplevel.fa.gz
```

The following command is used to fetch the annotation file (Arabidopsis_thaliana.TAIR10.57.gtf.gz):

```
$ wget ftp://ftp.ensemblgenomes.org/pub/plants/release-57/
gtf/arabidopsis_thaliana/Arabidopsis_thaliana.TAIR10.57.gtf.
gz
```

Using the most up-to-date genome version and the corresponding GTF file is crucial for accurate alignment. It is imperative that the chromosome naming in both the sequence file and the GTF file matches precisely for the commands to function properly. In cases where the sequence and annotation files are obtained from different sources, the users must verify the consistency in chromosome naming. To ensure consistency, the users can employ basic Linux command line tools like grep, sed, and others to check and potentially modify the chromosome names accordingly. This step will help prevent any potential alignment issues arising from mismatches in chromosome naming.

With the files procured, the next step is to generate the genome index. First, unzip the two files from Ensembl Plants and create a directory for the index:

```
$ gunzip Arabidopsis*gz
$ mkdir STAR_Index
```

The following command generates the genome index:

```
$ STAR --runThreadN 4 --runMode genomeGenerate –genomeDir
STAR_Index/ --genomeFastaFiles Arabidopsis_thaliana.TAIR10.
dna.toplevel.fa --sjdbGTFfile Arabidopsis_thaliana.TAIR10.57.
gtf
```

The genome indexing step is a critical prerequisite for the subsequent processes to run smoothly. The time required to complete this step is primarily dependent on two factors: the genome size of the organism under study and the available computational resources.

Once the genome index has been successfully created, the next step involves aligning the trimmed reads using STAR. Since we are dealing with paired-end reads here, it is crucial to use the STAR command that allows simultaneous incorporation of both the "left" and "right" reads for each sample. Additionally, as the read files are compressed in "*.gz" format, it is necessary to use the "--readFilesCommand zcat" option.

The command is as follows:

```
$ mkdir STAR_Alignment_1
$ STAR --runThreadN 10 --genomeDir STAR_Index –readFilesCom-
mand zcat --quantMode GeneCounts --readFilesIn CT1_1_Trimmed.
fastq.gz CT1_2_Trimmed.fastq.gz --outFileNamePrefix STAR_A-
lignment_1/
```

Please be aware that the command provided above is specifically for the first sample. To process all other samples, the users will need to run this command separately with the appropriate modifications to the input file names. To streamline this process and automate it for multiple samples, it is highly recommended to create a bash script. This script will enable efficient and consistent execution of the command for each sample, ensuring that all required alignments are performed and completed.

In the specified directory, running the alignment command generates six output files. Among them, three log files provide valuable information regarding the alignments and can offer insights into any potential failed runs, if they occur. For instance, the Log.final.out file for this dataset indicates the reads that were successfully aligned. The SJ.out.tab file contains filtered splice junctions detected during the mapping process, but it is not covered in this protocol. Among the most important output files from STAR is the Sequence Alignment/Map (SAM) file. It contains detailed information about every read in our dataset and its alignment to the indexed genome. For this tutorial, our primary focus is on the ReadsPerGene.out.tab file. This tab-delimited text file contains four columns for each gene in the genome. The first column represents the gene ID, followed by columns for unstranded counts, count for the first read of the paired-end reads, and count for the second read of the pair. For users dealing with stranded RNA-Seq data, it's important to refer to the STAR manual for more information on stranded alignment. Additional information on the SAM format specification can be found in the linked guide provided in Table 1.

After completing the alignment step, the next stage in the pipeline is the differential expression analysis. It is essential that all alignments are performed before proceeding to this step. This ensures that the data are appropriately processed and ready for assessment of differential gene expression between different conditions or samples under study.

2.4 Differential Gene Expression Analysis Using DESeq2

For conducting the differential expression analysis, we employ the DESeq2 R package. Various methods are available for importing data into DESeq2, even when using the same upstream pipeline. Users familiar with R can construct the required data tables using native functions or utilize dedicated software packages like tximport [6]. Additionally, in-house scripts written in scripting languages such as Perl or Python can also be used as an alternative approach. The choice of method depends on individual familiarity and preferences with the available tools. For convenience, scripts are included in the tutorial directory that can take a list of abundance files from STAR, and their associated experimental conditions, and generate the files necessary for import into DESeq2.

```
$ perl star2deseq_strand-specific_paired_end.pl
```

This will generate two tables containing raw counts ("counttable.txt") and experimental groups ("colddata.txt"). The next step will be to import them into R.

```
$ library("DESeq2")
$ cts <-
as.matrix(read.csv("counttable.txt",sep="\t",header=TRUE,
check.names=FALSE,row.names=1))
head(cts)
$ coldata <-
read.csv("coldata.txt",sep="\t",header=TRUE,check.name-
s=FALSE,row.names=1)
$ head(coldata)
$ coldata <- coldata[,c("condition","type")]
$ dds<-DESeqDataSetFromMatrix(countData=cts,colData=coldata,
design=~condition)
$ dds<-DESeq(dds)
```

It is highly recommended to consult the DESeq2 manual, as it serves as an excellent starting point for more detailed analysis and provides example codes for generating informative plots. These plots include assessments of RNA-Seq data quality, considerations for batch effects, handling time-series data, and exploring other variations to the standard workflow (*see* **Note 1**).

Once the "dds" object has been successfully generated, the next step is to conduct PCA analysis. The PCA plot allows for a comparison of each sample replicate within the dataset. It is important to note that a significant degree of separation among replicates belonging to the same experimental condition suggests poor reproducibility during the RNA extraction stage. The expectation is that replicates of a given sample cluster together, indicating better reproducibility and consistency.

The command is

```
$ library(ggplot2)
$ library(ggrepel)
$ rld<-rlog(dds,blind=FALSE)
$ data <- plotPCA(rld, returnData=TRUE)
$ percentVar <- round(100 * attr(data, "percentVar"))
$ qplot(data$PC1, data$PC2, color=data$group, data=data) +
xlab(paste0("PC1: ",percentVar[1],"% variance")) + ylab
(paste0("PC2: ",percentVar[2],"% variance")) + geom_point
(size=2) + geom_label_repel(aes(label = dds$condition),box.
padding = 0.35, point.padding = 0.5,segment.color = 'grey50')
```

This will generate a PCA plot with all the samples.

Moving forward with the analysis, the next step involves making comparisons between two conditions. To achieve this, we use a contrast table with the same annotation as the data tables originally imported into R. This contrast table allows to specify the specific conditions to be compared and facilitates the subsequent statistical comparisons and identification of differentially expressed genes between the chosen conditions. Additionally, during the comparison step, genes that have an adjusted p-value (p adjusted) less than 0.05 are listed. This yields a list of genes that exhibit statistically significant differential expression between the two conditions being compared. This step helps to prioritize genes that have the most substantial changes in expression and reduces the likelihood of false-positive results in the analysis.

```
$ comp1 <- results(dds, contrast = c("condition", "HL", "CT"))
$ resultsNames(comp1)
$ res1sig <- comp1[which(comp1$padj<0.05),]
$ write.table(res1sig, file = "Comp1_CT_vs_HL.txt", sep =
"\t", quote = FALSE)
```

The generated table contains valuable information for each gene or transcript represented in the indices used by the upstream aligner. It includes average expression values for both conditions, statistical test values, and the log2-transformed fold change values. These metrics determine the significance of differential expression between the conditions under investigation.

To make the most informed decisions during data interpretation, users are strongly encouraged to delve into the theoretical basis of DESeq2's significance metrics. Understanding the underlying statistical principles behind DESeq2's calculations will empower researchers to assess the results effectively and consider these metrics in the context of their specific experimental design. Such understanding is vital for ensuring reliable identification of genes with biologically meaningful changes in expression and also for further downstream analyses (*see* **Notes 2** and **3**).

3 Notes

1. This pipeline supplies a comprehensive starting point for conducting RNA-Seq analysis. Researchers are encouraged to review the manual or documentation of each tool, further explore them per their needs, and customize the commands to suit their specific biological questions. By adapting the pipeline to their unique research needs, researchers can derive more accurate and insightful results.
2. The subsequent steps in the pipeline will vary depending on the specific biological questions to be addressed. Some examples of additional analyses include Gene Ontology (GO) analysis or the construction of gene co-expression networks. These steps

can add further depth and context to the RNA-Seq data, providing valuable biological insights.

3. Throughout each stage of this pipeline, it is essential to be mindful of potential errors that may arise. However, due to the popularity of the tools used and the large user base, researchers can find substantial support within the active user community. This supportive community is a valuable resource for troubleshooting issues and obtaining guidance, making the RNA-Seq analysis process more accessible and efficient for researchers.

Availability

The sample data files used in the implementation of this protocol have been made available at the GitHub repository: https://github.com/sohamsg90/Protocol-for-RNAseq-analysis.

References

1. Love MI, Huber W, Anders S (2014) Moderated estimation of fold change and dispersion for RNA-seq data with DESeq2. Genome Biol 15(12):550
2. Andrews S (2010) FastQC: a quality control tool for high throughput sequence data
3. Dobin A, Davis CA, Schlesinger F, Drenkow J, Zaleski C, Jha S et al (2013) STAR: ultrafast universal RNA-seq aligner. Bioinformatics 29(1):15–21
4. Bolger AM, Lohse M, Usadel B (2014) Trimmomatic: a flexible trimmer for Illumina sequence data. Bioinformatics 30(15): 2114–2120
5. Balfagón D, Sengupta S, Gómez-Cadenas A, Fritschi FB, Azad RK, Mittler R et al (2019) Jasmonic acid is required for plant acclimation to a combination of high light and heat stress. Plant Physiol 181(4):1668–1682
6. Soneson C, Love MI, Robinson MD (2015) Differential analyses for RNA-seq: transcript-level estimates improve gene-level inferences. F1000Research, 4.

Chapter 2

Inferring Interaction Networks from Transcriptomic Data: Methods and Applications

Vikram Singh and Vikram Singh

Abstract

Transcriptomic data is a treasure trove in modern molecular biology, as it offers a comprehensive viewpoint into the intricate nuances of gene expression dynamics underlying biological systems. This genetic information must be utilized to infer biomolecular interaction networks that can provide insights into the complex regulatory mechanisms underpinning the dynamic cellular processes. Gene regulatory networks and protein–protein interaction networks are two major classes of such networks. This chapter thoroughly investigates the wide range of methodologies used for distilling insightful revelations from transcriptomic data that include association-based methods (based on correlation among expression vectors), probabilistic models (using Bayesian and Gaussian models), and interologous methods. We reviewed different approaches for evaluating the significance of interactions based on the network topology and biological functions of the interacting molecules and discuss various strategies for the identification of functional modules. The chapter concludes with highlighting network-based techniques of prioritizing key genes, outlining the centrality-based, diffusion- based, and subgraph-based methods. The chapter provides a meticulous framework for investigating transcriptomic data to uncover assembly of complex molecular networks for their adaptable analyses across a broad spectrum of biological domains.

Key words Transcriptome, Gene regulatory networks, Protein interaction networks, Bayesian model, Gaussian model, Interologous networks, Coexpression networks

1 Introduction

Rooted in the foundational discoveries of biological sciences is the central dogma of molecular biology, encapsulating the principle that DNA makes RNA, that further directs protein synthesis [1, 2]. Genetic information coded in DNA is transcribed into RNA through a complex, precise, and meticulously directed cellular process called transcription. The regulatory and housekeeping RNAs in concert with proteins intricately engage in spatiotemporally regulated interactions, finely tuned by specific environmental cues, to govern the multitude of life-sustaining processes [3–5]. In recent decades, the abundance of genomic sequences has markedly

Rajeev K. Azad (ed.), *Transcriptome Data Analysis*, Methods in Molecular Biology, vol. 2812,
https://doi.org/10.1007/978-1-0716-3886-6_2,

expanded, thanks to the successful completion of a number of genome projects and the sophistication of next-generation sequencing technologies [6]. Nevertheless, bridging the gap between gene expression data and the cellular function continues to stand as a formidable challenge [7]. Extensive transcriptome-wide studies need to be undertaken in order to effectively bridge this gap and determine the functional groupings among genetic elements. With the advent of high-throughput technologies, including microarrays, RNA-sequencing (RNA-seq), and the corresponding development of sophisticated data analysis methods, our ability to comprehend gene function and its relevance to diseases has undergone a major transformation. These cutting-edge techniques provide a systematic view of the functional status of genes, allowing us to see beyond the individual elements of the genetic circuitry and appreciate the unified whole [8].

Unraveling the molecular mechanisms of biological processes, and understanding the optimal utilization of intricate genetic circuitry as well as their potential implications in diseases, hinges on deciphering the complex interaction patterns among these genetic elements [9]. Thus, one of the primary objectives in biological research is to systematically identify all molecules within a living cell and the relationship among them to explain the phenotypic variability. Extensive research has demonstrated that the complex regulation governing gene expression is shaped by cis-acting DNA elements and complex protein interactions [10, 11]. Moreover, an evolving understanding of gene regulation has also unveiled the possibility of trans regulation, highlighting the nuanced interplay between DNA and transcribed RNA [12]. However, the functions of many genes are still less understood, a situation that has only become more complex with the recent identification of many novel noncoding genes. Recently, multiple genes were observed to be coexpressed, clustering together in patterns of expression. An observation led to the hypothesis that coexpression may be the result of the coregulation of genes, implicating complex regulatory networks [13]. These discoveries sparked the investigation of broader networks underlying gene coexpression, with the potential to shed light on the convoluted coordination of cellular processes. The coexpression networks represent the cooperative behavior of genes in response to different conditions. Subsequently, the constructed networks become the subject of in-depth examination and analyses, providing a comprehensive view of gene interactions and their implications.

As we delve deeper into the domain of network construction using transcriptomic data, we investigate how these networks reveal the symphony of molecular interactions within cells and the complexities of biological organization. In Subheading 2, we embark on a brief overview of the various platforms utilized for transcriptomic data generation followed by an overview of graph theory in

Subheading 3. Moving into Subheading 4, we delve into the core ideas associated with construction of interaction networks, exploring three pivotal methods: association-based methods, probabilistic methods, and interologous network construction techniques. In Subheading 5, we focus on analyzing the constructed network. Here, we uncover the significance of the predicted interactions within these networks, a critical step in deciphering their statistical and biological relevance. Subsequently, we discuss the methods to identify functional modules within these networks, elucidating the coordinated actions of genes or proteins in specific cellular processes. Finally, we elaborate various approaches to identify key drivers and molecular architects that govern the dynamics over these networks and ultimately control the cellular response.

2 A Primer on Transcriptomics

Almost all of the processes within a living organism, like gene expression, cellular regulation, and disease modulation, are substantially affected by the RNA molecules that include protein-coding mRNAs and noncoding RNAs, such as transfer RNAs (tRNAs), ribosomal RNAs (rRNAs), microRNAs (miRNAs), and long noncoding RNAs (lncRNAs) [14]. Thus, the transcriptome encompasses the entirety of RNA molecules present in a given cell or tissue in a specific temporal context [15]. This compilation reveals a multidimensional comprehension of cellular behavior by highlighting the complicated gene expression dynamics. Transcriptomics, a systematic investigation of the transcriptome, provides unprecedented insights into the genetic complexities that regulate cellular responses to developmental cues, environmental fluctuations, and disease manifestations [15]. Transcriptomics offers a panoramic perspective of the complex cellular dynamics, as it provides a comprehensive platform that transcends isolated gene analysis and gives a holistic understanding of the molecular dynamics driving biological systems. The rapid evolution of transcriptomics has been propelled by technological advancements [8]. Early attempts at understanding individual transcripts predated comprehensive transcriptomics approaches. Techniques like expressed sequence tags (ESTs) [16] and low-throughput Sanger sequencing [17] were utilized to sequence individual transcripts, providing insight into gene content. One of the pioneering sequencing-based methods, serial analysis of gene expression (SAGE) [18], emerged in 1995. However, the comprehensive expression and regulation of the entire transcriptome remained elusive until high-throughput techniques such as microarrays and RNA-Seq emerged [8].

Over the past few decades, two of the most commonly employed transcriptomics technologies have been microarray [19] and RNA-sequencing (RNA-seq) [20]. Microarrays utilize hybridization, involving thousands or even millions of short nucleotide oligomers known as probes strategically arrayed on a solid substrate (Fig. 1), typically glass. Fluorescently labeled transcripts are hybridized with these probes, resulting in fluorescence intensity at each probe location reflecting the corresponding transcript's abundance [19]. Microarrays generally fall into two main categories: low-density spotted arrays and high-density short probe arrays. Low-density arrays involve minute droplets of purified cDNAs placed on a glass slide. These arrays feature longer probes compared to high-density arrays, impacting their transcript resolution. Fluorescence ratios from different fluorophores are employed to calculate a relative measure of abundance in spotted arrays [21]. In contrast, high-density arrays employ single-channel detection, with each sample hybridized and detected individually [22]. Pioneered by the Affymetrix GeneChip array, high-density arrays utilize multiple short 25-mer probes to quantify individual transcripts,

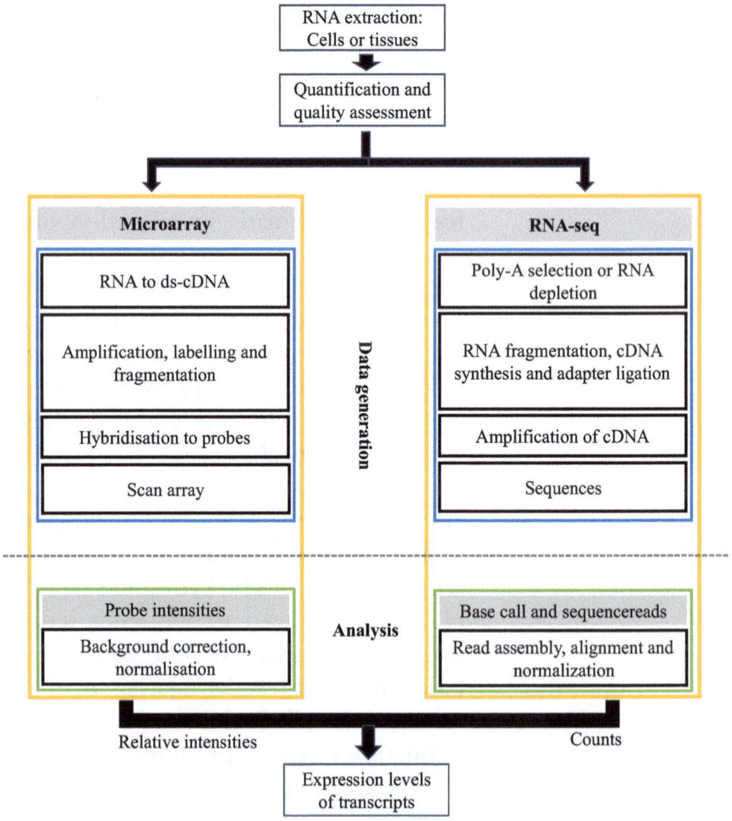

Fig. 1 An overview of the general steps involved in generating and preprocessing transcriptomic data using microarray and RNA-seq technologies

collectively assessing one gene. Microarrays are relatively less expensive and can be used to analyze a large number of genes simultaneously. However, they are less sensitive than RNA-seq and challenging to interpret. Another major drawback associated with microarrays is that these arrays necessitate a prior understanding of the organism under investigation, often derived from an annotated genome sequence or a library of expressed sequence tags (ESTs) that serve as the basis for generating the probes [8].

The advent of cutting-edge technologies, notably RNA-sequencing, has brought a revolutionary paradigm shift to transcriptomics [23]. This innovation enables the precise and comprehensive dissection of the RNA dynamics. RNA-Seq encompasses several crucial steps (Fig. 1) involving RNA isolation, library preparation, and sequencing. The process initiates with the extraction of RNA molecules from a biological sample. This step is pivotal to ensure the integrity and representation of the transcriptome. Subsequently, a cDNA library is synthesized that includes RNA fragmentation, reverse transcription to synthesize complementary DNA (cDNA), adapter ligation, and quality control using quantitative rt-PCR. Once the library is ready, the sequencing phase commences. High-throughput sequencing platforms decode the cDNA fragments, generating millions of short sequences or reads [20]. Sequenced nucleotide fragments are aligned to reference genomes or assembled de novo to reconstruct original RNA transcripts. RNA-Seq offers a broad dynamic range, enabling gene identification, real-time activity analysis, and accurate modeling of gene expression levels. Ongoing improvements in DNA-sequencing technology have enhanced RNA-Seq's throughput, accuracy, and read length, leading to its widespread adoption over microarrays. RNA-Seq requires lower RNA input than microarray, enabling finer examination of cellular structure at the single-cell or nuclear levels [24–26].

3 Biomolecular Interactions from Graph-Theoretic Perspectives

The biological world is organized into a hierarchical structure, spanning from the atomic and molecular level to ecosystems and the biosphere [27]. At each tier of this biological hierarchy, and even across different levels, interaction networks play crucial regulatory roles. Networks serve as a fundamental framework for understanding complex relationships within biological systems [28, 29]. Whether we are examining molecular interactions within a cell, the flow of energy and nutrients in ecosystems, or the intricate web of connections in the human brain, these systems have been aptly represented and analyzed using network frameworks [30]. In this context, an interaction network (Fig. 2a, b) is essentially a graph $G = (V, E)$ comprising of a set of nodes V and a

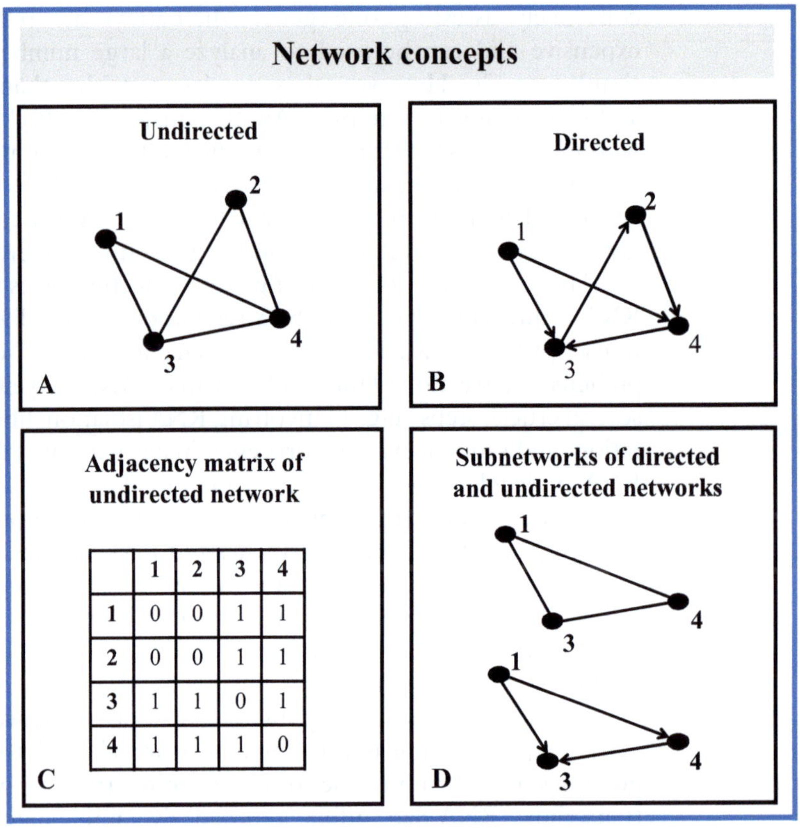

Fig. 2 Depiction of fundamental network-theoretic concepts. (**a**) introduces an undirected network featuring four nodes and five edges, each node boasting a degree of two.(**b**) A directed network comprising of four nodes and five edges. Notably, nodes 3 and 4 exhibit an in-degree of two and an out-degree of one, while node 2 possesses a balanced in-degree and out-degree of one, and node 1 showcases an out-degree of two alongside an in-degree of zero. (**c**) Adjacency matrix corresponding to the aforementioned undirected network, offering a visual representation of internode connections. (**d**) Subgraphs on nodes 1, 3, and 4 derived from the networks (a) and (b)

set of edges $E \subset V \times V$. Graph theory provides the formal foundation to explore and decipher the structural and functional properties of these networks, making it an indispensable tool for researchers. By treating biological entities as nodes and their interactions as edges, we can employ the rich toolbox of graph theory to gain insights into the organization, dynamics, and emergent properties of these systems [28]. These interactions underpin processes like signal transduction, gene regulation, and metabolism. However, deciphering these interactions, especially within the complex milieu of living cells, can be an immensely challenging endeavor. Fortunately, there exists an alternative approach, one that provides us with a computational lens into the intricate world of molecular interactions. This approach involves measuring the abundance of mRNA levels within cells. By quantifying the expression of thousands of genes simultaneously, researchers can generate detailed snapshots of gene expression patterns [31]. It is within these expression patterns that a wealth of information about underlying molecular interactions is concealed. Using mRNA expression data, computational methods can be employed to reconstruct the interaction structure (network topology) that governs these patterns [32].

Networks constructed from expression data would be directed (Fig. 2a) or undirected (Fig. 2b) graphs determined by the statistical model employed [33]. In graph theory, a directed network, often referred to as a directed graph or digraph, contains edges that are ordered pairs of vertices and thus have a specific orientation or direction associated with them. These edges represent causal relationships or information flow from one node (gene or protein) to another [34]. Conversely, an undirected network consists of edges that lack direction and typically represent symmetric associations, such as coexpression relationships between genes or proteins [32, 35]. The adjacency matrix A is a fundamental representation of networks, with rows and columns corresponding to nodes and entries indicating the presence or absence of edges (Fig. 2c). Each entry $a_{i,j}$ of the A is defined as

$$a_{i,j} = \begin{cases} x_{i,j} & \text{if } \{i,j \in E\} \\ 0 & \text{otherwise} \end{cases}$$

where $x_{i,j}$ would vary according to some weight function $w_{i,j}$ or have values 1 if the network is unweighted. The adjacency matrix for undirected networks is symmetric, i.e., if an edge exists between nodes i and j, the corresponding edge between nodes j and i is also present. In directed networks, the matrix can be asymmetric, reflecting the direction of interactions [36]. Some essential topological properties of a graph include adjacency, i.e., if nodes $i,j \in E(G)$, then node i is called adjacent to or neighbor of node j. All the nodes adjacent to node i constitute its neighborhood $N(i)$, which is defined as $N(i) = j \in V(G) | i,j \in E(G)$ [37]. The degree of a node i, denoted as $deg(i) = |N(i)|$, represents the total number of edges incident to a node i, offering insights into a node's centrality. In directed networks, we distinguish in-degree (incoming edges) and out-degree (outgoing edges), discerning the nature of interactions [37]. Similarly, several other properties describe the general structure, node rank, or local interconnectivity patterns between network nodes. These properties include the clustering coefficient, which mathematically quantifies how nodes tend to cluster, highlighting localized structures. Eigenvalues λ and eigenvectors unveil global network properties and central nodes, a key tool in identifying influential network components. The graph spectrum, derived from eigenvalues, encapsulates structural characteristics and stability [37].

4 Inferring Interaction Networks from RNA-seq Data

Differentially expressed genes (DEG) analysis often yields a substantial matrix, where each row represents a gene and the columns represent attributes such as fold change, *p-values*, or other relevant measures [38–40]. At this juncture, the sheer volume of data can be overwhelming for any scientist. Extracting meaningful insights and interpreting the biological significance necessitate a reduction in data complexity and the discovery of inherent structures within it. To tackle this challenge, several standard and widely applied methodologies come into play. These include clustering techniques [41] and principal component analysis [42], which help identify patterns and group genes with similar expression profiles. For more intricate analyses, researchers delve into inferring genome-scale networks, such as protein interaction networks [43], or leverage knowledge-based networks like pathway networks [44] and gene coexpression networks [45] derived from the expression data. These advanced approaches offer a deeper understanding of the interplay between genes and pathways, facilitating the extraction of biologically relevant information from the transcriptomic data. In the following, we discuss three different strategies (Fig. 3) for biomolecular interaction networks construction using transcriptomic data, including association-based networks, probabilistic networks, and interologous networks.

4.1 Association-Based Interaction Networks

Network approaches have been employed in molecular biology to explore the complex interaction patterns underlying expression data [43–45]. The nodes in these types of networks typically represent gene expression levels, and the edges that connect them represent interactions, which can be formulated either by correlation, mutual information, or some other mathematical model used to represent the system. It is crucial to remember that interactions in association networks based on correlation and mutual information do not identify the direction of effect between interacting genes and thus are undirected. Interactions in probabilistic networks, on the other hand, are directed. The initial step in building association networks is to create a distance matrix from expression data, a fundamental process for network analysis [46]. This step typically involves calculating numeric values that quantify the differences between gene expression patterns. Commonly, pairwise correlations or covariances between genes of interest are computed using expression values. There exists a number of correlation measures like Pearson's [47], Kendall's [48], and Spearman's [49] that have been commonly employed for this purpose [50, 51]. Alternatively, one can use information theory or more specifically mutual information score to measure the resemblance between pairs of expression vectors [50, 52]. The outcome of this step is an $m \times m$ matrix,

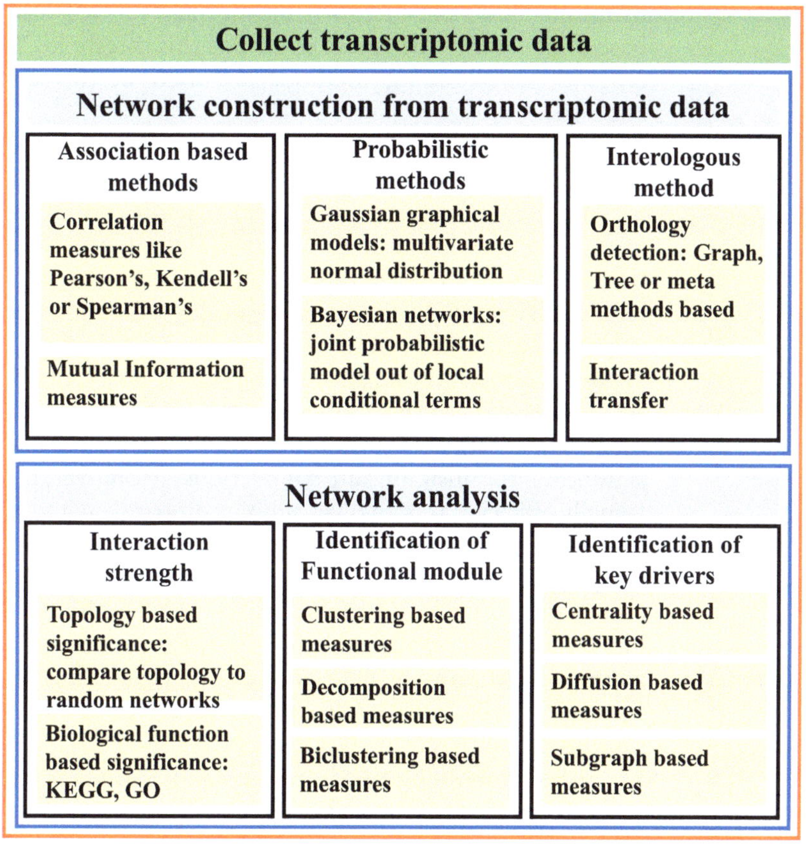

Fig. 3 An overview of (**a**) various methods used to infer interaction networks from transcriptomic data and (**b**) different network analysis methods

with m representing the number of analyzed genes. Each element within this $m \times m$ matrix signifies the distance between a pair of elements and is commonly referred to as a distance matrix, laying the groundwork for subsequent association network construction.

The distance matrix must then be converted to an adjacency matrix containing only those elements that survive specific measures of significance [50, 53]. Constructing an adjacency matrix from the initial distance matrix is a pivotal step in the creation of association networks. The purpose of this transformation is to identify significant relationships or connections between elements, effectively delineating the network's edges [50]. The resulting adjacency matrix is a binary representation, where surviving connections are marked as ones, signifying edges and nonsurviving entries are marked as zeros, indicating no connection. Converting the distance matrix into an adjacency matrix is of paramount importance as it profoundly influences the structure and interpretability of the network. Several approaches can be employed for this conversion; however, the most commonly employed approach is the threshold-based method [53]. Thresholding involves defining a threshold value and categorizing the values in the distance matrix based on whether they meet or exceed this threshold. Two primary types of thresholding methods are used in this context: hard and soft.

Hard thresholding is a straightforward method for converting a distance matrix into an adja- cency matrix, a critical step in coexpression network construction [54]. The process involves setting a predetermined threshold value, often determined through heuristics or statistical considerations. All values in the distance matrix below this threshold are assigned a value of zero, indicating insignificance, while those equal to or above the threshold are retained as connections in the adjacency matrix [55, 56]. One of the notable advantages of hard thresholding is its simplicity. It is easy to implement and interpret, providing a clear binary distinction between significant and insignificant connections. Moreover, it is computationally efficient, as it involves simple comparisons and assignments. However, this simplicity comes at a cost. Hard thresholding can lead to the loss of valuable information regarding the strength of connections below the threshold, potentially disregarding subtle but biologically relevant relationships. Further more, selecting an appropriate threshold value can be challenging, and the method is sensitive to this choice, with slight variations in threshold leading to significantly different network structures [56].

Soft thresholding in contrast is a more sophisticated approach to converting a distance matrix into an adjacency matrix. This method begins by computing correlation coefficients between elements in the distance matrix. These coefficients are then subjected to a power transformation, typically through squaring or applying other functions. This transformation accentuates stronger correlations while attenuating weaker ones. Subsequently, a hard threshold is applied to these transformed values, resulting in a weighted adjacency matrix [46]. In this weighted matrix, values below the threshold contribute to the network but with reduced weight, preserving information about the strength of relationships. One of the primary advantages of soft thresholding is its ability to retain information about the strength of connections, allowing for a more nuanced representation of the network. This feature can help reduce noise and capture subtle coexpression patterns that might be lost in a binary representation. It also provides flexibility in controlling the network's sparsity, as researchers can adjust the threshold to balance the number of retained connections [57]. However, this method introduces complexity due to the mathematical transformations involved, potentially making it more computationally intensive. Additionally, the selection of the power transformation exponent and threshold can be subjective, although they provide researchers with more options for fine-tuning the network's characteristics. The interpretation of weighted networks can also be more intricate than binary networks, often requiring additional analytical techniques. The final unweighted or weighted adjacency matrix thus obtained represents the network of interest, which is further analyzed using graph-theoretic approaches to explore coexpression patterns among different genes under different conditions.

4.2 Probabilistic Networks

Previously described data-driven methods rely on correlation or information-theoretic measures of dependence but do not explicitly define a probabilistic data model. In this section, we discuss two distinct classes of methods that explicitly begin with a probabilistic data model [33]. These methods employ global measures of fit, such as joint likelihood, or employ Bayesian approaches to discern the underlying network structure.

4.2.1 Gaussian Graphical Models

Gaussian graphical models (GGMs) are a powerful tool in statistical modeling that uses a graph to model the conditional dependence structure among continuous random variables [58]. GGMs are based on the assumption that the variables are jointly Gaussian, meaning their distribution is a multivariate Gaussian distribution. The graph consists of nodes representing random variables (gene expression measurements in our case) and edges representing conditional dependencies. Two variables are conditionally independent given a set of other variables in the model if their joint distribution is the same as the product of their conditional distributions. If an edge does not connect two nodes, they are conditionally independent, given all the other variables in the model. The graph can be either directed or undirected, depending on the type of conditional dependencies. The joint distribution of the random variables in a Gaussian graphical model is a multivariate normal distribution, which is given by

$$p(\mathbf{x}|\boldsymbol{\mu}, \Sigma) = \frac{1}{2\pi^{\frac{n}{2}}|\boldsymbol{\Sigma}|^{\frac{1}{2}}} \exp\left[-\frac{1}{2}(\mathbf{x}-\boldsymbol{\mu})^T\boldsymbol{\Sigma}^{-1}(\mathbf{x}-\boldsymbol{\mu})\right]$$

where $\mathbf{x}\in^{\mathbf{n}}$ is a multivariate normal vector, $\boldsymbol{\mu}$ is the mean vector, and $\boldsymbol{\Sigma}$ is the covariance matrix. The inverse of the covariance matrix, $\Omega=\Sigma^{-1}$, is called the precision matrix, and it encodes the partial correlations between the random variables, which measures the extent of correlation between these two variables after adjusting for the effect of the other variables. The precision matrix has a zero entry if and only if the corresponding nodes are conditionally independent in the graph. This means that the graph can be obtained from the sparsity pattern of the precision matrix and vice versa. The main problem in Gaussian graphical models is to estimate the precision matrix from data [59], which can be done by maximizing the log-likelihood function:

$$\ell(\boldsymbol{\Sigma}^{-1}) = \log|\boldsymbol{\Sigma}^{-1}| - \mathrm{tr}(\mathbf{S}\boldsymbol{\Sigma}^{-1}) - n\log(2\pi).$$

The term $-n\log(2\pi)$ is a constant that does not depend on $\boldsymbol{\Sigma}^{-1}$), and it is added to make the log-likelihood function consistent with the probability density function of the multivariate normal distribution. The term $\log|\boldsymbol{\Sigma}^{-1}|$ is the natural logarithm of the determinant of the precision matrix, which measures the information content of this matrix. The term $\mathrm{tr}(\mathbf{S}\boldsymbol{\Sigma}^{-1})$ is the trace of the product of the sample covariance matrix and the precision matrix, which measures how well the precision matrix fits the data. This is a convex optimization problem that can be solved by various methods, such as gradient descent, Newton's method, or coordinate descent. However, when the number of variables is large compared to the number of observations, the problem becomes ill-posed and the maximum likelihood estimator does not exist or is not unique. In this case, regularization techniques are needed to impose some constraints on the precision matrix, such as sparsity or low rank. Some common regularization methods include lasso that adds an $L1$ penalty on the entries of the precision matrix, which encourages sparsity and selects relevant edges in the graph. Graphical lasso applies lasso on each row (or column) of the precision matrix separately, which ensures positive definiteness and symmetry. Sparse inverse covariance estimation (SICE) applies graphical lasso on a penalized log-likelihood function that incorporates prior information on the graph structure [60].

4.2.2 Bayesian Interaction Networks

The methods discussed above range from analyzing marginal dependencies in coexpression measures to exploring conditional dependencies in partial correlation approaches and strive to unveil gene relationships based on various probabilistic dependencies. However, it is essential to note that these methods, consequently, cannot represent causal relationships between genes. In contrast, Bayesian networks provide a framework for modeling and studying the causal dependencies among genes [61]. It consists of nodes, which represent variables, and edges, which represent conditional dependencies [62]. A Bayesian network can be used to model gene expression data by assuming that each gene is a variable whose expression level depends on the expression levels of some other genes [34]. To generate a Bayesian network from gene expression data D, one needs to find the optimal network structure G that best fits the data, i.e., the set of nodes and edges that represent the actual gene interactions. The structure is a directed acyclic graph (DAG) where each node represents a variable, and each edge represents a conditional dependence. It represents the joint probability distribution of the genes as a product of local conditional probability distributions, each depending on the state of the parent nodes in the graph.

$$P(X_1, \ldots, X_p) = \prod_{i=1}^{p} P(X_i | Pa^G(X_i))$$

where Pa^G are all the parent nodes of X_i in the DAG G. This joint probability distribution represents a set of conditional independence relationships among the variables, meaning that each variable is independent of its nondescendants, given its parents. It is called the Markov assumption that, together with a restricted number of parents for each node, facilitates the definition of conditional probabilities and the effective execution of inference in a Bayesian network. Every directed acyclic graph G is assessed using a Bayesian score, represented as the posterior probability of G that is given by

$$S(G : D) = logP(G|D) = logP(D|G) + logP(G) + C.$$

Computing posterior probabilities in DAG modeling involves two key steps: learning the DAG structure from observational data and estimating conditional probabilities given this structure. This task is challenging due to the super-exponential growth in possible DAG topologies and the high dimensionality of expression data, leading to many equally scoring DAGs. To tackle this complexity, heuristic algorithms like greedy hill-climbing, simulated annealing, and genetic algorithms [63] are employed, iteratively adjusting the DAG topology. Biologically informed constraints and coexpression clustering help reduce the search space. Instead of selecting a single optimal DAG, the comparison of multiple DAGs with similar scores is a common practice, focusing on consistent topological features. The second step for computing the posterior probabilities involves parameter estimation. It depends on the form of the conditional probabilities (whether these are discrete, continuous, or involve mixture distributions) and the presence of missing data for any node. Various algorithms like the sum–product, maximum likelihood estimation (MLE), maximum a posteriori (MAP), and expectation–maximization (EM) are commonly applied to address this problem [34]. Additionally, prior information about parameter and graph distributions is incorporated into the scoring function calculation. It is crucial for the chosen scoring function to be decomposable into local scores for computational efficiency and to include features that guard against overfitting. Popular strategies to achieve this goal include the Bayesian Information Criterion (BIC) and the Bayesian Dirichlet equivalent (BDe) [64, 65].

4.3 Interologous Networks

The interologous network approach is a valuable strategy for constructing networks of contigs obtained from expression data, leveraging existing protein interaction information [66]. This method relies on the concept of orthology, which identifies orthologous proteins in a template organism and subsequently transfers interactions between these orthologous proteins to the query proteome, which consists of the proteins encoded by the contigs obtained after assembly. The procedure begins with the selection of one or more template organisms, typically those for which extensive protein interaction data is available. These template organisms serve as a reference, providing a foundation for inferring interactions in the target organism, where such data may be limited or unavailable [67].

4.3.1 Ortholog Identification

The first step involves identifying orthologous proteins between the template organism(s) and the target organism. Orthologs are genes that are related by vertical descent from a common ancestor and encode proteins with the same function in different species [68]. Orthologous relationships can be classified into three different types, namely, one-to-one orthologs, one-to-many orthologs, and many-to-many orthologs based on the evolutionary events that led to their divergence. Genes that have a single copy in each species and are derived from a single gene in the last common ancestor are called one-to-one orthologs. Genes having a single copy in one species, but multiple copies in another species, due to gene duplication events after speciation are called one-to-many orthologs. Finally genes that have multiple copies in both species, due to gene duplication events before and after speciation many-to-many orthologs.

In the quest to identify orthologous genes, several distinct approaches have emerged, each with its own set of tools and techniques. These methods can be broadly categorized into three main groups, namely, graph-based methods, phylogenetic tree-based methods, and meta or hybrid methods [69–71]. Graph-based approaches for orthology detection have evolved as a response to the growing availability of complete genome sequences and the need for efficient methods to discern orthologous genes. These methods construct graphs depicting genes or proteins as nodes, and edges signify their evolutionary relationships. Typically, these methods follow a two-phase process: the graph construction and clustering phases. The graph construction phase in orthology detection is based on the premise that orthologous genes are the homologous genes that have emerged due to speciation events in their most recent common ancestor. This implies they have undergone minimal divergence since they branched at the latest conceivable juncture in their evolutionary history. Graph-based approaches employ sequence similarity scores as a proxy for evolutionary divergence to determine orthologous genes by identifying the highest-scoring match or genome-wide best hit for each gene in the query genome [72]. Since the orthology relation is symmetric, reciprocal best hits, also known as bidirectional best hits (BBH), are mostly considered [73]. Maximum likelihood estimates computed based on amino acid substitutions are another commonly used measure of evolutionary distance between sequence pairs [74].

Following speciation events in the two genomes under consideration, there are instances where a gene might undergo duplication, giving rise to multiple orthologous genes or in-paralogs

[75]. The nature of the relationship, whether it is one-to-many or many-to-many, hinges on whether gene duplication occurred in one or both genomes. However, the conventional Bidirectional Best Hit (BBH) approach is limited to predicting only one-to-one relationships. To address this limitation, InParanoid [75, 76] identifies in-paralogous genes in the query genome that exhibit more significant similarity to each other than the BBH gene in the subject genome. Once pairs of orthologs have been identified, the next step involves clustering these genes into orthologous groups. Various clustering methods are employed for this purpose. Clusters of Orthologous Groups (COGs) method identifies triangles in the orthologous network and proceeds by merging triangles that share a common face, iteratively combining them until no more triangles can be added to the existing COGs [72]. OrthoMCL employs a Markov clustering technique by simulating random walks on the orthology graph and returns probabilities to pairs of genes belonging to the same cluster. Subsequently, the graph is partitioned into groups based on these probabilities [77]. OrthoDB [78] and EggNOG [79] employ a combination of hierarchical clustering and COGs-like clustering methods. In contrast, Hieranoid [80], OMAGETHOG's [81], and COCO-CL [82] generate hierarchical orthologous groups by utilizing a guide tree, taxonomic information, and correlations of similarity scores among homologous genes, respectively.

Tree-based methods for orthology inference primarily rely on reconstructing evolutionary history by combining the phylogenies of genes and species. A gene tree is initially constructed using multiple sequence alignments from candidate gene family sequences. The species tree and gene tree are then reconciled, and the internal nodes of the gene tree are annotated with various evolutionary processes, including speciation, loss, and duplication [71]. Predicting orthologous and paralogous relationships is simple once these events have been identified. A parsimony criterion is used by most tree reconciliation techniques, which favors reconciliations with the fewest gene duplications and deletions [83]. Many methods like RIO [84], Orthostrapper [85], PhylomeDB [86] implement tree reconciliation to generate accurate hierarchical ortholog groupings. Flat lists of orthologs lacking intragroup relationships are less informative than these hierarchical groups. Some significant disadvantages associated with phylogenetic tree-based methods, preventing their application to large-scale datasets, include computational complexity and dependence on accurate multiple alignments and trees [71].

Meta-methods amalgamate multiple orthology prediction approaches to generate a robust consensus output. They accomplish this by intersecting the results from several methods and assigning scores to each prediction based on the number of independent predictors that support that relationship. Consequently, predictions garner higher scores when they are corroborated by multiple predictors. Notably, methods like DIPOT [87], GET_HOMOLOGUES [88], COMPARE [89], and HCOP [90] adopt this approach to assign weights to their predictions. Certain methods incorporate postprocessing strategies to enhance their results. For instance, MOSAIC [91] conducts iterative graph-based optimization to incorporate missing orthologs effectively. MARIO [92] identifies common predictions from various orthology algorithms as seeds and incrementally adds additional sequences into different clusters using HMM profiles. OMA Hierarchical Orthologous Groups (HOG) employs an orthology graph to form groups progressively, beginning from specific taxonomic levels and merging toward the root of the species tree [93]. OrthoFinder2 [94] combines graph-based and tree-based approaches, first identifying orthogroups using the OrthoFinder graph-based method and then leveraging these to infer approximate gene trees and species trees, enhancing orthology and paralogy predictions. Methods such as WORMHOLE [95] employ machine learning algorithms to recognize patterns among multiple orthology prediction methods and subsequently leverage these patterns to identify novel orthologs.

4.3.2 Interaction Transfer and Network Construction

Once orthologs are identified, the known protein interactions from the template organism are transferred to the target organism. This transfer assumes that if two proteins in the template organism interact, their orthologous counterparts in the target organism are also likely to interact [67, 96]. Thus, transcript contigs in the target organism are connected based on the interactions between their corresponding orthologs in the template species [66]. This network represents potential functional associations and regulatory relationships among the transcript contigs. Several repositories collect, store, and annotate protein interaction data from various sources, such as experimental methods, computational methods, literature mining, and manual curation. Some of the commonly used databases of PPIs are BioGRID [97], IntAct [98], STRING [99], and MINT [100]. These databases can provide easy access and integration of protein interaction data, but they may have different standards, formats, and coverage. The interaction information for the template species is obtained from these publically available repositories.

5 Network Analysis

Building upon the foundational understanding of network construction from transcriptomic data, we now embark on the network analysis phase. Here, we will unravel the intricate relationships embedded within coexpression networks, ultimately shedding light on the functional dynamics of genes and proteins (Fig. 3). It aids in the interpretation of high-dimensional transcriptomic data, and we will explore how these networks can guide us in extracting meaningful biological insights, and advancing our knowledge across a spectrum of scientific domains, from fundamental research to the development of cutting-edge medical and biotechnological innovations.

5.1 Evaluation of Interaction Significance

In coexpression or protein interaction networks, each interaction is associated with a specific score, which is determined by the predictive method. This score gives an indication of the strength of the predicted interaction; however, this is approximate. The most authentic way to determine the significance of an interaction is through experimental validation. However, this process is often time-consuming, costly, and labor-intensive. As an alternative, computational methods are available to serve this purpose. Therefore, it is important to assess the significance of the predicted interactions using various methods, such as topological and biological function-based statistical methods [101].

5.1.1 Topology-Based Statistical Significance

Random networks are generated using null models such as Erdős–Rényi (ER) [102], scale-free (SF) [29], Configurational [103], etc., and are subsequently compared to the original network using selected topological properties. This allows for the computation of statistical measures like *z-scores*, *p-values*, or false discovery rates (FDR) to assess the significance of the predictions [96]. Another approach involves random sampling of a portion of the original network in each iteration, where the frequencies of each edge are recorded. The edges that are observed in a greater number of iterations are considered to be stronger or more significant [34]. Alternatively, mutual rank (MR) can be calculated, where various similarity or distance methods are employed to determine MR values for each prediction [104]. These statistical methods play a crucial role in evaluating the significance and confidence of predictions, but their effectiveness relies on the validity and robustness of the underlying models or assumptions. Nodes with associated *p-values* less than a predefined threshold, typically 0.01, in both the ER and SF models are considered key nodes or drivers.

5.1.2 Biological Function-Based Statistical Significance

Another approach to assess significance is by leveraging existing biological knowledge related to pathways, cellular localization, and gene ontology. It is a common belief that interacting proteins often participate in the same pathways and share cellular locations. To evaluate this, information about pathways and cellular locations for the interacting proteins is gathered, and their congruence is analyzed. Gene ontology provides valuable insights into cellular processes, molecular functions, and subcellular localization of proteins, making it a useful resource for assessing interaction significance [96]. Typically, quantitative scores for predictions are obtained using metrics like the Jaccard similarity index or various semantic similarity indices to measure the similarity [105] between the annotated attributes of the proteins.

5.2 Functional Modules Identification

Coexpression data provides the expression levels of genes or proteins under different conditions and gives insights into additional information to infer the functional modules, as proteins that are coexpressed are likely to be functionally related. Categorizing various module detection methods can be challenging due to the fine boundaries between them; for instance, matrix decomposition is an intermediate step in some clustering and biclustering algorithms. These methods are generally grouped into three categories: clustering, decomposition, and biclustering methods [106]. Clustering methods, among the oldest and still widely used, group genes based on their overall similarity in gene expression. Commonly applied clustering algorithms for module detection include K-medoids [107], K-means [108], Fuzzy-c-means [109], MCL [110], agglomerative hierarchical clustering [111], and WGCNA [46]. Decomposition methods aim to approximate the expression matrix through the product of smaller matrices. Within this approximation, two matrices hold the individual contributions of genes and samples to specific modules. By limiting the extent to which samples can contribute to a module, decomposition methods are effective at detecting local coexpression patterns. Commonly employed decomposition-based methods encompass Independent Component Analysis (ICA) [112], Principal Component Analysis (PCA) [113], and hybrid approaches that merge elements of both techniques [114]. Biclustering methods excel at identifying clusters of genes and samples that exhibit localized coexpression exclusively within the confines of the bicluster. Unlike decomposition methods, where all samples contribute to some extent, biclustering categorizes samples as either contributing or not to a specific module. Consequently, modules unveiled by biclustering methods are often more interpretable as they provide a more precise delineation of the exact source of local coexpression. In some instances, a biclustering approach can be considered an extension of an existing decomposition method, but with the added requirement that gene and sample contributions to a module are sparse, containing numerous zeros. Spectral biclustering [115], Iterative Signature Algorithm (ISA) [116], QUalitative BIClustering algorithm (QUBIC) [117], Bi-Force [118], and Factor Analysis for BIcluster Acquisition (FABIA) [119] are some of the commonly utilized biclustering techniques.

5.3 Identification of Key Drivers

Identifying key drivers within coexpression or protein interaction networks is a fundamental task in systems biology [56]. These identified key drivers within the network often correspond to proteins with critical roles in various pathways [120], or they may code for transcription factors that regulate gene expression [121]. Thus are crucial for understanding the underlying regulatory mechanisms and functional roles of genes or proteins. Several methodologies are employed for this purpose, ranging from straightforward topological metrics to more sophisticated statistical analyses.

5.3.1 Centrality-Based Methods

One of the simplest approaches involves using topological metrics to assess the importance of nodes within the network [122]. These metrics evaluate characteristics such as degree centrality [123], betweenness centrality [124], and closeness centrality [125], which respectively measure a node's connectedness, its position in facilitating information flow, and its proximity to other nodes. Nodes with high values of these metrics are considered key drivers, as they play pivotal roles in network connectivity and communication [122]. There exists a diverse array of centrality measures, numbering over 200 [126], each meticulously crafted to capture distinct facets of network topology. Consequently, a fusion of various centrality measures has been employed to unveil nodes of significance within networks. This amalgamation enables a more comprehensive understanding of node importance by considering a spectrum of network characteristics [127, 128]. For instance, by generating numerous realizations of these random networks, researchers compute topological metrics for the original network and the random ensembles. Statistical significance scores, often expressed as z-scores, are then calculated for each node within the original network. Nodes with associated *p-values* less than a predefined threshold, typically 0.01, are considered key nodes or drivers [96].

5.3.2 Diffusion-Based Methods

Diffusion-based methods for node prioritization are methods that simulate the spread of information or influence along paths within the network. They prioritize nodes based on their ability to propagate information to other nodes. Random-walk-based methods stand as a prominent class of diffusion-based node prioritization techniques, offering a means to gauge node importance in a network [129]. These methods leverage random walks, stochastic processes that emulate the meandering path of a random surfer navigating a graph. The surfer initiates the journey from a source node and, at each step, selects one of its neighbors to visit randomly. The likelihood of choosing a particular node hinges on the network's structural properties and edge weights. Node ranking in random-walk-based methods is contingent upon the probability of being visited by the random surfer, which effectively represents a node's proximity or significance concerning the source node.

Examples of such methods include Random Walk with Restart (RWR) [130], which gauges the likelihood of reaching each node from the source node through random walks with a restart probability. Personalized PageRank [131, 132], akin to RWR but customized to a specific set of query nodes, evaluates node importance based on their closeness to the query nodes. Diffusion Component Analysis (DCA) [133] identifies influential nodes by assessing their contributions to the diffusion process and utilizes matrix decomposition techniques to unearth significant nodes in information propagation.

5.3.3 Subgraph-Based Methods

Subgraph-based node prioritization methods are grounded in using subgraphs as fundamental units for network analysis and representation. These techniques assess the importance of nodes based on their roles and memberships within these subgraphs, which represent coherent structures in the network, such as communities, clusters, or motifs. Subgraph-based approaches possess the flexibility to capture both local and global patterns within the network, thereby excelling in the identification of critical nodes that facilitate connections both between subgraphs and within them. Common subgraph- based methodologies encompass module detection [106, 134], which identifies functionally related node groups, motif discovery [135, 136], which detects recurring structural patterns, community detection [137], which pinpoints densely connected node groups, and subgraph embedding, which maps nodes into vector spaces considering their subgraph structures [138, 139].

6 Multistep Protocol for Interaction Network Construction and Analysis

1. **Data Preprocessing:** Begin by collecting gene expression data from relevant experiments or databases. Clean this data and remove outliers and noise to ensure data integrity. Following this, apply normalization to reconcile the data, accounting for sample variations and setting the stage for consistent analysis (*see* Subheading 2 for details).
2. **Network Construction:** Follow one of the three abovementioned methodologies in Subheading 4 to construct the network. For association-based methods, calculate correlations or mutual information scores between gene expressions. Apply Gaussian graphical models or the Bayesian approach to infer the probabilistic network structure. Additionally, integrate protein–protein interaction data through interologous method, mapping gene expression onto the interactome to construct a comprehensive network.

3. **Reliability Assessment of Predicted Interactions:** Cross-validate the constructed network against known biological interactions or experimental data for reliability. Apply topological and biological function-based statistical significance assessment methods mentioned in Subheading 5.1 to ensure that predictions are significant.
4. **Network Visualization:** Use network visualization tools like Cytoscape [140] or Gephi [141] to visualize and interpret the network effectively. Highlight nodes based on expression levels, using distinct shapes or colors for key nodes or modules. This step ensures that the intricate network structures are not only analyzed but also visually accessible and understandable.
5. **Results Interpretation and Conclusion:** Now, synthesize and interpret the findings from network analysis. Summarize the modules (*see* Subheading 5.2), key nodes (*see* Subheading 5.3), and enriched functions, and discuss their biological implications and significance. This step provides a comprehensive understanding of the biological insights from the constructed and analyzed network.

7 Summary and Future Prospectives

In summary, this chapter has delved into the realm of network inference from transcriptomic data, highlighting the pivotal role of two primary datasets: microarrays and RNA-sequencing (RNA--seq). While microarrays were once the go-to technology, RNA-seq has surged ahead due to its increased sophistication and versatility. The construction of coexpression networks and interologous networks of contigs was thoroughly explored, offering insights into how these networks can reveal intricate biological relationships. Moreover, network analysis emerged as a central theme, encompassing the assessment of interaction significance, identifying functional modules, and discovering essential driver genes. These analytical processes shed light on the complex interplay within biological systems and provide a deeper understanding of gene regulation and function. Looking ahead, the field of network inference from transcriptomic data holds promising prospects. With the continuous advancements in high-throughput sequencing technologies, we can anticipate even more comprehensive and precise transcriptomic datasets. Integrative multiomics approaches that combine multiple data types, such as genomics and proteomics, are likely to become increasingly prevalent, enabling a more holistic understanding of cellular processes. Furthermore, the development of novel algorithms and computational techniques will play a pivotal role in refining network inference and analysis. These innovations will empower researchers to extract deeper insights from complex biological data, unraveling the intricacies of gene regulatory networks and contributing to breakthroughs in fields like systems biology and personalized medicine. In conclusion, as we navigate the ever-evolving landscape of transcriptomic data analysis, the future of network inference holds immense potential for uncovering the mysteries of cellular dynamics and advancing our knowledge of biological systems.

Acknowledgements

VS(first author) thanks Council of Scientific and Industrial Research (CSIR), India, for providing Senior Research Fellowship (SRF).

Funding:

Authors received no specific funding for this research work.

Authors Contributions:

VS(second author) conceptualized and designed the research framework. VS(first author) performed the computational experiments. VS(first author) and VS(second author) analyzed the data and interpreted results. VS(first author) and VS(second author) wrote and finalized the manuscript.

Competing Interests

The authors declare that they have no conflict.

Data and materials availability: NA.

References

1. Crick F (1970) Central dogma of molecular biology. Nature 227(5258):561–563
2. Bustamante C, Cheng W, Mejia YX (2011) Revisiting the central dogma one molecule at a time. Cell 144(4):480–497
3. Jacob F, Monod J (1961) Genetic regulatory mechanisms in the synthesis of proteins. J Mol Biol 3(3):318–356
4. Stark BC, Kole R, Bowman EJ, Altman S (1978) Ribonuclease P: an enzyme with an essential RNA component. Proc Natl Acad Sci 75(8):3717–3721
5. Morris KV, Mattick JS (2014) The rise of regulatory RNA. Nat Rev Genet 15(6): 423–437
6. Siva N (2008) 1000 Genomes project. Nat Biotechnol 26(3):256–257
7. Segal E, Battle A, Koller D (2002) Decomposing gene expression into cellular processes. In: Biocomputing 2003. World Scientific, Singapore, pp 89–100
8. Lowe R, Shirley N, Bleackley M, Dolan S, Shafee T (2017) Transcriptomics technologies. PLoS Comput Biol 13(5):e1005457
9. Emilsson V, Thorleifsson G, Zhang B, Leonardson AS, Zink F, Zhu J, Carlson S, Helgason A, Walters GB, Gunnarsdottir S et al (2008) Genetics of gene expression and its effect on disease. Nature 452(7186): 423–428
10. Maloy S, Stewart V (1993) Autogenous regulation of gene expression. J Bacteriol 175(2): 307–316
11. Maniatis T, Goodbourn S, Fischer JA (1987) Regulation of inducible and tissue-specific gene expression. Science 236(4806): 1237–1245
12. Killary A, Fournier R (1984) A genetic analysis of extinction: trans-dominant loci regulate expression of liver-specific traits in hepatoma hybrid cells. Cell 38(2):523–534
13. Wen X, Fuhrman S, Michaels GS, Carr DB, Smith S, Barker JL, Somogyi R (1998) Large-scale temporal gene expression mapping of central nervous system development. Proc Natl Acad Sci 95(1):334–339
14. Carninci P, Kasukawa T, Katayama S, Gough J, Frith M, Maeda N, Oyama R, Ravasi T, Lenhard B, Wells C et al (2005)

The transcriptional landscape of the mammalian genome. Science 309(5740):1559–1563
15. Jacquier A (2009) The complex eukaryotic transcriptome: unexpected pervasive transcription and novel small RNAs. Nat Rev Genet 10(12):833–844
16. Marra MA, Hillier L, Waterston RH (1998) Expressed sequence tags—ESTablishing bridges between genomes. Trends Genet 14(1):4–7
17. Sanger F, Nicklen S, Coulson AR (1977) DNA sequencing with chain-terminating inhibitors. Proc Natl Acad Sci 74(12): 5463–5467
18. Velculescu VE, Zhang L, Vogelstein B, Kinzler KW (1995) Serial analysis of gene expression. Science 270(5235):484–487
19. Schena M, Shalon D, Davis RW, Brown PO (1995) Quantitative monitoring of gene expression patterns with a complementary DNA microarray. Science 270(5235): 467–470
20. Ozsolak F, Milos PM (2011) RNA sequencing: advances, challenges and opportunities. Nat Rev Genet 12(2):87–98
21. Shalon D, Smith SJ, Brown PO (1996) A DNA microarray system for analyzing complex DNA samples using two-color fluorescent probe hybridization. Genome Res 6(7): 639–645
22. Lockhart DJ, Dong H, Byrne MC, Follettie MT, Gallo MV, Chee MS, Mittmann M, Wang C, Kobayashi M, Norton H et al (1996) Expression monitoring by hybridization to high-density oligonucleotide arrays. Nat Biotechnol 14(13):1675–1680
23. Wang Z, Gerstein M, Snyder M (2009) RNA-Seq: a revolutionary tool for transcriptomics. Nat Rev Genet 10(1):57–63
24. Van Verk MC, Hickman R, Pieterse CM, Van Wees SC (2013) RNA-Seq: revelation of the messengers. Trends Plant Sci 18(4):175–179
25. Huber W, Carey VJ, Gentleman R, Anders S, Carlson M, Carvalho BS, Bravo HC, Davis S, Gatto L, Girke T et al (2015) Orchestrating high-throughput genomic analysis with Bioconductor. Nat Methods 12(2):115–121
26. Amezquita RA, Lun AT, Becht E, Carey VJ, Carpp LN, Geistlinger L, Marini F, Rue-Albrecht K, Risso D, Soneson C et al (2020) Orchestrating single-cell analysis with Bioconductor. Nat Methods 17(2):137–145
27. MacMahon JA, Phillips DL, Robinson JV, Schimpf DJ (1978) Levels of biological organization: an organism-centered approach. Bioscience 28(11):700–704
28. Newman ME (2003) The structure and function of complex networks. SIAM Rev 45(2): 167–256
29. Albert R, Barabási AL (2002) Statistical mechanics of complex networks. Rev Mod Phys 74(1):47
30. Bascompte J (2009) Disentangling the web of life. Science 325(5939):416–419
31. Gardner TS, Di Bernardo D, Lorenz D, Collins JJ (2003) Inferring genetic networks and identifying compound mode of action via expression profiling. Science 301(5629): 102–105
32. D'haeseleer P, Liang S, Somogyi R (2000) Genetic network inference: from co-expression clustering to reverse engineering. Bioinformatics 16(8):707–726
33. Wang YR, Huang H (2014) Review on statistical methods for gene network reconstruction using expression data. J Theor Biol 362: 53–61
34. Friedman N, Linial M, Nachman I, Pe'er D (2000) Using Bayesian networks to analyze expression data. In: Proceedings of the Fourth Annual International Conference on Computational Molecular Biology, pp 127–135
35. Horvath S, Dong J (2008) Geometric interpretation of gene coexpression network analysis. PLoS Comput Biol 4(8):e1000117
36. Estrada E (2012) The structure of complex networks: theory and applications. Oxford University Press, Oxford
37. Assenov Y, Ramírez F, Schelhorn SE, Lengauer T, Albrecht M (2008) Computing topological parameters of biological networks. Bioinformatics 24(2):282–284
38. Wang L, Feng Z, Wang X, Wang X, Zhang X (2010) DEGseq: an R package for identifying differentially expressed genes from RNA-seq data. Bioinformatics 26(1):136–138
39. Kvam VM, Liu P, Si Y (2012) A comparison of statistical methods for detecting differentially expressed genes from RNA-seq data. Am J Bot 99(2):248–256
40. Slonim DK (2002) From patterns to pathways: gene expression data analysis comes of age. Nat Genet 32(4):502–508
41. Jain AK, Dubes RC (1988) Algorithms for clustering data. Prentice-Hall, London
42. Raychaudhuri S, Stuart JM, Altman RB (2000) Principal components analysis to summarize microarray experiments: application to sporulation time series. In: Biocomputing 2000. World Scientific, Singapore, pp 455–466

43. Pellegrini M, Haynor D, Johnson JM (2004) Protein interaction networks. Expert Rev Proteomics 1(2):239–249
44. Kanehisa M, Goto S (2000) KEGG: Kyoto encyclopedia of genes and genomes. Nucleic Acids Res 28(1):27–30
45. Van Dam S, Vosa U, van der Graaf A, Franke L, de Magalhaes JP (2018) Gene co-expression analysis for functional classification and gene–disease predictions. Briefings Bioinform 19(4):575–592
46. Langfelder P, Horvath S (2008) WGCNA: an R package for weighted correlation network analysis. BMC Bioinform 9(1):1–13
47. Pearson K (1895) VII. Note on regression and inheritance in the case of two parents. Proc R Soc Lond 58(347-352):240–242
48. Kendall MG (1938) A new measure of rank correlation. Biometrika 30(1/2):81–93
49. Spearman C (1961) The proof and measurement of association between two things.
50. Song L, Langfelder P, Horvath S (2012) Comparison of co-expression measures: mutual information, correlation, and model based indices. BMC Bioinform 13(1):1–21
51. Kumari S, Nie J, Chen HS, Ma H, Stewart R, Li X, Lu MZ, Taylor WM, Wei H (2012) Evaluation of gene association methods for coexpression network construction and biological knowledge discovery. PLoS One 7(11):e50411
52. Butte AJ, Kohane IS (1999) Mutual information relevance networks: functional genomic clustering using pairwise entropy measurements. In: Biocomputing 2000. World Scientific, Singapore, pp 418–429
53. López-Rozo N, Romero M, Finke J, Rocha C (2022) A Network-based Approach for Inferring Thresholds in Co-expression Networks. In: International Conference on Complex Networks and Their Applications. Springer, Berlin, pp 265–276
54. Paci P, Colombo T, Fiscon G, Gurtner A, Pavesi G, Farina L (2017) SWIM: a computational tool to unveiling crucial nodes in complex biological networks. Sci Rep 7(1):44797
55. Butte AJ, Ye J, Häring H, Stumvoll M, White M, Kohane I (2000) Determining significant fold differences in gene expression analysis. In: Biocomputing 2001. World Scientific, Singapore, pp 6–17
56. Carter SL, Brechbühler CM, Griffin M, Bond AT (2004) Gene co-expression network topology provides a framework for molecular characterization of cellular state. Bioinformatics 20(14):2242–2250
57. Zhang B, Horvath S (2005) A general framework for weighted gene co-expression network analysis. Stat Appl Genet Mol Biol 4(1):1–45
58. Schäfer J, Strimmer K (2005) An empirical Bayes approach to inferring large-scale gene association networks. Bioinformatics 21(6): 754–764
59. Edwards D (2012) Introduction to graphical modelling. Springer Science & Business Media
60. Friedman J, Hastie T, Tibshirani R (2008) Sparse inverse covariance estimation with the graphical lasso. Biostatistics 9(3):432–441
61. Pearl J (1988) Probabilistic reasoning in intelligent systems: networks of plausible inference. Morgan Kaufmann, Los Altos
62. Needham CJ, Bradford JR, Bulpitt AJ, Westhead DR (2007) A primer on learning in Bayesian networks for computational biology. PLoS Comput Biol 3(8):e129
63. Yu J, Smith VA, Wang PP, Hartemink AJ, Jarvis ED (2002) Using Bayesian network inference algorithms to recover molecular genetic regulatory networks. In: International Conference on Systems Biology, vol 2002
64. Yoo C, Thorsson V, Cooper GF (2001) Discovery of causal relationships in a gene-regulation pathway from a mixture of experimental and observational DNA microarray data. In: Biocomputing 2002. World Scientific, Singapore, pp 498–509
65. Cooper GF, Herskovits E (1992) A Bayesian method for the induction of probabilistic networks from data. Mach Learn 9:309–347
66. Singh G, Singh V, Singh V (2019) Construction and analysis of an interologous protein–protein interaction network of Camellia sinensis leaf (TeaLIPIN) from RNA-Seq data sets. Plant Cell Rep 38:1249–1262
67. Matthews LR, Vaglio P, Reboul J, Ge H, Davis BP, Garrels J, Vincent S, Vidal M (2001) Identification of potential interaction networks using sequence-based searches for conserved protein-protein interactions or "interologs". Genome Res 11(12): 2120–2126
68. Fitch WM (1970) Distinguishing homologous from analogous proteins. Systematic Zool 19(2):99–113
69. Kuzniar A, van Ham RC, Pongor S, Leunissen JA (2008) The quest for orthologs: finding the corresponding gene across genomes. Trends Genet 24(11):539–551
70. Kristensen DM, Wolf YI, Mushegian AR, Koonin EV (2011) Computational methods

for Gene Orthology inference. Briefings Bioinform 12(5):379–391

71. Tekaia F (2016) Inferring orthologs: open questions and perspectives. Genom Insights 9:GEI–S37925
72. Tatusov RL, Koonin EV, Lipman DJ (1997) A genomic perspective on protein families. Science 278(5338):631–637
73. Overbeek R, Fonstein M, D'souza M, Pusch GD, Maltsev N (1999) The use of gene clusters to infer functional coupling. Proc Natl Acad Sci 96(6):2896–2901
74. Wall D, Fraser H, Hirsh A (2003) Detecting putative orthologs. Bioinformatics 19(13): 1710–1711
75. Remm M, Storm CE, Sonnhammer EL (2001) Automatic clustering of orthologs and in-paralogs from pairwise species comparisons. J Mol Biol 314(5):1041–1052
76. O'brien KP, Remm M, Sonnhammer EL (2005) Inparanoid: a comprehensive database of eukaryotic orthologs. Nucleic Acids Res 33 (suppl_1):D476–D480
77. Li L, Stoeckert CJ, Roos DS (2003) OrthoMCL: identification of ortholog groups for eukaryotic genomes. Genome Res 13(9): 2178–2189
78. Waterhouse RM, Zdobnov EM, Tegenfeldt F, Li J, Kriventseva EV (2011) OrthoDB: the hierarchical catalog of eukaryotic orthologs in 2011. Nucleic Acids Res 39(suppl_1): D283–D288
79. Jensen LJ, Julien P, Kuhn M, von Mering C, Muller J, Doerks T, Bork P (2007) eggNOG: automated construction and annotation of orthologous groups of genes. Nucleic Acids Res 36(suppl_1):D250–D254
80. Schreiber F, Sonnhammer EL (2013) Hieranoid: hierarchical orthology inference. J Mol Biol 425(11):2072–2081
81. Train CM, Glover NM, Gonnet GH, Altenhoff AM, Dessimoz C (2017) Orthologous Matrix (OMA) algorithm 2.0: more robust to asymmetric evolutionary rates and more scalable hierarchical orthologous group inference. Bioinformatics 33(14):i75–i82
82. Jothi R, Zotenko E, Tasneem A, Przytycka TM (2006) COCO-CL: hierarchical clustering of homology relations based on evolutionary correlations. Bioinformatics 22(7): 779–788
83. Goodman M, Czelusniak J, Moore GW, Romero-Herrera AE, Matsuda G (1979) Fitting the gene lineage into its species lineage, a parsimony strategy illustrated by cladograms constructed from globin sequences. Systematic Biol 28(2):132–163
84. Zmasek CM, Eddy SR (2002) RIO: analyzing proteomes by automated phylogenomics using resampled inference of orthologs. BMC Bioinform 3(1):1–19
85. Storm CE, Sonnhammer EL (2002) Automated ortholog inference from phylogenetic trees and calculation of orthology reliability. Bioinformatics 18(1):92–99
86. Huerta-Cepas J, Capella-Gutierrez S, Pryszcz LP, Marcet-Houben M, Gabaldon T (2014) PhylomeDB v4: zooming into the plurality of evolutionary histories of a genome. Nucleic Acids Res 42(D1):D897–D902
87. Hu Y, Flockhart I, Vinayagam A, Bergwitz C, Berger B, Perrimon N, Mohr SE (2011) An integrative approach to ortholog prediction for disease-focused and other functional studies. BMC Bioinform 12:1–16
88. Contreras-Moreira B, Vinuesa P (2013) GET_HOMOLOGUES, a versatile software package for scalable and robust microbial pangenome analysis. Appl Environ Microbiol 79(24):7696–7701
89. Salgado D, Gimenez G, Coulier F, Marcelle C (2008) COMPARE, a multi-organism system for cross-species data comparison and transfer of information. Bioinformatics 24(3): 447–449
90. Eyre TA, Wright MW, Lush MJ, Bruford EA (2007) HCOP: a searchable database of human orthology predictions. Briefings Bioinform 8(1):2–5
91. Maher MC, Hernandez RD (2015) Rock, paper, scissors: harnessing complementarity in ortholog detection methods improves comparative genomic inference. G3: Genes, Genomes Genet 5(4):629–638
92. Pereira C, Denise A, Lespinet O (2014) A meta-approach for improving the prediction and the functional annotation of ortholog groups. BMC Genom 15(6):1–8
93. Altenhoff AM, Gil M, Gonnet GH, Dessimoz C (2013) Inferring hierarchical orthologous groups from orthologous gene pairs. Plos One 8(1):e53786
94. Emms DM, Kelly S (2019) OrthoFinder: phylogenetic orthology inference for comparative genomics. Genome Biol 20:1–14
95. Sutphin GL, Mahoney JM, Sheppard K, Walton DO, Korstanje R (2016) WORMHOLE: novel least diverged ortholog prediction through machine learning. PLoS Comput Biol 12(11):e1005182
96. Singh V, Singh G, Singh V (2019) TulsiPIN: an interologous protein interactome of Ocimum tenuiflorum. J Proteome Res 19(2): 884–899

97. Oughtred R, Stark C, Breitkreutz BJ, Rust J, Boucher L, Chang C, Kolas N, O'Donnell L, Leung G, McAdam R, et al. (2019) The BioGRID interaction database: 2019 update. Nucleic Acids Res 47(D1):D529–D541
98. Kerrien S, Aranda B, Breuza L, Bridge A, Broackes-Carter F, Chen C, Duesbury M, Dumousseau M, Feuermann M, Hinz U, et al. (2012) The IntAct molecular interaction database in 2012. Nucleic Acids Res 40(D1): D841–D846
99. Szklarczyk D, Gable AL, Nastou KC, Lyon D, Kirsch R, Pyysalo S, Doncheva NT, Legeay M, Fang T, Bork P et al (2021) The STRING database in 2021: customizable protein–protein networks, and functional characterization of user-uploaded gene/measurement sets. Nucleic Acids Res 49(D1):D605–D612
100. Licata L, Briganti L, Peluso D, Perfetto L, Iannuccelli M, Galeota E, Sacco F, Palma A, Nardozza AP, Santonico E, et al. (2012) MINT, the molecular interaction database: 2012 update. Nucleic Acids Res 40(D1): D857–D861
101. Peng X, Wang J, Peng W, Wu FX, Pan Y (2017) Protein–protein interactions: detection, reliability assessment and applications. Briefings Bioinform 18(5):798–819
102. Erdős P, Rényi A et al (1960) On the evolution of random graphs. Publ Math Inst Hung Acad Sci 5(1):17–60
103. Newman M (2018) Networks. Oxford University Press, Oxford
104. Obayashi T, Hibara H, Kagaya Y, Aoki Y, Kinoshita K (2022) ATTED-II v11: a plant gene coexpression database using a sample balancing technique by subagging of principal components. Plant Cell Physiol 63(6): 869–881
105. Yu G, Li F, Qin Y, Bo X, Wu Y, Wang S (2010) GOSemSim: an R package for measuring semantic similarity among GO terms and gene products. Bioinformatics 26(7): 976–978
106. Saelens W, Cannoodt R, Saeys Y (2018) A comprehensive evaluation of module detection methods for gene expression data. Nat Commun 9(1):1090
107. Schubert E, Rousseeuw PJ (2021) Fast and eager k-medoids clustering: O (k) runtime improvement of the PAM, CLARA, and CLARANS algorithms. Inform Syst 101: 101804
108. Hartigan JA, Wong MA (1979) Algorithm AS 136: a k-means clustering algorithm. J R Stat Soc Series C 28(1):100–108
109. Bezdek JC, Ehrlich R, Full W (1984) FCM: The fuzzy c-means clustering algorithm. Comput Geosci 10(2-3):191–203
110. Van Dongen SM (2000) Graph clustering by flow simulation. PhD thesis
111. Kaufman L, Rousseeuw PJ (2009) Finding groups in data: an introduction to cluster analysis. John Wiley & Sons, London
112. Hyvarinen A (1999) Fast and robust fixed-point algorithms for independent component analysis. IEEE Trans Neural Netw 10(3): 626–634
113. Pedregosa F, Varoquaux G, Gramfort A, Michel V, Thirion B, Grisel O, Blondel M, Prettenhofer P, Weiss R, Dubourg V et al (2011) Scikit-learn: machine learning in Python. J Mach Learn Res 12:2825–2830
114. Yao F, Coquery J, Lê Cao KA (2012) Independent principal component analysis for biologically meaningful dimension reduction of large biological data sets. BMC Bioinform 13: 1–15
115. Kluger Y, Basri R, Chang JT, Gerstein M (2003) Spectral biclustering of microarray data: coclustering genes and conditions. Genome Res 13(4):703–716
116. Ihmels J, Friedlander G, Bergmann S, Sarig O, Ziv Y, Barkai N (2002) Revealing modular organization in the yeast transcriptional network. Nat Genet 31(4):370–377
117. Li G, Ma Q, Tang H, Paterson AH, Xu Y (2009) QUBIC: a qualitative biclustering algorithm for analyses of gene expression data. Nucleic Acids Res 37(15):e101–e101
118. Sun P, Speicher NK, Röttger R, Guo J, Baumbach J (2014) Bi-Force: large-scale bicluster editing and its application to gene expression data biclustering. Nucleic Acids Res 42(9): e78–e78
119. Hochreiter S, Bodenhofer U, Heusel M, Mayr A, Mitterecker A, Kasim A, Khamiakova T, Van Sanden S, Lin D, Talloen W, et al. (2010) FABIA: factor analysis for bicluster acquisition. Bioinformatics 26(12):1520–1527
120. Wisecaver JH, Borowsky AT, Tzin V, Jander G, Kliebenstein DJ, Rokas A (2017) A global coexpression network approach for connecting genes to specialized metabolic pathways in plants. Plant Cell 29(5):944–959
121. Haynes BC, Maier EJ, Kramer MH, Wang PI, Brown H, Brent MR (2013) Mapping functional transcription factor networks from gene expression data. Genome Res 23(8): 1319–1328

122. Koschützki D, Schreiber F (2008) Centrality analysis methods for biological networks and their application to gene regulatory networks. Gene Regul Syst Biol 2:GRSB–S702
123. Jeong H, Mason SP, Barabási AL, Oltvai ZN (2001) Lethality and centrality in protein networks. Nature 411(6833):41–42
124. Joy MP, Brock A, Ingber DE, Huang S (2005) High-betweenness proteins in the yeast protein interaction network. J Biomed Biotechnol 2005(2):96
125. Wuchty S (2002) Interaction and domain networks of yeast. Proteomics 2(12):1715–1723
126. Jalili M, Salehzadeh-Yazdi A, Asgari Y, Arab SS, Yaghmaie M, Ghavamzadeh A, Alimoghaddam K (2015) CentiServer: a comprehensive resource, web-based application and R package for centrality analysis. PLoS One 10(11):e0143111
127. Ma HW, Zeng AP (2003) The connectivity structure, giant strong component and centrality of metabolic networks. Bioinformatics 19(11):1423–1430
128. Lareau CA, White BC, Oberg AL, McKinney BA (2015) Differential co-expression network centrality and machine learning feature selection for identifying susceptibility hubs in networks with scale-free structure. Biodata Mining 8:1–17
129. Ata SK, Wu M, Fang Y, Ou-Yang L, Kwoh CK, Li XL (2021) Recent advances in network-based methods for disease gene prediction. Briefings Bioinform 22(4):bbaa303
130. Köhler S, Bauer S, Horn D, Robinson PN (2008) Walking the interactome for prioritization of candidate disease genes. Am J Hum Genet 82(4):949–958
131. Brin S (1998) The PageRank citation ranking: bringing order to the web. Proc Asis 1998 98: 161–172
132. Wang H, Wei Z, Gan J, Wang S, Huang Z (2020) Personalized pagerank to a target node, revisited. In: Proceedings of the 26th ACM SIGKDD International Conference on Knowledge Discovery & Data Mining, pp 657–667
133. Cho H, Berger B, Peng J (2015) Diffusion component analysis: unraveling functional topology in biological networks. In: International Conference on Research in Computational Molecular Biology, Springer, Berlin, pp 62–64
134. Ideker T, Ozier O, Schwikowski B, Siegel AF (2002) Discovering regulatory and signalling circuits in molecular interaction networks. Bioinformatics 18(suppl_1):S233–S240
135. Milo R, Shen-Orr S, Itzkovitz S, Kashtan N, Chklovskii D, Alon U (2002) Network motifs: simple building blocks of complex networks. Science 298(5594):824–827
136. Pržulj N, Corneil DG, Jurisica I (2004) Modeling interactome: scale-free or geometric? Bioinformatics 20(18):3508–3515
137. Newman ME, Girvan M (2004) Finding and evaluating community structure in networks. Phys Rev E 69(2):026113
138. Alsentzer E, Finlayson S, Li M, Zitnik M (2020) Subgraph neural networks. Adv Neural Inform Process Syst 33:8017–8029
139. Adhikari B, Zhang Y, Ramakrishnan N, Prakash BA (2018) Sub2vec: feature learning for subgraphs. In: Advances in Knowledge Discovery and Data Mining: 22nd Pacific-Asia Conference, PAKDD 2018, Melbourne, VIC, Australia, June 3-6, 2018, Proceedings, Part II 22. Springer, Berlin, pp 170–182
140. Shannon P, Markiel A, Ozier O, Baliga NS, Wang JT, Ramage D, Amin N, Schwikowski B, Ideker T (2003) Cytoscape: a software environment for integrated models of biomolecular interaction networks. Genome Res 13(11):2498–2504
141. Bastian M, Heymann S, Jacomy M (2009) Gephi: an open source software for exploring and manipulating networks. In: Proceedings of the International AAAI Conference on Web and Social Media, vol 3, pp 361–362

Chapter 3

EMPathways2: Estimation of Enzyme Expression and Metabolic Pathway Activity Using RNA-Seq Reads

Filipp Rondel, Hafsa Farooq, Roya Hosseini, Mark Grinshpon, and Alex Zelikovsky

Abstract

In this chapter, we outline an approach to analyzing metatranscriptomic data, focusing on the assessment of differential enzyme expression and metabolic pathway activities using a novel bioinformatics software tool, EMPathways2. The analysis pipeline commences with raw data originating from a sequencer and concludes with an output of enzyme expressions and an estimate of metabolic pathway activities. The initial step involves aligning specific transcriptomes assembled from RNA-Seq data using Bowtie2 and acquiring gene expression data with IsoEM2. Subsequently, the pipeline proceeds to quality assessment and preprocessing of the input data, ensuring accurate estimates of enzymes and their differential regulation. Upon completion of the preprocessing stage, EMPathways2 is employed to decipher the intricate relationships between genes, enzymes, and pathways. An online repository containing sample data has been made available, alongside custom Python scripts designed to modify the output of the programs within the pipeline for diverse downstream analyses. This chapter highlights the technical aspects and practical applications of using EMPathways2, which facilitates the advancement of transcriptome data analysis and contributes to a deeper understanding of the complex regulatory mechanisms underlying living systems.

Key words RNA-Seq data analysis, Alignment, Enzyme expression, Metabolic pathway expression analysis, Metagenomics

1 Introduction

Understanding metabolic pathways is crucial for elucidating the complex regulatory mechanisms underlying cellular functions, as these pathways represent the interconnected series of biochemical reactions that maintain and modulate the dynamic balance of a living system. In this chapter, we present a comprehensive guide to EMPathways2, a novel bioinformatics software tool designed to unravel the mysteries of metabolic pathways by analyzing metatranscriptomic data and quantifying differential enzyme expressions.

Metabolic pathways play a vital role in cellular processes such as energy production, biosynthesis of biomolecules, and

Rajeev K. Azad (ed.), *Transcriptome Data Analysis*, Methods in Molecular Biology, vol. 2812,
https://doi.org/10.1007/978-1-0716-3886-6_3,

detoxification [1–5]. Accurate assessment of differential expression in these pathways provides invaluable insights into an organism's response to environmental changes, gene regulation, and potential therapeutic targets for various diseases. Traditional methods, such as Gene Set Enrichment Analysis (GSEA), have been widely employed to estimate gene expression and activities under multiple conditions [6]. However, GSEA has limitations in its ability to estimate activity levels, offering only a binary indication of gene presence or absence.

In this protocol, we describe a pipeline titled EMPathways2, which has been developed to address such limitations and offer a more comprehensive view of metabolic pathway activities. In contrast to GSEA, EMPathways2 enables biologists to calculate metabolic pathway activity levels and infer enzyme expression across multiple conditions with greater accuracy and detail. This enhanced capacity allows researchers to delve deeper into the functional roles of genes and their products within the context of the entire metabolic network, ultimately facilitating a more thorough understanding of the intricate relationships between genes and enzymes. EMPathways2 can be found in the repository referenced in Table 1.

In this chapter, we will provide a step-by-step guide on using EMPathways2 for metatranscriptomic data analysis and discuss the technical aspects of the software. By the end of this chapter, readers will gain a solid understanding of the advantages EMPathways2 offers over traditional methods like GSEA and how this powerful tool can be employed to advance the field of transcriptome data analysis.

Table 1
Software and URLs

Software	URL
DOE-JGI	`https://jgi.doe.gov/`
KEGG	`https://www.kegg.jp/`
STAR	`https://github.com/alexdobin/STAR`
HISAT2	`http://daehwankimlab.github.io/hisat2/`
IsoEM2	`https://github.com/mandricigor/isoem2`
EMPathways2	`https://github.com/frondel/EMPathways2`
SAM Format Specification	`https://samtools.github.io/hts-specs/SAMv1.pdf`
Windows Subsystem for Linux	`https://docs.microsoft.com/en-us/windows/wsl/install-win10`
Linux	`https://itsfoss.com/install-software-from-source-code/`

2 Software and Data

Before starting the tasks outlined in each section of this protocol, a few initial steps need to be accomplished. These involve acquiring the data from the designated repository and installing the required software listed in Table 1. To use the tools in this pipeline, a GNU/Linux command line environment is necessary, which can be achieved via a standalone installation or a virtual machine (VM) as long as adequate resources are allocated to the VM. For Windows users, the Windows Subsystem for Linux is a viable option, as it offers a complete GNU/Linux command line environment within Windows without the need for virtualization and resource allocation.

Common distributions, such as Ubuntu and Debian, may already include many essential tools in their repositories for effortless installation. However, these versions could be outdated, potentially leading to workflow complications. Although several programs are natively available for Mac OS X, users might have to compile them from source packages. In contrast, binaries are typically accessible for GNU/Linux.

Table 1 contains a general guide on software installation and modifying the $PATH. Nonetheless, the steps needed for compiling software from the source can differ among programs. It is advised that users download the most recent available binary and add it to the /usr/local/bin directory whenever possible.

The commands featured in this pipeline are designed for a standard desktop with four CPU cores and 8 GB of RAM. Handling larger datasets might require more RAM, and users with extra cores and memory can modify the relevant parameters by referring to the manual for each particular program.

This estimation of enzyme expression and pathway activity level requires procuring the metatranscriptomic dataset of a plankton community of the surface waters of the Northern Gulf from the repository. RNA-Seq data was sampled along with six different environmental parameters, and samples were gathered from two different depths, 2 meters and 18 meters [4]. An initial annotation of RNA-Seq data was acquired through the DOE-JGI Metagenome Annotation Pipeline (MAP v.4; JGI portal) [7]. It is available in the JGI Genome Online Database through GOLD ID Gs0110190.

3 Methods

3.1 Data Preparation and Quality Assessment

The sample dataset used to run EMPathways2 (Fig. 1) is a metatranscriptomic dataset of a plankton community from the surface waters of the Northern Gulf of Mexico. The RNA-Seq data and respective environmental parameters were sampled in July 2015 at 2 depths—2 m and 18 m—every 4 h throughout 48 h totaling in 13 samples per depth. Six environmental parameters—including PAR (photosynthetic active radiation) and seawater dissolved oxygen concentration, density, salinity, temperature, and chlorophyll concentration—were measured for each sample. Out of 26 samples, 4 samples (Day 1, 12:00, 18 m; Day 2, 20:00, 2 m; Day 3, 08:00, 2 m; Day 4, 12:00, 18 m) were discarded as they did not contain enough reads.

The MAP pipeline comprises feature prediction, including the identification of protein-coding genes. Initially, the MEGAHIT metagenome assembler assembles RNA-Seq reads into scaffolds [8]. Subsequently, various software suites (GeneMark.hmm, MetaGeneAnnotator, Prodigal, and FragGeneScan) predict genes on the assembled scaffolds [9–12]. The MAP pipeline employs enzyme commission (EC) numbers to annotate genes, which are a necessary

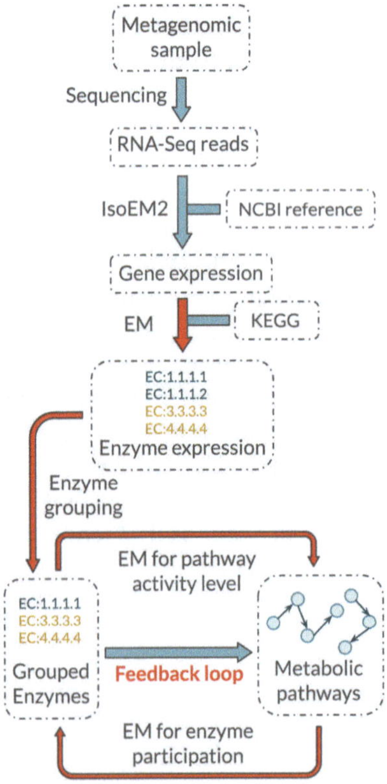

Fig. 1 EMPathways2 pipeline

input for the model. These annotations are acquired through homology searches (using USEARCH) within a nonredundant protein-sequence database (maxhits = 50, e-value = 0.1), where each protein is assigned to a KEGG Orthology (KO) group. For each KO, the top five hits, provided that the identity score is at least 30% and 70% of the protein length is matched, are utilized. The KO IDs are then converted into EC numbers using the KEGG KO to EC mapping [13].

3.2 Align RNA-Seq Read

The first step in our pipeline involves aligning RNA-Seq reads to a reference genome using an aligner tool such as HISAT2 or STAR [14, 15]. In this particular case, we went with STAR. The command for this process is as follows:

```
$ mkdir STAR_Output
$ STAR --runThreadN NumberOfThreads --genomeDir STAR_Gen-
ome_Index --readFilesCommand zcat --readFilesIn Read1.fq.gz
Read2.fq.gz --outFileNamePrefix STAR_Output/
```

In this command, `NumberOfThreads` specifies the number of threads for parallel processing, while `STAR_Genome_Index` refers to the directory containing the genome index files.

`Read1.fq.gz` and `Read2.fq.gz` denote the input read files in FASTQ format, which can be compressed using gzip. The output files will be saved in the STAR Output directory.

3.3 Produce Gene Expression Data Using IsoEM2

After the alignment step, the IsoEM2 is used to estimate gene expression levels from the aligned reads [16]. Before running IsoEM2, input files need to be prepared, which include a set of known isoforms in GTF format and a file with aligned reads in SAM format. To ensure proper analysis, the aligned reads must be sorted by read name. If you are uncertain whether your reads are sorted, you can run the following command to sort the file:

```
sort -k 1,1 aligned_reads.sam > aligned_reads_sorted.sam
```

Once the files are prepared, you can execute IsoEM2 using the command line as follows:

```
isoem2 -G genes.gtf -m 200 -d 20 aligned_reads_sorted.sam
```

In this example command, `-G genes.gtf` specifies the known genes and isoforms in GTF format, `-m 200` indicates the fragment length mean, and `-d 20` sets the fragment length standard deviation. IsoEM2 also offers various optional global and library options to fine-tune the analysis and accommodate specific sequencing techniques, such as directional sequencing, quality score-based read weighting, and disabling length normalization for data generated using 5′ or 3′ end-sequencing protocols. The output files will be saved in the specified output directory.

3.4 Calculate Enzyme and Metabolic Pathway Expression with EMPathways2

Finally, the EMPathways2 tool calculates enzyme and metabolic pathway expression based on the gene expression data obtained from IsoEM2. The command for this process is

```
$ python empathways2.py -ge gene_enzyme_file -epd enzyme_-
pathway_dictionary -gexp gene_expression_file -eo enzyme_out-
put -po pathway_output --theta_e 0.0001 --theta_p 0.0001
```

In this command, `gene_enzyme_file` is the input file containing gene–enzyme relationships, and `enzyme_pathway_dictionary` refers to the Enzyme Pathway Dictionary (EPD) file. The `gene_expression_file` contains the gene expression data, such as FPKM values, derived from IsoEM output. The optional `-eo` and `-po` flags allow users to specify output filenames for enzymes and pathways, respectively. The `--theta_e` and `--theta_p` parameters represent the convergence thresholds for enzyme and pathway calculations, with default values set to 0.000001.

EMPathways2 pipeline generates two output files as a result of the analysis: the enzyme expression output file and the pathway activity output file. These files provide valuable information on the enzyme expression levels and the metabolic pathway activity levels, respectively. The enzyme expression output file contains enzyme expression levels for each enzyme in the dataset. Each line in the file represents the expression level of a single enzyme, formatted as `enzyme_id:expression`. This file enables researchers to examine the expression levels of individual enzymes and identify those that may play crucial roles in the biological processes being studied. The pathway activity output file provides information on the activity levels of metabolic pathways. Each line in this file represents the activity level of a metabolic pathway, formatted as `pathway_id:activity` level. By analyzing these pathway activity levels, researchers can gain insights into the overall metabolic landscape of the organisms under investigation and identify key pathways that may be involved in the response to specific conditions or stimuli.

Both the enzyme and pathway IDs use KEGG nomenclature, which offers several benefits. Firstly, KEGG provides accurate and experimentally verified mappings, ensuring reliable results in the analysis. Secondly, the use of KEGG streamlines the analysis process by integrating well-established biological knowledge, allowing for easier interpretation of the results. Furthermore, KEGG nomenclature facilitates comparison and integration with other studies that utilize KEGG data, promoting consistency and reproducibility in the research community. Lastly, relying on KEGG mappings saves time and resources compared to creating custom mappings, allowing researchers to focus on the biological interpretation and potential implications of their findings.

By utilizing the EMPathways2 pipeline and its output files, researchers can gain valuable insights into the enzyme expression and metabolic pathway activity levels in their datasets, paving the way for a deeper understanding of the complex interactions and regulatory mechanisms that govern biological processes.

4 Conclusion

In this chapter we have described an improved maximum likelihood-based pipeline, EMPathways2, for the estimation of metabolic pathway activity and enzyme expression levels. This pipeline is designed to be highly adaptable and user-friendly, making it compatible with a wide range of RNA-Seq analyses. Although no software can replace the expertise of a skilled bioinformatician, this pipeline serves as an excellent starting point for researchers who are new to RNA-Seq analysis or those who want to delve deeper into the computational aspects of their work. EMPathways2 also provides a solid foundation for those interested in building their own customized pipelines.

We have optimized the pipeline for typical total mRNA-derived RNA-Seq datasets. However, it is important to recognize that certain situations may necessitate adjustments to the default parameters. EMPathways2 is particularly valuable for researchers working with organisms lacking established genomic references, a challenge frequently encountered in plant research.

Ultimately, the quality of the results produced by any pipeline or bioinformatician is contingent upon the accuracy of the experimental data. Researchers must carefully consider the experimental conditions being compared and employ appropriate pipelines for reliable and robust conclusions.

References

1. Quijano C, Trujillo M, Castro L, Trostchansky L (2016) Interplay between oxidant species and energy metabolism. Redox Biol 8:28–42
2. Peng X, Chen Z, Farshidfar Z et al (2018) Molecular characterization and clinical relevance of metabolic expression subtypes in human cancers. Cell Rep 23(1):255–269.e4
3. Gianoulis TA, Raes J, Patel PV et al (2009) Quantifying environmental adaptation of metabolic pathways in metagenomics. Proc Natl Acad Sci USA 106(5):1374–1379
4. Rondel F, Hosseini R, Sahoo B et al (2020) Estimating enzyme participation in metabolic pathways for microbial communities from RNA-seq data. In: Bioinformatics research and applications. Springer, Heidelberg, pp 335–343
5. Taverne YJ, Merkus D, Bogers AJ et al (2018) Reactive oxygen species: Radical factors in the evolution of animal life: a molecular timescale from earth's earliest history to the rise of complex life. BioEssays 40(3)
6. Clark NR, Ma'ayan A (2011) Introduction to statistical methods for analyzing large data sets: gene-set enrichment analysis. Sci Signal 4(190):tr4
7. Huntemann M, Ivanova NN, Mavromatis K et al (2016) The standard operating procedure of the DOE-JGI metagenome annotation pipeline (MAP v.4). Stand Genomic Sci 11:17

8. Li D, Liu CM, Luo R, Sadakane K, Lam TW (2015) MEGAHIT: an ultra-fast single-node solution for large and complex metagenomics assembly via succinct de Bruijn graph. Bioinformatics 31(10):1674–1676
9. Rho M, Tang H, Ye Y (2010) FragGeneScan: predicting genes in short and error-prone reads. Nucleic Acids Res 38(20):e191
10. Hyatt D, Chen GL, Locascio PF et al (2010) Prodigal: prokaryotic gene recognition and translation initiation site identification. BMC Bioinformatics 11:119
11. Noguchi H, Taniguchi T, Itoh T (2008) MetaGeneAnnotator: detecting species-specific patterns of ribosomal binding site for precise gene prediction in anonymous prokaryotic and phage genomes. DNA Res 15(6):387–396
12. Lukashin V, Borodovsky M (1998) GeneMark.hmm: new solutions for gene finding. Nucleic Acids Res 26(4):1107–1115
13. Kanehisa M (2000) KEGG: Kyoto encyclopedia of genes and genomes
14. Dobin A, Davis CA, Schlesinger F et al (2013) STAR: Ultrafast universal RNA-seq aligner. Bioinformatics 29(1):15–21
15. Kim D, Paggi JM, Park C, Bennett C, Salzberg SL (2019) Graph-based genome alignment and genotyping with HISAT2 and HISAT-genotype. Nat Biotechnol 37(8):907–915
16. Mandric I, Temate-Tiagueu Y, Shcheglova T et al (2017) Fast bootstrapping-based estimation of confidence intervals of expression levels and differential expression from RNA-Seq data. Bioinformatics 33(20):3302–3304

Chapter 4

Efficient and Powerful Integration of Targeted Metabolomics and Transcriptomics for Analyzing the Metabolism Behind Desirable Traits in Plants

Christopher R. Johnston, Cintia L. Arias, and Ana Paula Alonso

Abstract

Through current mass spectrometry methods and multiple RNA-Seq technologies, large metabolomics and transcriptomics datasets are readily obtainable, which provide a powerful and global perspective on metabolism. Indeed, one "omics" method is often not enough to draw strong conclusions about metabolism. Combining and interpreting multiple "omics" datasets remains a challenging task that requires careful statistical considerations and pre-planning. Here we describe a protocol for obtaining high-quality metabolomics and transcriptomics datasets in developing plant embryos followed by a robust approach to integration of the two. This protocol is readily adjustable and scalable to any other metabolically active organ or tissue.

Key words Multi-omics, Metabolomics, Transcriptomics, RNA-Seq, Plant metabolism, Differential expression, LC-MS/MS, Overrepresentation analysis

1 Introduction

Understanding the metabolism in any given plant organ at any given developmental stage requires the combination of multiple data types. Indeed, targeted global metabolite content (the metabolome) alone gives powerful phenotypic information but little to none on the mechanisms giving rise to such phenotypes, while global transcript content alone (the transcriptome) only provides information on gene expression that may or may not be the cause of these phenotypic differences [1]. However, metabolic states in plant tissues are better described when transcriptomics and metabolomics are properly combined, as they allow for corroboration of the metabolome composition with the gene expression that (a) directly gives rise to the actual enzymes catalyzing metabolic reactions and (b) determines a large part of the regulatory framework controlling signaling, hormone responses, and responses to external stimuli, among other processes [2]. While this

Rajeev K. Azad (ed.), *Transcriptome Data Analysis*, Methods in Molecular Biology, vol. 2812,
https://doi.org/10.1007/978-1-0716-3886-6_4, © The Author(s), 2024

methodology is broadly applicable when studying metabolism, it requires a good deal of methodological rigor to ensure the procedures are coordinated from start to finish.

First, experiments must be designed to collect enough material for both metabolite and transcript quantification, ideally from relatively homogeneous individuals at the same developmental stage under the same environmental conditions. In addition to the experimental design, careful consideration of a precise and reliable extraction method needs to take place. These methods differ for each type of "omics" experiment; while one can typically extract all RNA with only one extraction procedure for transcriptome analysis, for metabolomic analysis there is not one solvent or solvent mix to extract and dissolve all metabolites due to their great chemical diversity [3]. Often an experimental design must incorporate even more extra replicates for metabolite quantification, as a separate extraction may be required for each class of metabolite (i.e., organic acids, phosphorylated compounds, sugars and sugar alcohols, amino acids, phenolics, etc.).

Second, appropriate use of statistics is required to appropriately integrate both metabolomic and transcriptomic datasets. Transcriptomic datasets almost always contain many orders of magnitude more features than targeted metabolomics datasets, and because of this, care must be taken to ensure that transcript trends do not overpower metabolite data. This can be accomplished by using different strategies of assigning weights to each dataset but must be done so in accordance with the desired analysis [4]. Additionally, results from these analyses can be obtained to study features on a global scale (e.g., full coexpression networks, cluster analyses) or from a pathway perspective to address more specific questions.

The following protocol outlines a workflow for generating both a transcriptomic and metabolomic dataset comparing two experimental groups—using plant embryos as an example—followed by a method of integrating results from each dataset. The plant embryo example followed in this protocol compares two varieties of a hypothetical oilseed species, one producing a high amount of oil (variety "A"), which serves as the experimental group displaying a desirable trait, and then a low-oil variety (variety "B"). The targeted metabolomics protocol allows for the quantification of 100+ water-soluble metabolites: amino acids, sugars, sugar alcohols, organic acids, and phosphorylated compounds from freeze-dried plant embryos. However, it can be used for any freeze-dried biological tissue/sample by simply adjusting the starting material amount as necessary. The transcriptomics section of the protocol describes an efficient RNA extraction procedure. Similar to the metabolite extraction method, the RNA one is effective for any tissue (from either plants, animals, or microorganisms); however it strictly requires that tissues remain frozen following collection. After the RNA extraction is a walkthrough for how to perform a differential gene expression analysis, which is valuable in and of itself. Finally, the transcriptomic data is integrated with the metabolomics dataset for a final multi-omics analysis, providing a powerful and global perspective on metabolism. To interpret the results and understand the metabolism behind a desirable trait in plant, an example of the association of metabolomic and transcriptomic data with seed/embryo oil content is provided.

2 Materials

2.1 Sample Collection

In transcriptomic and metabolomic studies, fast processing and cooling of the samples is key to obtaining accurate data.

1. Dissecting microscope.
2. Forceps.
3. 2 mL Sarstedt screw cap tubes (Sarstedt AG & Co. KG, Germany).
4. Ice-cold double distilled water (ddH_2O).
5. 2 mL Sarstedt tubes with perforated lids for freeze-drying (lids can be perforated by puncturing two to three times with a disposable syringe needle, etc.).
6. Liquid nitrogen.
7. Lyophilizer with a chamber that cools up to −80 °C and allows a vacuum of 0.070 mBar.

2.2 Metabolomics

2.2.1 Metabolite Extraction

1. Refrigerated microcentrifuge at 4 °C.
2. Refrigerated benchtop centrifuge at 4 °C.
3. Retsch mill MM400 (Retsch, USA, Verder Scientific, Inc., Newtown, PA) or 2010 Geno/Grinder® (SPEX™ Sample-Prep, Metuchen, NJ) (or thorough tissue grinder).
4. Water bath at 100 °C.
5. Ice-cold ddH_2O in a 50 mL falcon tube (in a bucket of ice).
6. 100 °C ddH_2O in a 50 mL falcon tube (in the water bath).
7. A mixture of the compounds that will be used as internal standards (IS). It is recommended to use at least one IS for each family of metabolite to be quantified. The analytes used need to be absent from the samples and have a similar chemical structure than the compounds that will be analyzed, for example, [U-^{13}C]-glycine, for amino acids, [U-^{13}C]-mannose, for sugar and sugar alcohols, and [U-^{13}C]-fumarate for organic acids and phosphorylated compounds. The final concentration of the IS needs to be in the same order of magnitude as the concentration of the metabolites that are going to be quantified. For example, 10 μL of an aqueous solution containing

4 mM [U-^{13}C]-glycine, 10 mM [U-^{13}C]-mannose, and 1 mM [U-^{13}C]-fumarate could be used for extractions in which 5 mg of freeze-dried embryos is extracted and resuspended in a final volume of 500 μL.

8. Ice-cold, labeled 5 mL syringes, one for each sample and one for every IS control.
9. Ice-cold 15 mL falcon tubes, one for each sample and one for every IS control.
10. 0.22 μm syringe filters (Cat. No. SLGVR33RS, Millipore-Sigma, Burlington, MA), one for each sample and one for every IS control.
11. 5 mm tungsten beads, one for each sample and one for every IS control.
12. Perforated lids of 15 mL falcon tubes, one for each sample and one for every IS control.
13. Ice-cold, labeled 3 kDa Amicon Ultra 0.5 mL column filtering devices, one for each sample and one for every IS control.
14. Ice-cold, labeled 0.2 μm Nanosep filtering devices, one for each sample and one for every IS control.
15. Liquid nitrogen.
16. Lyophilizer with a chamber that cools to −80 °C and maintains a vacuum of ~0.070 mBar.
17. Heat-resistant tube rack for incubation in water bath.

2.2.2 Metabolite Quantification

Prepare all solutions using ddH_2O and LC-MS-grade solvents at room temperature (unless indicated otherwise). The model of columns and LC-MS/MS systems are interchangeable with proper method development.

1. Agilent amber screw vials and caps.
2. 10 mM HCl.
3. LC-MS-grade acetonitrile, water, and acetic acid.
4. Acetonitrile/water (60:40, v/v).
5. External standards (ES) mix. For each family of compounds, a specific number of standard mixtures are prepared considering the impurities that are present in some "pure standards." For each of the metabolites that will be quantified, a pure standard is needed. The final dilution of the ES is performed with the same solvents as the sample final dilution. The exact final concentration of each standard needs to be known and preferably be in the same order of magnitude as the levels in the samples.
6. 0.1% acetic acid in acetonitrile, volume dependent on the number of samples to run.

7. 0.1% acetic acid in ddH_2O, volume dependent on the number of samples to run.
8. 0.5 and 75 mM of potassium hydroxide (KOH), volume dependent on the number of samples to run. For more accurate and reproducible preparation of these solutions, it is imperative to use degassed LC-MS-grade water and to measure the volume of the KOH solution (45% w/w) by weight (do not use KOH pellets), as previously described [3].
9. LC columns (Table 1).
10. Agilent 1290 Infinity II coupled to an AB SCIEX QTRAP 6500+ LC-MS/MS system.
11. Analyst software—SCIEX (https://sciex.com/products/software/analyst-software).
12. MetaboAnalyst v. 5.0 (browser-based) (https://www.metaboanalyst.ca/) [5].

2.3 Transcriptomics

Solutions must be made using RNase-free reagents. Furthermore, any tools and containers/bottles used must be autoclaved for 30–60 min at 121 °C. The service end of the probes used for pH measurement can be placed in 70% ethanol (EtOH) for 1 min and then in 1 M sodium hydroxide for 5 min. Rinse by dunking the probe's end in RNase-free water repeatedly, until a stable pH can be read using the clean probe.

2.3.1 RNA Extraction

1. 1.5 and 2 mL snap-top sample tubes, autoclaved.
2. Deionized water or 18.2 MΩ-cm water.
3. Diethyl dicarbonate (DEPC).
4. Ethanol (EtOH).
5. Sodium hydroxide (NaOH).
6. 2,2′,2″,2‴-(Ethane-1,2-diyldinitrilo)tetraacetic acid (EDTA).
7. 2-Amino-2-(hydroxymethyl)propane-1,3-diol (Tris).
8. Hydrochloric acid (HCl).
9. Chloroform.
10. Isoamyl alcohol.
11. Sodium acetate (NaOAc).
12. Glacial acetic acid.
13. Lithium chloride (LiCl).
14. Glycogen (we used RNase-free, Thermo Fisher Scientific Cat. No. R0551).
15. N,N,N-Trimethylhexadecan-1-aminium bromide (CTAB).
16. Polyvinylpolypyrrolidone (PVPP).
17. Sodium chloride (NaCl).

Table 1
Liquid chromatograph parameters

	Amino acids			Sugars and sugar alcohols			Organic acids and phosphorylated compounds		
Sample preparation									
Dilution solvent	1 mM HCl			Acetonitrile/H_2O (60:40)			H_2O		
Columns									
Name	Symmetry C18 (4.6 × 75 mm; 3.5 μm) [6]			Shodex Asahipak NH2P-50 2D (2 mm × 150 mm; 5 μm) [7]			Dionex IonPac AS11 (250 × 2 mm; 13 μm) [3]		
Precolumn	Symmetry C18 (3.9 × 20 mm; 5 μm) [6]			Shodex Asahipak NH2P-50 2A (2 mm × 10 mm; 5 μm) [7]			Dionex IonPac AG11 (50 × 2 mm; 13 μm)		
Trap-column	None			None			Dionex IonPac ATC-3 (35 × 4 mm; 55 μm)		
Suppressor	None			None			Dionex AERS 500e (2 mm)		
Solutions									
Solvent A	100% acetonitrile + 0.1% acetic acid			100% acetonitrile			0.5 mM KOH		
Solvent B	H_2O + 0.1% acetic acid			H_2O			75 mM KOH		
Needle wash solution	Acetonitrile/isopropanol/methanol/H_2O (25:25:25:25) + 0.1% formic acid [6]			50% methanol			50% methanol		
Seal of pump wash solution	50% methanol			50% methanol			H_2O		
Pump(s) used	Binary			Binary			Binary for column Isocratic for suppressor		
Temperature									
Column	30 °C			30 °C			23 °C		
Autosampler	15 °C			20 °C			4 °C		
Liquid chromatography parameters									
Flow and gradient	*Time (min)*	*%A*	*Flow (mL/min)*	*Time (min)*	*%A*	*Flow (mL/min)*	*Time (min)*	*%A*	*Flow (mL/min)* [3]
	0	0	0.8	0	85	0.4	0	100	0.35
	2	0	0.8	8.5	85	0.4	2	100	0.35
	6.5	20	0.8	8.6	78	0.4	13	95.2	0.35
	6.6	80	0.8	16	72	0.4	20	95.2	0.35
	8.5	80	0.8	19	72	0.4	30	87.3	0.35

(continued)

Table 1
(continued)

	Amino acids			Sugars and sugar alcohols			Organic acids and phosphorylated compounds		
	8.6	0	0.8	20	68	0.4	45	73.8	0.35
	10	0	0.8	25	68	0.4	65	33.6	0.35
				26	85	0.4	70	0	0.35
				30	85	0.4	75	0	0.35
							75.1	100	0.35
							80	100	0.35
Total running time (min)									
Time	10			30			80		
Column storage solution									
Solvent	60% acetonitrile			80% acetonitrile			15.5% of 75 mM KOH		

Symmetry C18, Waters Corp., Milford, MA; Shodex Asahipak NH2P-50 2D, Showa Denko K.K., Tokyo, Japan; Dionex IonPac AS11, AG11, Thermo Fisher Scientific, Waltham, MA

18. Spermidine.
19. 2-Mercaptoethanol.
20. Aluminum foil.
21. Calibrated pH meter.
22. Magnetic stirrer.
23. Autoclave.
24. Liquid nitrogen.
25. Access to gel capillary electrophoresis instrument or service.

2.3.2 Linux

A Linux distribution such as Ubuntu (or Ubuntu Server) is required. Installing these on your current operating system requires some specific instructions, so be sure to follow carefully. The programs used here will work on Ubuntu v. 22.01.1 and later. If using a high-performance computer remotely, WinSCP (https://winscp.net/eng/index.php) is an easy way to navigate your directories. The command line attached to WinSCP is powered by a local installation of PuTTY (https://www.putty.org).

1. HISAT2 v. 2.2.1 (http://daehwankimlab.github.io/hisat2/) [8].
2. samtools v. 1.16.1 (www.htslib.org) [9].
3. StringTie v. 2.2.0 (https://ccb.jhu.edu/software/stringtie/) [10].

4. gffread v. 0.12.7 (https://github.com/gpertea/gffread) [11].
5. Trinity v. 2.14.0 (https://github.com/trinityrnaseq/trinityrnaseq) [12].
6. Trans-ABySS v. 2.0.1 (https://github.com/bcgsc/transabyss) [13].
7. AGAT v. 1.0.0 (https://github.com/NBISweden/AGAT).
8. EvidentialGene (http://arthropods.eugenes.org/EvidentialGene/evigene/).
9. Salmon v. 1.9.0 (https://combine-lab.github.io/salmon/) [14].
10. bedtools v. 2.30.0 (https://github.com/arq5x/bedtools2/releases) [15].
11. minimap2 v 2.24 (https://github.com/lh3/minimap2) [16].

2.3.3 R

The R statistical package (www.r-project.org) needs to be installed (v. 4.1.3 or later). A convenient way to use R is via RStudio (v. 2022.07.2+576) (https://posit.co/download/rstudio-desktop/) which allows for easy navigation of R terminals and directories and streamlines maintaining script files. The local installation of R is accessed by RStudio.

1. DESeq2 v. 1.38.1 (https://bioconductor.org/packages/release/bioc/html/DESeq2.html) [17].
2. tximport v. 1.26.0 (https://bioconductor.org/packages/release/bioc/html/tximport.html) [18].
3. tximportData v. 3.16 (https://bioconductor.org/packages/release/data/experiment/html/tximportData.html).

2.3.4 Windows/PC

1. FastQC v. 0.11.9 (https://www.bioinformatics.babraham.ac.uk/projects/fastqc/) [19].
2. Blast2GO v. 6.0 (https://www.biobam.com/blast2go-previous-versions/). This is currently available as a stand-alone and is free for academic use after registering at https://www.biobam.com/blast2go-basic/ [20]. It is also part of the proprietary OmicsBox suite (https://www.biobam.com/omicsbox/).
3. PANTHER v. 17.0 (browser-based) (http://www.pantherdb.org) [21, 22].
4. MetaboAnalyst v. 5.0, as used previously.

3 Methods

1. Peel immature seeds using forceps and the dissecting microscope; discard the seed coat and endosperm. Cool down the tissue immediately after dissection in 2 mL Sarstedt tubes

3.1 Targeted Metabolomics

3.1.1 Plant Embryo Collection and Preparation

placed on ice. The number of embryos to be collected for one biological replicate should be equivalent to at least 1.5 and 3 mg of dry weight (DW) for smaller embryos and larger embryos, respectively. Once the required number of embryos is collected, freeze the 2 mL Sarstedt tube in liquid nitrogen. Keep the tubes at −80 °C until all the biological replicates have been collected.

2. Dry the frozen embryos for 3 days in a lyophilizer, capping the tubes with perforated lids to allow the gaseous water to be released from the samples.
3. Grind each replicate separately using a 5 mm tungsten bead in a Retsch mill/tissue grinder for 5 min at 30 Hz.
4. For larger embryos, weigh 5 ± 0.1 mg of the powder of each replicate into a 2 mL Sarstedt tube. For smaller embryos, the same 2 mL Sarstedt tube containing the sample can be used for the extraction, as long as the DW of the material is between 1.5 and 10 mg.

3.1.2 Metabolite Extraction with Boiling Water

We recommend starting the processing of the samples 2 days before LC-MS/MS runs to avoid long periods of storage.

1. Spin down the 2 mL Sarstedt tube containing the dried powder and the 5 mm tungsten bead using a microcentrifuge set to 4 °C at 17,000 × *g* for 1 min. This step is used to pull the dried material to the bottom of the tube.
2. Remove tubes from microcentrifuge and place them immediately on ice.
3. At this point, label 1 empty 2 mL Sarstedt tube (IS-1, IS-2, . . ., IS-*n*) for every 12 samples. These tubes are processed exactly the same way as the sample tubes but will contain only IS and no biological samples.
4. Start a timer and process one tube at a time.
5. Add 10 μL of the IS mix (4 mM [U-^{13}C]-glycine, 10 mM [U-^{13}C]-mannose, 1 mM [U-^{13}C]-fumarate) to the first tube.
6. Immediately add 1 mL of boiling ddH_2O to that first tube.
7. Screw the lid tightly, vortex, and transfer the tube immediately to the boiling water bath (a heat-resistant floating rack is helpful if water level is high). Record exact stopwatch time when the incubation started for that first tube.
8. Proceed with the following tubes (one at a time) in the same way as the first tube, following **steps 5**–7. Try to space them about 30 s apart; this way the addition of boiling water to samples can be staggered and processed more efficiently.
9. After 5 min of incubation, remove each tube from the water bath, vortex, and put it back for another 5 min.

10. After a total of 10 min incubation, place the first tube quickly on ice, and do so, until the last sample is removed from the water bath. Once the last tube has been placed on ice, wait for 3–5 min before continuing with the following step.
11. Centrifuge the tubes at 17,000 × *g* and 4 °C for 5 min.
12. Remove tubes carefully from the microcentrifuge to avoid disrupting the pellet.
13. Transfer each supernatant to a pre-labeled 5 mL syringe with a 0.22 μm filter.
14. Push supernatant through the filter by using the pre-labeled plunger that comes with the 5 mL syringe, and collect it in a 15 mL falcon tube previously placed on ice.
15. Add 1 mL of ice-cold ddH_2O to the 2 mL Sarstedt screw cap tubes containing the pellets, vortex until pellet is resuspended, and centrifuge at 17,000 × *g* and 4 °C for 5 min.
16. Filter each supernatant as explained before using the same syringes and filters for each sample and collecting the eluent in the same 15 mL falcon tubes. After this step each falcon tube will contain 2 mL of ice-cold extract.
17. Rinse syringes with filters twice with 1 mL of ice-cold ddH_2O.
18. Discard syringes and filters.
19. Cap the 15 mL falcon tubes each containing approximately 4 mL of ice-cold extracts, and centrifuge at 1000 × *g* and 4 °C for 60 s.
20. Remove caps of the 15 mL falcon tubes and replace them with perforated lids.
21. Freeze the 15 mL falcon tubes in liquid nitrogen for 15–20 s.
22. Lyophilize the frozen extracts for 24 h at −80 °C.

3.1.3 Resuspension of Lyophilized Samples

1. Add 500 μL of ice-cold ddH_2O to the lyophilized samples, change the lids of the tubes to new ones, and vortex until complete dissolution.
2. Centrifuge the resuspended samples at 1000 × *g* at 4 °C for 60 s.
3. Split samples as follows:
4. 150 μL is loaded on a 0.22 μm Nanosep filter (use for sugars and sugar alcohols).
5. The remaining volume is added to a 3 kDa Amicon Ultra 0.5 mL column filtering device.
6. Centrifuge the filtering devices at 14,000 × *g* at 4 °C for 60 min. At the end of the centrifugation, a maximum of only 5 μL should be lost via retention in the μm Nanosep filter and only 50–75 μL in the 3 kDa Amicon filters. Do not exceed

14,200 × *g* for the 3 kDa Amicon filtering device because the membrane will break and macromolecules (e.g., proteins) will be released in the eluate.

7. Eluates are placed on ice and the filters are discarded.
8. The filtered samples can be stored in the refrigerator until LC-MS/MS processing. If storage time is over 2 days, it is highly recommended to freeze the extracts and keep them at −20 °C until further analysis.

3.1.4 Metabolite Quantification

The water-soluble metabolites are quantified using an Agilent 1290 Infinity II coupled to an AB SCIEX QTRAP 6500+ LC-MS/MS system through multiple reaction monitoring (MRM). Three specific procedures previously described in the literature are used for quantifying (1) amino acids and their derivatives, (2) soluble sugars and sugar alcohols, and (3) phosphorylated compounds and organic acids [3, 7, 23, 24]. These metabolites are of particular interest because they are key players in the central metabolism of many species. The preparation of the samples, chromatographic columns, mobile phases, gradients, flow, and remaining LC parameters are described in Table 1, while the MS specifications for each method and metabolite are shown in Table 2. The length of the run is 10, 30, and 80 min for steps (**1**), (**2**), and (**3**), respectively. In each case, a calibration curve with each pure standard is required as the area/mol ratio is specific for each compound and allows the identification of linearity range and the limits of detection and quantification. In general, the calibration graphs show excellent linearity in the low fmol to high pmol range and detection limits varied from 0.1 to 122.5 fmol. General steps for each method are as follows:

1. Create an acquisition method in the Analyst software (*see* **Note 1**) under "Acquisition Mode" tab for each family of compounds using the information provided in Tables 1 and 2. For the LC, define Solvents A and B, percentage of each solvent at each given time, flow, autosampler temperature, and column temperature. For the MS sections, set the polarity, Q1 and Q3 masses, dwell time, entrance potential (EP), declustering potential (DP), collision energy (CE), and the collision cell exit potential (CXP), as well as the source parameters: curtain gas (CUR), collision gas (CAD), ion spray voltage (IS), temperature (TEM), and ion source gas 1 and 2 (GS1 and GS2). Note that each of these parameters was optimized for each of the metabolites and depends on the specific instrumentation used.
2. Purge and prime the LC section with 50% of each mobile phase.

3. Install the corresponding column, carefully considering the direction of the flow and making sure that the system does not show leakages, especially in the connections of the tubing entering and exiting the column.
4. Equilibrate the column for 20 min using the percentage of Solvents A and B and flow with which the chromatography run starts (Table 1). In the case of metabolites from step (**3**) that are separated in anion exchange column, a suppressor needs to be connected downstream of the column to remove the salts from the mobile phase before reaching the MS using isocratic pump, as previously described [3].
5. Create a sample submission batch in "Acquisition Mode"; select one of the acquisition methods created in **step 1**; add three samples; label them as blanks; select the path where the data from the runs will be saved, the rack type, the vial position, and the direction in which the vials will be used (columns or rows); and select an injection volume of 5 μL. We advise to include, in the name of the run, the name of the sample, the dilution (for the samples), and the injection volume. For example, the blanks will be Blankl_5ul_001, Blank2_5ul_002, and Blank3_5ul_003. Note that mu in μL is replaced with "u," as special characters can cause problems with sample/batch submissions; using spaces should also be avoided.
6. Submit the blanks, and check if they appear in the queue, and verify that all the information was added correctly.
7. Prepare a blank vial and insert in rack 1 position 1. The blank will consist of 1 mL of the same solvent that will be used to dilute the samples (Table 1).
8. Equilibrate the MS for 60 s, connect the tubing coming out of the column to the MS, and check if the shape of the ionization spray is adequate (the stream should be in a "V" shape); if not, adjust the protuberance of the capillary from the electrospray probe to not more than 1 mm.
9. Start the run of the blanks, and check the chromatograms in real time ("Explore Mode" in the Analyst software), where flat curves for total ion count (TIC) along the duration of the run for each of the transitions are expected. If peaks appear, run more blanks until they are no longer seen.
10. Add the ES run in the sample submission batch ("Acquisition Mode" in the Analyst software). We suggest the following dilutions for the mix of ES (six vials): 1000 μM, 100 μM, 10 μM, 1 μM, 100 nM, and 1 nM. Inject each ES concentration using different injection volumes: 2, 5, and 10 μL. When entering the information in the sample submission batch, add one row for each dilution and for each injection volume, assigning a vial position.

Table 2
Mass spectrometer parameters

Column	Metabolite	Retention time (min)	Polarity (*see* Note 3)	Q1 (Da)	Q3 (Da)	Dwell time (ms)	EP (V)	DP (V)	CE (V)	CXP (V)	CUR	IS	TEM	GS1	GS2	CAD
Amino acids	Alanine	0.88	+	89.9	44.0	20	10	20	13	20	40	3000	600	40	40	Medium
	Arginine	0.62	+	175.0	70.0	20	10	45	27	8	40	3000	600	40	40	Medium
	Asparagine	0.88	+	133.0	87.0	20	10	20	13	10	40	3000	600	40	40	Medium
	Aspartate	1.40	+	133.9	88.0	20	10	20	11	14	40	3000	600	40	40	Medium
	Citrulline	0.97	+	176.0	159.0	20	10	10	13	10	40	3000	600	40	40	Medium
	Cysteine	0.97	+	122.0	76.0	20	10	40	17	34	40	3000	600	40	40	Medium
	Cystine	0.86	+	241.0	152.0	20	10	20	17	18	40	3000	600	40	40	Medium
	GABA	0.67	+	104.0	87.0	20	10	5	13	10	40	3000	600	40	40	Medium
	Glutamate	1.14	+	147.9	84.0	20	10	20	21	10	40	3000	600	40	40	Medium
	Glutamine	0.91	+	147.0	130.0	20	10	20	13	14	40	3000	600	40	40	Medium
	Glycine	0.86	+	76.2	30.1	20	10	20	19	14	40	3000	600	40	40	Medium
	Histidine	0.61	+	156.0	110.0	20	10	20	19	12	40	3000	600	40	40	Medium
	Homoserine	0.90	–	118.0	100.0	50	−10	−20	−14	−11	40	−4500	600	40	40	Medium
	Hydroxyproline	0.93	+	132.0	86.0	20	10	17	14	10	40	3000	600	40	40	Medium
	Isoleucine	2.37	+	132.0	86.0	20	10	17	14	10	40	3000	600	40	40	Medium
	Leucine	2.57	+	132.0	86.0	20	10	17	14	10	40	3000	600	40	40	Medium
	Lysine	0.61	+	147.0	130.0	20	10	20	13	14	40	3000	600	40	40	Medium
	Methionine	1.81	+	149.9	104.0	20	10	15	13	12	40	3000	600	40	40	Medium
	Nicotinamide	4.22	+	122.9	80.0	20	10	45	27	10	40	3000	600	40	40	Medium
	Nicotinic acid	3.07	+	123.9	80.0	20	10	15	27	10	40	3000	600	40	40	Medium
	Ornithine	0.60	+	133.0	116.0	20	10	5	13	12	40	3000	600	40	40	Medium
	Phenylalanine	4.92	+	166.0	120.0	20	10	10	17	14	40	3000	600	40	40	Medium
	Proline	1.09	+	115.9	70.0	20	10	10	19	8	40	3000	600	40	40	Medium
	Serine	0.87	+	106.0	60.0	20	10	25	15	8	40	3000	600	40	40	Medium
	Threonine	0.92	–	118.0	74.0	50	−10	−20	−14	−11	40	−4500	600	40	40	Medium
	Tryptophan	5.90	+	205.0	188.0	20	10	15	13	10	40	3000	600	40	40	Medium
	Tyrosine	3.23	+	182.0	165.0	20	10	10	13	10	40	3000	600	40	40	Medium
	Valine	1.28	+	117.9	72.0	20	10	10	15	10	40	3000	600	40	40	Medium
	^{13}C-Glycine (IS)	0.86	+	78.2	31.1	20	10	20	19	14	40	3000	600	40	40	Medium

(continued)

Table 2 (continued)

Column	Metabolite	Retention time (min)	Polarity (*see* Note 3)	Q1 (Da)	Q3 (Da)	Dwell time (ms)	EP (V)	DP (V)	CE (V)	CXP (V)	CUR	IS	TEM	GS1	GS2	CAD
Sugars and sugar alcohols	Arabitol	4.65	–	150.9	89.0	150	−10	−40	−16	−37	30	−4500	350	60	60	Medium
	Arabinose	4.10	–	148.9	89.0	150	−10	−35	−10	−15	30	−4500	350	60	60	Medium
	Chiro-inositol	11.10	–	179.0	161.0	150	−10	−40	−16	−27	30	−4500	350	60	60	Medium
	Erythritol/threitol	3.15	–	121.1	89.0	150	−10	−20	−13	−11	30	−4500	350	60	60	Medium
	Fructose	5.40	–	179.0	89.0	150	−10	−40	−10	−9	30	−4500	350	60	60	Medium
	Galactitol	7.45	–	181.0	89.0	150	−10	−40	−10	−9	30	−4500	350	60	60	Medium
	Galactose	7.75	–	179.0	89.0	150	−10	−40	−10	−9	30	−4500	350	60	60	Medium
	Glucose	8.25	–	179.0	89.0	150	−10	−40	−10	−9	30	−4500	350	60	60	Medium
	Glycerol	2.16	–	91.0	59.0	150	−10	−30	−12	−7	30	−4500	350	60	60	Medium
	Inositol	11.55	–	178.9	87.0	150	−10	−65	−22	−9	30	−4500	350	60	60	Medium
	Maltose	13.83	–	341.1	161.0	150	−10	−50	−10	−7	30	−4500	350	60	60	Medium
	Mannitol	7.21	–	181.0	89.0	150	−10	−40	−10	−9	30	−4500	350	60	60	Medium
	Mannose	6.85	–	185.0	92.0	150	−10	−40	−10	−9	30	−4500	350	60	60	Medium
	Pinitol	5.30	–	193.0	161.0	150	−10	−35	−14	−9	30	−4500	350	60	60	Medium
	Raffinose	18.10	–	503.2	178.9	150	−10	−140	−32	−11	30	−4500	350	60	60	Medium
	Ribitol/xylitol	4.52	–	150.9	89.0	150	−10	−40	−16	−37	30	−4500	350	60	60	Medium
	Ribose	3.00	–	148.9	89.0	150	−10	−35	−10	−15	30	−4500	350	60	60	Medium
	Scyllo-inositol	12.30	–	179.0	161.0	150	−10	−40	−16	−27	30	−4500	350	60	60	Medium
	Sorbitol	6.60	–	181.0	89.0	150	−10	−50	−20	−9	30	−4500	350	60	60	Medium
	Stachyose	21.80	–	665.1	383.1	150	−10	−220	−48	−25	30	−4500	350	60	60	Medium
	Sucrose	12.20	–	341.1	89.0	150	−10	−90	−23	−21	30	−4500	350	60	60	Medium
	Verbascose	24.60	–	828.1	545.1	150	−10	−280	−52	−25	30	−4500	350	60	60	Medium
	Xylose	4.42	–	148.9	89.0	150	−10	−35	−10	−15	30	−4500	350	60	60	Medium
	^{13}C-Mannose (IS)	6.85	–	185.0	92.0	150	−10	−40	−10	−9	30	−4500	350	60	60	Medium

Organic acids and phosphorylated compounds	2- and 3-Phosphoglycerate	39.7	–	184.9	79.0	10	−10	−30	−18	−9	30	−4500	650	40	60	High
	2-Ketoglutarate	24.30	–	145.1	101.0	10	−10	−40	−12	−11	30	−4500	650	40	60	High
	2-Phosphoglycolate	40.66	–	155.1	79.0	10	−10	−30	−16	−11	30	−4500	650	40	60	High
	6-Phosphogluconate	37.58	–	274.9	97.0	10	−10	−35	−20	−13	30	−4500	650	40	60	High
	Adipic acid	17.47	–	144.9	101.0	10	−10	−25	−16	−15	30	−4500	650	40	60	High
	ADP	49.53	–	425.9	79.0	10	−10	−80	−70	−35	30	−4500	650	40	60	High
	ADP-glucose	30.15	–	587.9	346.0	10	−10	−135	−34	−17	30	−4500	650	40	60	High
	AMP	28.90	–	345.9	79.0	10	−10	−20	−104	−37	30	−4500	650	40	60	High
	Ascorbate	18.66	–	175.0	115.0	10	−10	−60	−16	−17	30	−4500	650	40	60	High
	ATP	59.90	–	505.9	159.0	10	−10	−130	−34	−17	30	−4500	650	40	60	High
	Azelaic acid	27.70	–	187.0	125.0	10	−10	−90	−20	−15	30	−4500	650	40	60	High
	CDP	41.90	–	401.9	79.0	10	−10	−70	−88	−11	30	−4500	650	40	60	High
	Cis-aconitate	43.84	–	172.9	85.0	10	−10	−30	−16	−13	30	−4500	650	40	60	High
	Citrate	41.64	–	190.9	87.0	10	−10	−40	−22	−13	30	−4500	650	40	60	High
	CMP	19.20	–	321.9	79.0	10	−10	−60	−72	−3	30	−4500	650	40	60	High
	CTP	55.00	–	481.9	159.0	10	−10	−135	−38	−17	30	−4500	650	40	60	High
	Deoxyxylulose 5-P	15.60	–	212.9	97.0	10	−10	−20	−16	−11	30	−4500	650	40	60	High
	Erythrose 4-P	55.80	–	198.9	97.0	10	−10	−15	−12	−9	30	−4500	650	40	60	High
	Fructose 6-P	26.70	–	258.9	79.0	10	−10	−40	−60	−9	30	−4500	650	40	60	High
	Fructose-1,6-bisP	54.05	–	168.9	97.0	10	−10	−15	−16	−11	30	−4500	650	40	60	High
	Fumarate	25.73	–	114.9	71.0	10	−10	−40	−10	−9	30	−4500	650	40	60	High
	Galactose 1-P	14.80	–	258.9	79.0	10	−10	−40	−60	−9	30	−4500	650	40	60	High
	GDP	62.93	–	441.9	79.0	10	−10	−100	−80	−35	30	−4500	650	40	60	High
	GDP-fucose	50.77	–	587.9	442.0	10	−10	−145	−36	−55	30	−4500	650	40	60	High
	GDP-glucose	52.30	–	603.9	362.0	10	−10	−160	−36	−17	30	−4500	650	40	60	High
	GDP-mannose	52.60	–	603.9	424.0	10	−10	−150	−40	−19	30	−4500	650	40	60	High
	Gluconate	2.70	–	194.9	75.0	10	−10	−120	−22	−9	30	−4500	650	40	60	High
	Glucosamine 1-P	14.00	–	257.9	79.0	10	−10	−70	−22	−39	30	−4500	650	40	60	High
	Glucosamine 6-P	25.15	–	257.9	97.0	10	−10	−20	−22	−9	30	−4500	650	40	60	High
	Glucose 1-P/mannose 1-P	15.35	–	258.9	79.0	10	−10	−40	−60	−9	30	−4500	650	40	60	High
	Glucose 6-P	25.53	–	258.9	79.0	10	−10	−40	−60	−9	30	−4500	650	40	60	High
	Glutarate	17.47	–	130.9	87.0	10	−10	−30	−16	−13	30	−4500	650	40	60	High
	Glycerol-Ps	15.20	–	170.9	79.0	10	−10	−25	−24	−11	30	−4500	650	40	60	High
	GMP	52.12	–	361.9	79.0	10	−10	−25	−70	−9	30	−4500	650	40	60	High
	GTP	69.35	–	521.9	159.0	10	−10	−135	−36	−23	30	−4500	650	40	60	High

(continued)

Table 2 (continued)

Column	Metabolite	Retention time (min)	Polarity (*see* Note 3)	Q1 (Da)	Q3 (Da)	Dwell time (ms)	EP (V)	DP (V)	CE (V)	CXP (V)	CUR	IS	TEM	GS1	GS2	CAD
	IDP	62.20	–	426.9	79.0	10	−10	−80	−102	−9	30	−4500	650	40	60	High
	IMP	51.25	–	346.9	79.0	10	−10	−75	−68	−35	30	−4500	650	40	60	High
	Isocitrate	42.60	–	190.9	173.0	10	−10	−35	−12	−11	30	−4500	650	40	60	High
	ITP	69.00	–	506.9	159.0	10	−10	−75	−32	−19	30	−4500	650	40	60	High
	Malate	18.64	–	132.9	114.9	10	−10	−75	−14	−55	30	−4500	650	40	60	High
	Maleic acid	22.70	–	114.9	71.0	10	−10	−20	−12	−9	30	−4500	650	40	60	High
	Malonic acid	20.35	–	102.9	59.0	10	−10	−30	−12	−15	30	−4500	650	40	60	High
	Mannose 6-P	28.15	–	258.9	79.0	10	−10	−40	−60	−9	30	−4500	650	40	60	High
	Mevalonate	2.65	–	146.9	59.0	10	−10	−35	−14	−9	30	−4500	650	40	60	High
	N-Acetylglucosamine 1-P	13.85	–	299.9	97.0	10	−10	−30	−20	−13	30	−4500	650	40	60	High
	N-Acetylglucosamine 6-P	26.90	–	299.9	97.0	10	−10	−30	−20	−13	30	−4500	650	40	60	High
	Phosphoenolpyruvate	43.70	–	166.9	79.0	10	−10	−30	−34	−37	30	−4500	650	40	60	High
	Pimelic acid	18.82	–	158.9	97.0	10	−10	−60	−18	−11	30	−4500	650	40	60	High
	Pyruvate	4.45	–	86.9	43.0	10	−10	−80	−10	−19	30	−4500	650	40	60	High
	Quinic acid	2.50	–	190.9	85.0	10	−10	−50	−26	−9	30	−4500	650	40	60	High
	Ribose 1-P	15.65	–	229.0	79.0	10	−10	−35	−26	−35	30	−4500	650	40	60	High
	Ribose 5-P/ribulose 5-P	28.50	–	229.0	97.0	10	−10	−35	−26	−35	30	−4500	650	40	60	High
	Ribulose-1,5-bis-P	37.80	–	229.0	97.0	10	−10	−35	−26	−35	30	−4500	650	40	60	High
	Sebacic acid	33.10	–	201.0	139.0	10	−10	−105	−22	−7	30	−4500	650	40	60	High
	Sedoheptulose 7-P	30.15	–	288.9	97.0	10	−10	−30	−28	−11	30	−4500	650	40	60	High
	Shikimate	3.05	–	172.9	111.0	10	−10	−30	−12	−5	30	−4500	650	40	60	High
	Suberic acid	22.65	–	172.9	111.0	10	−10	−90	−18	−15	30	−4500	650	40	60	High
	Succinate	18.30	–	116.9	73.0	10	−10	−80	−18	−9	30	−4500	650	40	60	High
	Sucrose 6-P	17.60	–	420.9	79.0	10	−10	−70	−92	−9	30	−4500	650	40	60	High
	Tartaric acid	20.45	–	148.9	87.0	10	−10	−25	−18	−7	30	−4500	650	40	60	High
	Trans-aconitate	46.29	–	172.9	85.0	10	−10	−30	−16	−13	30	−4500	650	40	60	High
	Trehalose 6-P	14.90	–	420.9	79.0	10	−10	−70	−92	−9	30	−4500	650	40	60	High
	UDP	57.61	–	403.0	79.0	10	−10	−80	−90	−35	30	−4500	650	40	60	High
	UDP-arabinose	44.12	–	534.9	323.0	10	−10	−100	−30	−17	30	−4500	650	40	60	High
	UDP-galactosamine	42.10	–	605.9	385.0	10	−10	−160	−38	−23	30	−4500	650	40	60	High

UDP-galactose	44.50	–	564.9	323.0	10	−10	−130	−34	−15	30	−4500	650	40	60	High
UDP-galacturonic acid	57.60	–	578.9	403.0	10	−10	−220	−32	−21	30	−4500	650	40	60	High
UDP-glucosamine	42.63	–	605.9	385.0	10	−10	−160	−38	−23	30	−4500	650	40	60	High
UDP-glucose	45.50	–	564.9	323.0	10	−10	−130	−34	−15	30	−4500	650	40	60	High
UDP-glucuronic acid	58.75	–	578.9	403.0	10	−10	−220	−32	−21	30	−4500	650	40	60	High
UDP-rhamnose	47.70	–	548.9	323.0	10	−10	−130	−30	−17	30	−4500	650	40	60	High
UDP-xylose	46.05	–	534.9	323.0	10	−10	−100	−30	−17	30	−4500	650	40	60	High
UMP	45.00	–	322.9	79.0	10	−10	−40	−70	−1	30	−4500	650	40	60	High
UTP	65.40	–	482.9	159.0	10	−10	−200	−34	−17	30	−4500	650	40	60	High
Xylulose 5-P	29.25	–	229.0	97.0	10	−10	−35	−26	−35	30	−4500	650	40	60	High
^{13}C-Fumarate (IS)	25.73	–	118.9	73.9	10	−10	−40	−10	−9	30	−4500	650	40	60	High

Q1 quadrupole 1, *Q3* quadrupole 3, *EP* entrance potential, *DP* declustering potential, *CE* collision energy, *CXP* collision exit potential, *CUR* curtain gas, *IS* ion spray voltage, *TEM* temperature, *GS1* ion source gas 1, *GS2* ion source gas 2, *CAD* collision gas

11. Insert the ES vials in the rack according to the positions assigned in the sample submission.
12. Run the different dilutions of ES mix to generate a calibration curve.
13. Prepare a quantitative method in the Analyst software ("Quantitation Mode") to integrate the area under each peak. First, select one sample to be used as an example for the automatic integration of the peaks; typically, this is done by running the ES so that each metabolite is obtained in one injection. For each metabolite determine the correct transition (parent and daughter ion masses, Q1 and Q3), using one row for each metabolite. In the case of isomers (i.e., several metabolites in the same transition but at different retention times), duplicate the transition as many times as the number of isomers. Then, for each analyte, set the correct RT, smoothing factor (normally we used a factor between 1 and 2 to avoid an automatic integration of half of the peak, but it depends on the peak's shape), and the zone of the chromatogram to measure the noise (i.e., the zone with no peaks, even for the sample chromatograms).
14. Apply the quantitation method to the ES samples and check manually if each automatic integration was done correctly; if not, manually correct it.
15. Graph the calibration curves (area vs pmol) for each metabolite and determine linearity range, limits of detection, and quantification. According to the sensitivity obtained for each metabolite (slope of the graph), determine an ideal ES mix for which the concentration of each of the metabolites is approximately in the middle of the calibration curve using only one injection volume. This dilution of ES will be run together with the samples. In Table 3, the concentration and injection volumes for the three different methods are shown, establishing one ideal ES mix for amino acids and derivatives (named White, injection volume of 6 μL), two for soluble sugars and sugar alcohols (named Yellow and Red, injection volume of 10 μL), and four for organic acids and phosphorylated compounds (named Yellow, Red, Blue, and Orange, injection volume of 10 μL).
16. Prepare one representative sample for each "type" of sample to determine the level of dilution needed and the injection volume to have all the metabolites in the linear range of the curves. In Table 1, the solvents used for the preparation of the samples in each method can be found. In general, an initial trial would be as follows:

Table 3
External standard mixes

Column	Standard mix	Metabolite	Concentration (μM)	IV (μL)
Amino acids	White	Alanine	1.25	6
		Arginine	0.25	6
		Asparagine	1.25	6
		Aspartate	1.25	6
		Citrulline	0.25	6
		Cysteine	1.25	6
		Cystine	1.00	6
		GABA	0.25	6
		Glutamate	1.25	6
		Glutamine	1.25	6
		Glycine	1.25	6
		Histidine	0.25	6
		Homoserine	1.00	6
		Hydroxyproline	0.25	6
		Isoleucine	0.25	6
		Leucine	0.25	6
		Lysine	1.25	6
		Methionine	0.25	6
		Nicotinamide	0.25	6
		Nicotinic acid	0.25	6
		Ornithine	1.25	6
		Phenylalanine	0.25	6
		Proline	0.25	6
		Serine	0.25	6
		Threonine	0.25	6
		Tryptophan	0.25	6
		Tyrosine	0.25	6
		Valine	0.25	6
Sugars and sugar alcohols	Yellow	Chiro-inositol	1.00	10
		Erythritol/threitol	1.00	10
		Fructose	1.00	10
		Galactitol	0.50	10
		Glucose	1.00	10
		Glycerol	100.00	10
		Inositol	1.00	10
		Mannose	10.00	10
		Raffinose	1.00	10
		Ribitol/xylitol	1.00	10
		Ribose	10.00	10
		Scyllo-inositol	1.00	10
		Sorbitol	1.00	10
		Stachyose	1.00	10
		Sucrose	1.00	10
		Trehalose	1.00	10
		Verbascose	100.00	10
		Xylose	10.00	10

(continued)

Table 3
(continued)

Column	Standard mix	Metabolite	Concentration (μM)	IV (μL)
	Red	Arabinose	10.00	10
		Arabitol	0.50	10
		Galactose	10.00	10
		Maltose	10.00	10
		Mannitol	0.50	10
		Pinitol	1.00	10
Organic acids and phosphorylated compounds	Yellow	2-Ketoglutarate	0.10	10
		Cis-aconitate	0.25	10
		Citrate	0.10	10
		Fructose 6-P	0.05	10
		Fumarate	0.25	10
		Galactose 1-P	0.025	10
		Glucose 1-P/mannose 1-P	0.025	10
		Glucose 6-P	0.025	10
		Isocitrate	0.10	10
		Malate	0.25	10
		Mannose 6-P	0.025	10
		Pyruvate	10.00	10
		Ribose 1-P	0.025	10
		Shikimate	0.10	10
		Succinate	0.25	10
		Trans-aconitate	0.50	10
		Xylulose 5-P	1.00	10
	Blue	2- and 3-Phosphoglycerate	0.025	10
		Adipic acid	1.00	10
		Ascorbate	10.00	10
		Azelaic acid	0.10	10
		Erythrose 4-P	2.50	10
		GDP-glucose	0.50	10
		Gluconate	0.01	10
		Glucosamine 1-P	0.05	10
		Glutarate	0.10	10
		Maleic acid	0.10	10
		Malonic acid	0.25	10
		Mevalonate	0.05	10
		N-Acetylglucosamine 6-P	0.05	10
		Pimelic acid	0.10	10
		Quinic acid	0.05	10
		Ribulose 1,5 bis-P	0.25	10
		Sebacic acid	0.25	10
		Suberic acid	0.10	10
		Tartaric acid	0.10	10
		UDP-glucosamine	0.25	10
		UDP-glucose	0.50	10
		UDP-glucuronic acid	10.00	10
		UDP-xylose	0.50	10

(continued)

Table 3 (continued)

Column	Standard mix	Metabolite	Concentration (μM)	IV (μL)
	Green	6-Phosphogluconate	0.05	10
		ADP	0.05	10
		ADP-glucose	0.25	10
		AMP	0.025	10
		ATP	0.05	10
		CDP	0.05	10
		CMP	0.10	10
		CTP	0.05	10
		Deoxyxylulose 5-P	0.10	10
		Fructose-1,6-bisP	0.05	10
		GDP	0.10	10
		Glycerol-Ps	0.25	10
		GMP	0.025	10
		GTP	0.10	10
		IDP	0.10	10
		IMP	0.10	10
		ITP	0.10	10
		Phosphoenolpyruvate	0.10	10
		Ribose 5-P/ribulose 5-P	0.50	10
		Sedoheptulose 7-P	0.025	10
		Sucrose 6-P	0.25	10
		Trehalose 6-P	0.25	10
		UDP	0.05	10
		UDP-glucose	0.25	10
		UMP	0.025	10
		UTP	0.05	10
	Orange	2-Phosphoglycolate	0.10	10
		GDP-fucose	0.25	10
		GDP-mannose	0.25	10
		Glucosamine 6-P	0.10	10
		N-Acetylglucosamine 1-P	0.10	10
		UDP-arabinose	2.50	10
		UDP-galactosamine	0.25	10
		UDP-galactose	0.25	10
		UDP-galacturonic acid	10.00	10
		UDP-rhamnose	5.00	10

IV injection volume

(1) *For amino acids*:

- 50× dilution in the vial → 5 μL injection
- 880 μL of ddH_2O + 100 μL of 10 mM HCl + 20 μL of sample after Amicon filtration

(2) *For sugars*:

- 50× dilution in the vial → 5 μL injection
- 980 μL of acetonitrile/water solution (60:40; v/v) + 20 μL of sample after Nanosep filtration

(3) *For organic acids and phosphorylated compounds*:

- 10× dilution in the vial → 5 μL injection
- 900 μL of ultrapure water + 100 μL of sample after Amicon filtration

(*see* **Note 2**)

17. Run the trial and identify the level of dilution needed for each type of sample. Multiple runs with different dilutions and/or injection volumes may be required for the samples where the metabolites to be quantified are present in a large range of concentrations. Furthermore, adjusting the DP for some of the metabolites in a run may avoid the need to run the samples multiple times.
18. Prepare the rest of the samples following the best dilution identified in the trial.
19. Add the samples to the sample submission batch in "Acquisition Mode." If the samples are run on a different day than the ES calibration curves, plan runs of the ideal ES mix every 12–15 samples. We recommend distributing samples, ES mix, IS, and blanks in the following order: blank (using the same injection volume than the samples), ES mix, IS (using the same injection volume than the samples), 12–15 samples; blank, ES mix, IS, 12–15 samples; etc. It is important to run ideal ES mix, IS, and blanks before and after every 12–15 samples to account for run variations that may occur over long running times.
20. After the runs, apply the same quantification method created in **step 13** to the full batch of samples, ES mix, IS, and blanks. Check the automatic integrations, and if considerable variations in the retention time occurred between the runs of the ES calibration curves and the sample batch, adjustments to the quantitative method will need to be performed for effective automatic integration. Small variations in the preparation of the mobile phase could lead to shifts in the RT. It is also possible when dealing with a big number of samples and especially for the organic acids and phosphorylated compounds (80 min runs) that shifts in the RT and/or area of the standards are seen between the first samples and the last ones. In those cases, the normalization and the quantification will be done only considering the flanking IS and ES mix for every 12–15 samples, respectively.
21. For the calculation of each metabolite concentration in pmol/mg DW, first we calculate a correction factor for each sample

between the area of the IS in the samples and the area obtained in the IS tubes (where only the IS was present); this will be applied to the areas of each metabolite to account for sample loss throughout extraction and preparation or extract concentration during filtration:

$$\text{Correction factor} = \frac{\text{Original peak area of 13C} - \text{mannose in biological sample}}{\text{Average peak area of 13C} - \text{mannose across all IS}} \quad (1)$$

$$\text{Corrected peak area} = \frac{\text{Original peak area}}{\text{Correction factor}} \quad (2)$$

Correction factors closer to 1 indicate that no significant losses nor concentration occurred throughout sample preparation. Then, we calculate the quantity (mol) of each metabolite by comparing the area of a given metabolite in the sample with the area and known concentration of this same metabolite in the ideal ES mix (or the calibration curves if processed simultaneously with the samples):

$$\text{mol} = \frac{\text{mol of ES injected}}{\text{average peak area of ES}} \times \text{corrected area} \times \text{dilution factor} \quad (3)$$

The dilution factor is calculated as follows:

$$\text{Dilution factor} = \frac{\text{sample extraction volume} \times \text{total volume in LCMS vial}}{\text{volume injected} \times \text{volume of sample added to LCMS vial}} \quad (4)$$

After considering the total dilution and the mg of DW used for each sample, we obtain the concentration in pmol/mg DW of each metabolite in each sample:

$$\text{pmol/mg DW} = \frac{\text{mol} \times 1 \times 10^{12}}{\text{mg DW of sample}} \quad (5)$$

3.1.5 Metabolome Visualization and Analysis

There are different free software packages that allow the visualization and analysis of metabolomic data. Here, we suggest the use of MetaboAnalyst, which not only allows the analysis of the metabolomic data (targeted and untargeted) but also contains tools to integrate the results with transcriptomic data [5]. Some characteristics that made this software our first choice are the user-friendly design, the frequent updates and generation of new tools, and the availability of tutorials, manuals, examples, and even personal technical support to help the user navigate through the page. For each of the tools offered in MetaboAnalyst, the R commands are available for export to allow the user to have some freedom to redefine some variables, which can be especially important for image processing. We describe a series of analyses to explore, understand, and visualize the data using the "Statistical Analysis [one factor]" option in the module overview page, which one can navigate to

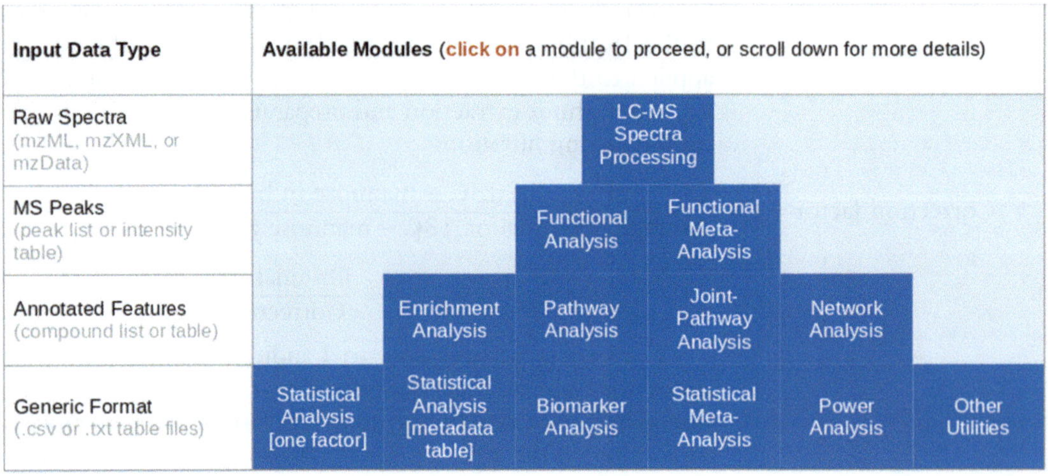

Fig. 1 Different analysis modules available in MetaboAnalyst 5.0

by clicking the "Click here to start" button on the home page (Fig. 1).

Data Entry

There are different ways to enter the data into MetaboAnalyst. For targeted metabolomics, we recommend the use of .csv files with metabolite concentrations expressed in the same units for each biological replicate and sample type. Using columns or rows for metabolites or samples is inconsequential as the software allows one to select the way data is organized. Different strategies to deal with missing values, data normalization, data transformation, and scaling are offered, but the best approach needs to be evaluated in a case-by-case manner. As many of the statistical analyses offered follow the assumption that the data is distributed normally, the user is responsible to select the best data transformation that satisfies that requirement. While navigating through the tools, the options "Data editor" and "Data filter" are available to reduce or modify the samples and metabolites that are being analyzed. A common way of looking at this type of data is to perform a log transformation and autoscaling.

Statistical Analysis Module

1. First, assess the quality and variation of the data to evaluate how the biological replicates group together; for this use the "hierarchical clustering dendrogram." Replicates from the same sample are expected to group together. The output also allows one to visualize the general arrangement of the samples/conditions. A complementary visualization of the data is given through "principal component analysis (PCA)." This analysis reduces the large multivariate data into a few "components" that are constructed as linear combinations of the initial variables (metabolites) and still containing most of the information of the large set. Furthermore, by the interpretation of the

individual coefficients for each variable from a component, information regarding which metabolites contribute the most to the separation of the samples can be discovered. These two exploratory tools allow one to evaluate the replicates, detecting outliers and the degree of variation in the data.

2. Run a significance test (ANOVA or t-test depending on your data), and identify the metabolites that are significantly different between the conditions being analyzed. Reducing the number of variables to only those metabolites that are significantly different could help to reduce the noise and facilitate the understanding of the general metabolic changes that most differentiate the conditions analyzed.
3. When analyzing a large set of conditions, evaluate the relationship between the metabolite's changes through a "Correlations" and/or "PatternHunter" analysis from the sidebar menu. This can be done for all metabolites against all or just the associations with one metabolite or trait, respectively. This is especially powerful if one has multiple agronomic traits in their data, such as oil content, protein content, harvesting index, etc. The direction of the correlation and its strength can be used for identifying processes that respond together under particular circumstances and opens further research to understand the type of relationship that linked them together.

Pathway Analysis Module

Although this protocol will address joint-pathway analysis using both transcriptomics and metabolomics data later, it is helpful to look at just metabolomics data alone to see patterns and inspect results [4, 5]. One must replace the metabolite names by the KEGG C number in the .csv file and perform a pathway analysis in MetaboAnalyst (Fig. 1). Since targeted metabolomics data is being used, make sure to upload a reference metabolome listing all the compounds that are being detected in the LC-MS/MS method (this is important for statistical reasons, as false significance would come from using an entire database of compounds in such a case). The results will indicate if the significantly different metabolites detected above are common members of a pathway. Keep notes on these results, as it is useful to see if trends with metabolomics data match the integrated analysis later on.

Biomarker Analysis Module

This tool is useful for identifying metabolite changes that are associated with a certain condition. This analysis allows the discovery of biomarkers based on receiver operating characteristic (ROC) curves, or manually specifying biomarker models to perform new sample predictions. This is a particularly powerful method if using very large metabolite sets and/or population-level sample data, for which it can display which metabolites are most important for explaining the data trends as a whole.

3.2 Transcriptomics

3.2.1 Embryo Collection and Preparation

1. When collecting samples, be sure to use tissue containing appropriate amounts of RNA for enough of a yield to obtain a good library. In general, plant samples have higher concentrations of RNA in younger tissues including leaves and seeds. Preliminary extractions may need to be performed to determine what amount of tissue is required for no less than ~150 ng/μL. A maximum of 1000 ng/μL is recommended, as RNA concentration instruments often give distorted results above this concentration. Serial dilutions generally will be necessary to prepare for RNA-Seq, so an excess concentration of RNA gives more control for dilution.
2. Vials or containers used for collecting tissue must be autoclaved for 30–60 min at 121 °C to remove RNases. In addition, any tool or surface that may come into contact with samples (forceps, spatulas, liquid nitrogen containers, etc.) must be RNase-free. If these items cannot be autoclaved, this is ideally done with a commercial solution such as RNaseZAP™ (Sigma-Aldrich, St. Louis, MO); however success has been achieved by performing a thorough scrubbing with EtOH (70% using RNase-free water). Be sure to use RNaseZAP or 70% EtOH on any working surfaces, nitrogen buckets, ice trays, etc.
3. Collected tissue should be immediately placed in liquid nitrogen in vials that are not fully sealed so as to prevent implosion. The time between collection and freezing should be as minimal as possible; if thawing occurs at any point, the sample should be discarded and recollected using fresh tissue. If collection of samples requires significant time, sample tubes can be placed on dry ice or in a rack in liquid nitrogen. Furthermore, samples can be stored at −80 °C following collection if not ready for processing.
4. Samples should be pulverized into as fine of a powder as possible while in liquid nitrogen. This can be done by placing sample vials into a rack seated in liquid nitrogen and carefully grinding by hand with compatible RNase-free plastic pestles. When working with tissue that is harder, such as seeds, it is important to begin the grinding slowly to ensure that tissue does not eject out when under pressure from the pestle. RNA yield increases exponentially with greater surface area for extraction, so the finest powder that can be obtained under frozen conditions is highly desirable. Samples can be placed in a −80 °C freezer if not being immediately used or kept in liquid nitrogen if extraction is to take place right away. If placing in a −80 °C freezer, wait several hours after placing in freezer before closing the vial lids to allow temperature to equilibrate and avoid implosion of the vial.

3.2.2 Reagent Preparation

1. Add 1 mL of 0.1% diethyl dicarbonate (DEPC) solution per 1000 mL of total (untreated) water, mix well, then cover bottle with foil, and allow to incubate overnight at ~35 °C. Vent bottle by loosely placing cap on and autoclave with a liquid cycle at 121 °C for 30 min.
2. Prepare 0.5 M 2,2′,2″,2‴-(ethane-1,2-diyldinitrilo)tetraacetic acid (EDTA) by adding 116.0 g EDTA to 500 mL of deionized water and place on a magnetic stirrer. Adjust pH to 8.0 using sodium hydroxide and wait until dissolved. Autoclave at 121 ° C for 30 min.
3. Prepare 1 M 2-amino-2-(hydroxymethyl)propane-1,3-diol (Tris) by adding 60.6 g Tris to 500 mL RNase-free water. Place on magnetic stirrer and adjust pH to 8.0 with hydrochloric acid.
4. Prepare a 24:1 (v/v) chloroform/3-methylbutan-1-ol (isoamyl alcohol) by adding 30 mL of isoamyl alcohol to 720 mL of chloroform in a 1 L bottle. Cover in foil to keep out of light.
5. Prepare 500 mL of 70% EtOH (v/v) in RNase-free water, and another 500 mL of 100% EtOH.
6. Commercial RNA-grade glycogen is the best option for this procedure, but RNase-free glycogen of any sort will work. Dilute glycogen to aliquots of 500 μg in autoclaved 1.5 or 2 mL snap-top centrifuge tubes.
7. Prepare 200 mL of 3.2 M sodium acetate (NaOAc) by adding 52.5 g to 200 mL RNase-free water. Place on magnetic stirrer and adjust pH to 5.5 using glacial acetic acid.
8. Prepare 200 mL of 8 M lithium chloride (LiCl) by adding 67.8 g LiCl to 200 mL RNase-free water.
9. Create the N,N,N-trimethylhexadecan-1-aminium bromide (CTAB) extraction buffer stock solution by adding 8.0 g of CTAB, 46.8 g of sodium chloride (NaCl), and 12 g of 1-ethenylpyrrolidin-2-one (polyvinylpolypyrrolidone, PVPP) to a 500 mL glass bottle. Add 400 mL of RNase-free water to bring the concentration of CTAB to 2% w/v, NaCl to 2 M, and PVPP to 3% w/v.
10. Add 40 mL of 1 M Tris–HCl and 20 mL of 0.5 M EDTA to the CTAB buffer to obtain a final concentration of 100 mM Tris–HCl and 25 mM EDTA. Add 0.5 mL of 0.1% DEPC to the buffer and incubate overnight at ~35 °C before autoclaving at 121 °C for 30 min. This will serve as the CTAB stock solution.
11. Prepare the working extraction buffer. Determine the volume necessary by multiplying each sample +2 for overage by 1050 μL. For 12 samples, 15 mL is enough and will be used in this example (all concentrations can be adjusted if more or less volume of extraction buffer is needed). Add 14.5 mL of the

CTAB stock solution to a RNase-free tube; then add 8.1 μL of N1-(3-aminopropyl)butane-1,4-diamine (spermidine) and 450 μL of 2-sulfanylethan-1-ol (2-mercaptoethanol).

3.2.3 RNA Extraction (See Note 4)

RNA extraction was adapted from Horn et al. [25].

1. For a 2 mL volume sample vial containing 1000 μg of tissue, a 600 μL volume CTAB extraction buffer heated to 65 °C is added. Ensure that PVPP is resuspended in buffer prior to adding to samples. This mass to volume ratio of sample to vial is recommended to be used; thus if using a larger or smaller amount, adjust reagent volumes in protocol accordingly. Density of sample tissue matters; however as long as ground tissue does not take up more than ~1/4 of the vial used for extraction, there should be enough contact of the tissue's surface area with the extraction buffer for adequate yield.
2. Vortex samples at 500–1000 rpm for 10 min at 65 °C.
3. Add 600 μL of 24:1 chloroform/isoamyl alcohol (v/v) to each sample and vortex at 500–1000 rpm for 2 min at 65 °C.
4. Centrifuge samples at 13,000 × *g* for 10 min at room temperature, and transfer upper aqueous phase to a 2 mL autoclaved microcentrifuge tube seated on ice.
5. Add 450 μL of CTAB extraction buffer to organic phase remaining in sample tubes for a sequential extraction, and vortex at 500–1000 rpm for 2 min at 65 °C.
6. Transfer upper aqueous phase to the same tube on ice used for **step 4**. Briefly vortex to combine and return to ice.
7. Add 800 μL of 24:1 chloroform/isoamyl alcohol (v/v), briefly vortex to mix, and then centrifuge at 12,000 × *g* for 15 min at 4 °C.
8. Transfer 300 μL of upper aqueous phase into a fresh autoclaved 1.5 mL microcentrifuge tube on ice; then transfer another 300 μL of the remaining upper aqueous phase to a second autoclaved 1.5 mL microcentrifuge tube. The division of sample into two separate tubes allows for the following precipitation to be carried out at a lower volume, thus resulting in a faster reaction with the RNA extract. Ensure that tubes are labeled as both subsamples will be recombined into one sample in further steps following precipitation.

3.2.4 Precipitations and Resuspension

All discarded supernatants/solutions should be placed appropriately in chemical waste.

1. Add 30 μL of 3.2 M NaOAc (pH = 5.5) to each subsample.
2. Add 75 μL of RNA-grade glycogen (0.5 mg/mL) to each subsample.

3. Add 600 μL of 100% EtOH to each tube. Vortex well for 5–10 s; then incubate for 12 h at −20 °C.
4. Centrifuge subsamples for 1 h at 13,000 × *g* and 4 °C. A small white RNA pellet may be visible at this point, but its absence does not indicate an unsuccessful extraction thus far. Carefully decant supernatant into waste without disturbing RNA pellet by slowly pouring out. This and further decanting should be done instead of removing supernatant with a pipette to reduce degradation/contamination and damage to the pellet.
5. Rinse subsamples by adding 1 mL of 70% EtOH; then vigorously flick sample to detach pellet from sidewall of microcentrifuge tube (the pellet should be seen floating in the solution, if a pellet was observed). If no pellet is present, flick and agitate sample for 1 min. Vortex subsamples at 500–1000 rpm for 15 min at 4 °C.
6. Centrifuge at 21,100 × *g* and 4 °C for 15 min. Again, carefully decant supernatant into waste.
7. Repeat **steps 2** and **3** two more times, resulting in a total of three washes of the RNA pellet.
8. Open tubes and place laterally in a laminar sterile flow hood for ~15 min, with the opening of the tube facing the flow to dry pellet. This is recommended instead of drying on ice because it takes less time, thus reducing the chance for contaminants to be introduced to the pellets.
9. Resuspend pellets in 103 μL of commercial-grade RNase-free water; then recombine the two corresponding, resuspended subsamples into a fresh, autoclaved 1.5 mL microcentrifuge tube. At this point there should once again be only one tube per sample.
10. Add 94 μL of 8 M LiCl to each sample. Vortex thoroughly for 5–10 s; then incubate for 12 h at −20 °C.
11. Centrifuge samples for 1 h at 13,000 × *g* and 4 °C. Carefully decant the small amount of LiCl supernatant into waste.
12. Rinse samples by adding 1 mL of 70% EtOH; then vigorously flick sample to detach pellet from sidewall of microcentrifuge tube (the pellet should be seen floating in the solution, if a pellet was observed). If no pellet is present, flick sample for 1 min. Vortex samples at 500–1000 rpm for 15 min at 4 °C.
13. Centrifuge at 21,100 × *g* and 4 °C for 15 min. Carefully decant supernatant into waste.
14. Repeat **steps 2** and **3** two more times, resulting in a total of three washes of the RNA pellet.

15. Open tubes and place laterally in a laminar sterile flow hood for ~15 min, with the opening of the tube facing the flow to dry pellet.
16. Dissolve pellet in commercial-grade RNase-free water. We recommend starting with 20–30 μL. Vigorously pipette the water up and down into the pellet until fully dissolved. If insoluble portions of the pellet are observed, remove liquid RNA solution, transferring to a new autoclaved 1.5 mL microcentrifuge tube. Discard insoluble pellet.

3.2.5 Quality Confirmation and Library Prep

1. An initial rough estimate of RNA quality can be determined via a spectrophotometer, with absorbance at 260/absorbance at 280 nm (A260/A280) in the range of 1.8–2.2. If the concentration exceeds 1000 ng/μL, dilute until it falls below this concentration, as A260/A280 can be distorted when analyzed at such a high concentration.
2. Run capillary gel electrophoresis to determine RNA integrity according to instrument manufacturer settings. The integrity is determined by an RNA integrity number (RIN, on a scale of 1–10), usually directly reported by the instrument. In plant tissue, a RIN of 8–10 can be confidently used for RNA-Seq. However, since plants can contain significant amounts of other ribosomal RNAs including those from the chloroplast as well as other small RNAs, the original algorithm used for RIN calculation may cause distortions. In the case where at least one significant band is observed other than the 18S and 28S ribosomal RNA, a returned RIN value of 7 is considered high-quality RNA that is suitable for RNA-Seq. Ensure at this step no genomic DNA is present; else a DNA cleanup kit may need to be used for further purification.

This protocol will work with RNA-Seq data using any type of read length, paired- or single-end reads, or strand specificity. However, it is very important to keep track of these parameters before proceeding to further steps as any downstream usage of RNA-Seq libraries will rely heavily on accounting for these. For instance, if using a typical Illumina TruSeq protocol, the reads are generally going to be strand-specific short reads that are paired-end, where the first strand is the reverse complement of the original cDNA template. The program FastQC is a good option for determining read quality, determining adapter or contaminant presence, among other information concerning usability of reads [19]. Instructions for interpreting FastQC data based on different read types are included. The remaining protocol assumes the reads are ready to be used.

3.2.6 Transcript Assembly: Reference Genome Available (Skip to Subheading 3.2.7 if There Is No Available Reference)

Open the Linux client; from here forward, all lines of computer code for Linux are prefaced by "$." First, all experimental reads must be combined for this step, because we need to use all the evidence we can for accurate construction of transcripts. This can be done using the "cat" function in this way, as appropriate depending on your library:

```
$ cat Read_fileA.fastq.gz Read_fileB.fastq.gz > Reads_com-
bined.fastq.gz
```

1. If a reference genome is available, download the full FASTA file along with any GFF/GTF (genome feature format) files if available. This can be done on NCBI (https://www.ncbi.nlm.nih.gov/), Phytozome (https://phytozome-next.jgi.doe.gov/), or any number of different databases. For this example, we will use a FASTA file called "example_genome.fasta" and an example GFF file called "example_genomeGFF.gff."
2. Reads obtained from RNA-Seq first need to be aligned (aka mapped) to the genome using a short-read mapper such as HISAT2 or STAR (https://github.com/alexdobin/STAR). This will allow for the transcripts we are ultimately trying to quantify to be oriented on the reference genome, which will help assist in the assembly of the transcripts in subsequent steps. We will perform this mapping using HISAT2. First, an index needs to be created from the reference genome that will allow HISAT2 to efficiently map reads to it using the following command:

```
$ hisat2-build example_genome.fasta Ref_genome -p 8
```

   This will build the index files using a base name "Ref_genome" (*see* **Note 5**).

3. We will then map reads to the reference. For paired-end strand-specific short reads (common with current Illumina protocols), the script needs to be tailored to include right- and left-end reads and the orientation of strand specificity too:

```
$ hisat2 -x Ref_genome -1 Reads_left.fastq.gz -2
Reads_right.fastq.gz –S
Illumina_paired_alignment.sam -p 8 –dta
--rna-strandness RF
```

   The --dta option tells the program to map in a way that will allow for downstream assembly to be performed. The --rna-strandedness RF tells HISAT2 that the reads are strand-specific and that the first strand corresponds to the reverse complement ("R") of the original cDNA template.

If performing using single-end short reads:

```
$ hisat2 -x Ref_genome -U Reads.fastq.gz –S
Illumina_single_alignment.sam -p 8 --dta
```

Finally, if performing using long reads, the parameters need to be relaxed a bit when mapping:

```
$ hisat2 -x Ref_genome -U LongReads.fastq.gz -S
Illumina_long_alignment.sam -p 8 --dta
--score-min L,0.0,-2
```

4. Once mapped, a SAM file that is generally very large (on the order of hundreds of GB) is generated. Compress this significantly by converting the alignment into a BAM file using the "view" function in samtools:

```
$ samtools view -S -b -@ 8
Illumina_paired_alignment.sam >
Illumina_paired_alignment.bam
```

5. The alignment then needs to be sorted using the "sort" function in samtools. This allows the entire alignment to be sorted by coordinates of the aligned reads in the reference genome:

```
$ samtools sort –o
Illumina_paired_alignment_sorted.bam -O bam -@ 8
Illumina_paired_alignment.bam
```

Finally, if it was necessary to perform several alignments due to computing limitations, they can be merged like so:

```
$ samtools merge -@ 8 -o alignments_merged.bam
alignment1.bam alignment2.bam alignment3.bam
```

6. The program StringTie will be used to assemble transcripts from the alignment file since we have a reference genome (and potentially a corresponding GFF file). If a GFF file is available with the reference genome, it is strongly recommended to be used as StringTie will allow the features included (i.e., UTRs, exons, start and stop sites, genes) to guide the process. In this case one would perform the assembly like this for the paired-end strand-specific case (leave off --rf if paired-end reads are not strand-specific or if using single-end reads):

```
$ stringtie -o StringTie_output.gtf –G
```

```
example_genomeGFF.gff -p 8
Illumina_paired_alignment_sorted.bam --rf
```

In the case of long reads:

```
$ stringtie -L -o StringTie_output.gtf -G
example_genomeGFF.gff -p 8
Illumina_long_alignment_sorted.bam
```

The output of StringTie is a GTF file that contains the positions of each transcript within the reference.

7. The resulting GTF file can then be fed into gffread to extract the FASTA sequences for each transcript into one multi-FASTA file. First, create an index to speed this step up using the "faidx" command in samtools:

```
$ samtools faidx StringTie_output.gtf -@ 8
```

Then feed the GTF file into gffread:

```
$ gffread -w stringtie_transcripts.fasta
-g example_genome.fasta StringTie_output.gtf
```

The resulting transcript FASTA will be used for calculating expression in later steps.

3.2.7 Transcript Assembly: No Reference Genome Available

Trinity requires a lot more computing power compared to StringTie and will potentially take a very long time if not using a high-performance computer. Be sure to utilize all available computing power in the script, as insufficient memory can cause a premature crash of the program (*see* **Note 6**).

1. If a reference is unavailable, a de novo assembly can be generated, where reads themselves are used to assemble transcript sequences via overlap based on a k-mer length. Trinity is an effective program for this and has been shown to surpass benchmarks desired for transcriptome assemblies. Trinity can be initiated like so for paired-end, strand-specific reads:

```
$ Trinity --seqType fq --left Reads_left.fastq.gz
--right Reads_right.fastq.gz --max_memory 16G
 --CPU 8 --full_cleanup --SS_lib_type RF --output./trini-
ty_out_paired
```

For our single-end reads:

```
$ Trinity --seqType fq --single Reads.fastq.gz
--max_memory 16G --CPU 8 --full_cleanup
--SS_lib_type RF --output ./trinity_out_single
```

Finally, if using long reads, one doesn't need any additional parameters with the single-end example:

```
$ Trinity --seqType fq --single LongReads.fastq.gz
--max_memory 16G --CPU 8 --full_cleanup
--SS_lib_type RF --output ./trinity_out_long
```

2. The primary output from Trinity is a .fasta file, so gffread is not needed. If one needs to merge several assemblies, the "transabyss-merge" command from Trans-ABySS is an effective tool. "transabyss-merge" will find redundant transcripts and keep the longer version and remove the shorter transcript. Since Trinity uses a constant k-mer size of 25, this needs to be accounted for in the merge command (if changed to any other setting, be sure to adjust --mink and --maxk accordingly):

```
$ transabyss-merge assembly1.fasta assembly2.fasta
assembly3.fasta --mink 25 --maxk 25
```

3.2.8 Gene to Transcript Matrix Preparation and Assembly Processing

It is important to note that at this point, all of the sequences in the FASTA file are at the transcript level, not gene level. For a true differentially expressed gene (DEG) analysis, we need to obtain a tab-separated text file (.TSV) with the first column listing all transcripts and the second column corresponding to each transcript's parent gene (it must be in this order). Trinity itself will output a file with suffix ".gene_trans_map," which will correspond transcripts to a parent "gene," which Trinity defines as a cluster of transcripts. The GTF output of the StringTie assembly will also give this information; the value in the "gene_id" field for each transcript feature is the gene, and the child transcript ID is in the "transcript_id" field. A quick way to format this is to use the AGAT package (*see* **Note 7**):

```
$ agat_sp_extract_attributes.pl –gff
 StringTie_output.gtf --att transcript_id,gene_id-t transcript
-o gene_to_transcripts.output -m
```

1. It is extremely important at this point to make sure that the assembly's .fasta file has correctly formatted headers corresponding to the "gene_to_transcripts.output" file. If the assembly .fasta file sequence IDs corresponding the sequences to the transcripts in the "gene_to_transcripts.output" or ". gene_trans_map" files are after any white space (tabs or spaces), they may get erased by filtering, and the way to correlate genes with the filtered transcripts will be lost. Particularly, if StringTie was used, the "gene_id" and "transcript_id" in the StringTie_output.gtf that were extracted above may get cut out of

the .fasta file after processing. If we have a "gene_to_transcripts.output" or ".gene_trans_map" formatted like the following:

```
Transcript_1_name Gene_1_name
Transcript_2_name Gene_1_name
Transcript_3_name Gene_2_name
Transcript_4_name Gene_3_name
Transcript_5_name Gene_4_name
```

then headers in the corresponding fasta file should look like this:

```
>Transcript_1_name
```

and not this:

```
>NonamEVm013594t1 Any; other; text; oid=Transcript_1_name;
```

If this latter example is the case, one option is to export the .fasta file to any text editor and delete anything before "`Transcript_1_name`" with the find and replace function; this is useful if it is formatted unusually. However, in this more common specific situation, everything in the header needs to be removed before and including the "`oid=`" character and then the "`;`" at the end, which can be done using the `sed` command like this:

```
$ sed 's/>.*oid=* */>/' -i
stringtie_transcripts.fasta
$ sed 's/;//' -i stringtie_transcripts.fasta
```

then check the headers to confirm they are correct:

```
$ head stringtie_transcripts.fasta
```

2. Assemblies generated via de novo programs such as Trinity can often contain many more isoforms than a guided assembly, and even after merging there may be significant amounts of fragments, duplicates, and chimeras. There are many programs available for filtering these assemblies. In this case we will use "tr2aacds" from EvidentialGene to filter our raw assembly (either "stringtie_transcripts.fasta" or the Trinity output .fasta file shown here as "assembly.fasta"):

```
$ tr2aacds.pl -log -cdna assembly.fasta -tidy
-NCPU 8 -MAXMEM 8000
```

(*see* **Note 8**).

Navigating the output folders, find the file named something like "assembly.okay.mrna," usually in the "okayset" folder. This will be your filtered assembly. For the purpose of this protocol, we will rename "assembly.okay.mrna" to "assembly.fasta." If a StringTie assembly is being used, it is generally less important to do this step as StringTie is more conservative in its approach to constructing transcripts (however it should be considered depending on the assembly).

3.2.9 Transcript Quantification

1. With the final assembly in hand, the reads can be mapped directly to it using Salmon. Note that many quantification programs require that the reads be mapped to a reference genome, but Salmon is convenient because the reads are mapped directly to the transcriptome assembly. First, align reads for each biological sample again using HISAT2 but this time to the assembly (indexed first), in order to obtain a separate BAM file for each biological sample. It is important to note that the resulting BAM alignments should not be sorted, which is required by Salmon for quantification:

```
$ hisat2-build assembly.fasta assembly -p 8
$ hisat2 -x assembly -1 Sample1_Reads_left.fastq.gz
-2 Sample2_Reads_right.fastq.gz –S
Sample1_alignment.sam -p 8 --rna-strandness RF
$ samtools view -S -b -@ 8 Sample1_alignment.sam > Sample1_a-
lignment.bam
```

Then, perform the quantification on the resulting BAM file:

```
$ salmon quant -t assembly.fasta -l A –a
Sample1_alignment.bam -o ./sample1_algn_quant
```

The resulting output will be a file with suffix "quant.sf," which contains the transcripts per million (TPM) and the number of reads mapped for each transcript, as well as some other information (*see* **Note 9**).

2. Since quantification is being done, it is time to switch to R; from here forward all lines of computer code for R are prefaced by ">." The data has to be imported in a specific way for the DEG analysis to work. A convenient pipeline for this whole process is mediated by the "tximport" package. This will allow the "quant.sf" files generated by Salmon to be directly imported into R for DEG analysis. First, convert the

"gene_to_transcripts.output" file into .csv format to read into R. Then, in R, load tximport and import it:

```
> library(tximport))
> gene.and.trans <- read.csv("/path/to/gene_to_transcripts.
out put.csv")
```

Then, create a working directory in the location containing the "salmon" output folders:

```
> workdir <- "/path/to/my/salmon/output"
```

Next, import into R a list of the samples that will be included in the analysis. The list should be a .txt file ("sample_-names.txt" here, with control and treated samples) formatted so that each sample listed corresponds to the folder name containing each sample's "quant.sf" file in the working directory:

```
sample_name
Control1
Control2
Control3
Control4
Treated1
Treated2
Treated3
Treated4
```

taking care to have the header on line 1. You can then read this list into an R object along with each sample's corresponding folder in the working directory:

```
> sample_names <-
read.table(file.path(workdir,"sample_names.txt"),
header = TRUE)
> samples <- file.path(workdir,sample_names
$sample_name, "quant.sf")
> names(samples) <- c("Control1","Control2","Control3",
"Control4","Treated1","Treated2","Treated3","Treated4")
```

Now, to put all expression data in one place using tximport, with the transcripts and corresponding genes also taken into account:

```
> tx.imported.data <- tximport(samples,
type = "salmon", tx2gene = gene.and.trans,
```

```
dropInfReps = TRUE)
```

3. Next load "DESeq2" (> library(DESeq2)) and create a data frame associating each sample to its corresponding treatment factor, and assign the raw counts from each "quant.sf" to the samples:

```
> DESeq2.dataframe <- data.frame(condition =
factor(rep(c("Control", "Treated"), each = 4)))
```

Note that the "`each`" argument indicates how many replicates there are per treatment factor.

```
> rownames(DESeq2.dataframe) <- colnames
(tx.imported.data$counts)
> DESeq2.DEG.input <- DESeqDataSetFromTximport
(tx.imported.data, DESeq2.dataframe, ~condition)
```

4. It is a good idea to filter out lowly expressed transcripts, since these may either be misassemblies from StringTie/Trinity or transcripts that are so lowly expressed that any change would be considered biologically irrelevant to the question. For this example, we will consider that to be any transcript in which the sum of all samples' counts is under 20:

```
> filtered.data <-
rowSums(counts(DEseq2.DEG.input)) >= 20
> DESeq2.DEG.input <-
DESeq2.DEG.input[filtered.data,]
```

5. Finally, run the DESeq2 analysis and collect the results into a . csv file:

```
> DESeq2.DEG.analysis <- DESEq(DESeq2.DEG.input)
```

Then, make a results object:

```
> DESeq2.output <- results(DESeq2.DEG.analysis,
contrast=c("condition","Treated","Control"))
```

Note that when setting the "contrast" argument, the treatment factor that is desired to be the reference is placed last. To get our results out at the significance level desired ($p_{\text{adj}} = 0.01$ here), sort and write to a .csv file:

```
> DESeq2.output.significant <- subset
(DESeq2.output, padj < 0.01)
```

```
> write.csv(DESeq2.output.significant,
"DEG_results.csv")
```

Otherwise, to get the results for all genes regardless of significance, disregard the "`subset`" command.

3.2.10 Annotation

1. After quantification, one may want to annotate the transcript sequences so that one can get an idea of the function of each transcript (and thus corresponding gene). This will be performed in the Linux console. A simple way to do this is to obtain the fasta of proteomes (on NCBI they contain the format ".faa") of several related annotated species and merge into one file:

```
$ cat species1_aas.fasta species2_aas.fasta
species3_aas.fasta > database.fasta
```

2. Blast2GO is a simple to use desktop program for identifying homologs in the assembly via homology to the transcripts in the database just created. Once installed, import the final assembly by clicking the arrow in the "start" box and selecting "Load Sequences (fasta)." Then, click the arrow by the "blast" box; select "Make Blast Database" to convert the fasta database "database.fasta" to the formats needed by Blast2GO to perform a search. Once completed, click the "blast" box and select "Local Blast" and click "Next >." Use the "blastx-fast" option in "Blast Program" so that the transcripts assembled are searched based on their expected protein product(s). Leave all other options at default and click "Next >" until you can hit "Run."
3. Export the annotation as a fasta and/or as a table for later use as needed by going to "File > Export>" and selecting the appropriate option in the menu.

If interested in the identity of the gene (not transcript) sequences themselves and using the StringTie procedure, the locus coordinates of the transcript features that are associated with each gene must be first extracted. Basically, the coordinates of the genome are being isolated from which the transcript (s) originate(s), obtaining the fasta sequences inside these genome coordinates, and doing a homology search on those genome sequences to predict the protein product encoded by their child transcript(s). The following script will take the StringTie GTF file and format as a tab-separated file (TSV):

```
$ agat_convert_sp_gff2tsv.pl –gff
StringTie_output.gtf -o StringTie_output.tsv
```

Then extract only the information necessary for creation of a BED file. This can be done by rearranging the format with awk:

```
$ awk -v OFS='\t' '$3=="gene"' StringTie_output.tsv
 > stringtie_coords.1.tsv
$ awk -v OFS='\t' '{print $1,$4,$5,$10,$6,$7}'
 stringtie_coords.1.tsv > stringtie_coords.2.tsv
$ awk -v OFS='\t' '$4 != "N/A"' strin
 gtie_coords.2.tsv > stringtie_coords.3.tsv
$ awk -v OFS='\t' '{sub("-1","-",$6);print}'
 stringtie_coords.3.tsv > stringtie_coords.4.tsv
$ awk -v OFS='\t' '{sub("1","+",$6);print}'
 stringtie_coords.4.tsv > stringtie_coords.bed
```

This BED file will then be used by bedtools to obtain the genome sequences, using the original genome reference fasta as the sequence source:

```
$ bedtools getfasta -fi example_genome.fasta -bed
 stringtie_coords.bed -s -name -fo
 stringtie_genes.fasta
```

These fasta sequences can then be put into Blast2GO as above to predict identity (*see* **Note 10**).

The identity of Trinity gene sequences cannot be extracted for two reasons, primarily because no reference genome sequence was available and also because the definition of "gene" in Trinity is a "cluster" of transcripts with similar sequence content and is thus not confirmed to originate from the same locus of a genome. If a quality annotation was done and the right databases were used, one can infer the function of the gene itself based on what each transcript of the gene (i.e., cluster) was identified as, assuming consistency across all transcripts of that gene.

3.2.11 Overrepresentation Analysis

Since a typical RNA-Seq dataset contains a large sample size of potentially tens of thousands of transcripts, including an overrepresentation on transcriptome data alone is recommended (there may be strong trends that can only be observed when using this dataset alone). There are numerous programs and methods for performing gene set overrepresentation analysis as well as characterization. This protocol uses the Gene Ontology Project database, which groups genes based on three types of categorization domains: biological process, molecular function, and the cellular component the gene/gene function is associated with. Each of these domains consists of a large hierarchy of further categories (or "terms") that fall under their scope. For the purpose of this protocol, the assumption is that the list of DEGs is what is being analyzed. The use of this analysis will allow for the user to determine if specific pathways, reactions, functions, cell compartments, etc. are statistically overrepresented by the DEG gene list, which can identify what kind of processes are altered between treatment factors.

1. The PANTHER knowledgebase will be used for GO analysis. Go to the PANTHER website (pantherdb.org) and click on the "Tools" tab on the main page, and select the "Gene List Analysis" on the tools tab. This tool will allow the following to be put as input, according to the boxes pictured (Fig. 2): (1) a text box to enter a set of genes (DEGs or any genes of interest) and options for the input format, (2) an option to select organism (most closely related to your species if your specific species isn't listed), and (3) an option to select what kind of analysis to perform.
2. To determine if the DEGs are associated with any GO term (s) in particular, only the list of IDs of the DEGs are needed (no quantification data is necessary here). PANTHER allows for multiple kinds of IDs to be used, but the most common are RefSeq, UniProt, and Ensembl accession numbers/IDs, and for plants TAIR IDs work as well. Simply use the Blast2GO output to assign these homolog IDs to the corresponding DEGs. For this example, multiple TAIR IDs (all DEGs in this example) were entered into box 1, selected *Arabidopsis thaliana* in box 2, and "statistical over-representation test" in box 3. After selecting for the overrepresentation test in box 3, there are numerous options for which databases to analyze. GO domains are often returned as massive lists of categories at every hierarchical level, making conclusions a bit obscure. There is a "slim" version of these databases in the PANTHER knowledgebase which includes only the most common subsets of GO terms of interest put together; these are shown as "PANTHER GO-Slim." For this example, we are interested in biological processes associated with our DEG list, so we will select "PANTHER GO-Slim Biological Process." Click on "submit."
3. The next form allows for submission of a reference list using many different options, which is useful if one wants to search DEGs against a custom database. For this purpose, we will simply select "Arabidopsis thaliana genes" under the "Default whole-genome lists" box. Finally, at the analysis summary screen (Fig. 3), parameters can be selected to be used for the test.
4. Select "Fisher's Exact" test type and use the "calculate false discovery rate" option. The false discovery rate (FDR) will allow for a greater power in estimation of what GO terms are

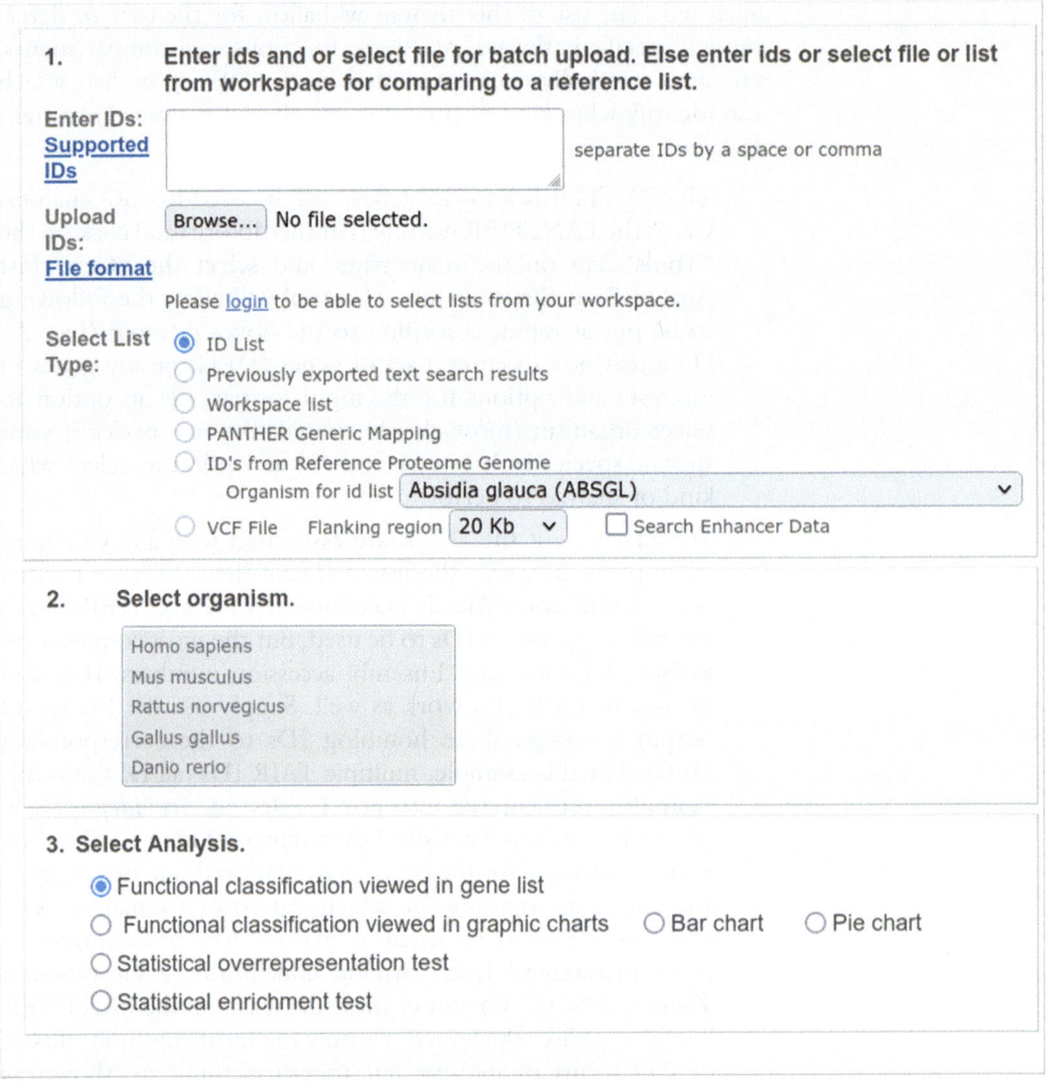

Fig. 2 Input form for gene ontology analysis on PANTHER website

overrepresented but may increase the rate of falsely significant hits; if a more conservative approach is desired, select the Bonferroni correction. Finally, click "Launch analysis" and you will see a result similar to Fig. 4 if any significant hits were found.

In this example, the most specific GO terms that are significant are "response to osmotic stress," "response to heat," "response to gibberellin," "response to wounding," and "regulation of defense response." The parent terms of these significant child GO terms are also listed if they are significant, which will give a more bird's-eye view about which biological processes are significantly overrepresented by our DEG list. The correct interpretation of these data is

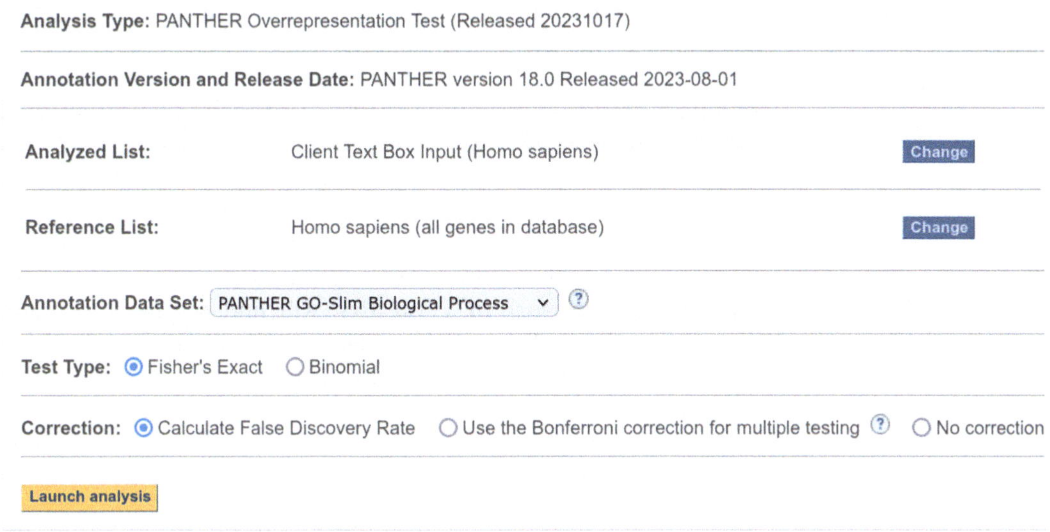

Fig. 3 Summary of analysis and options using PANTHER-Go Slim overrepresentation test

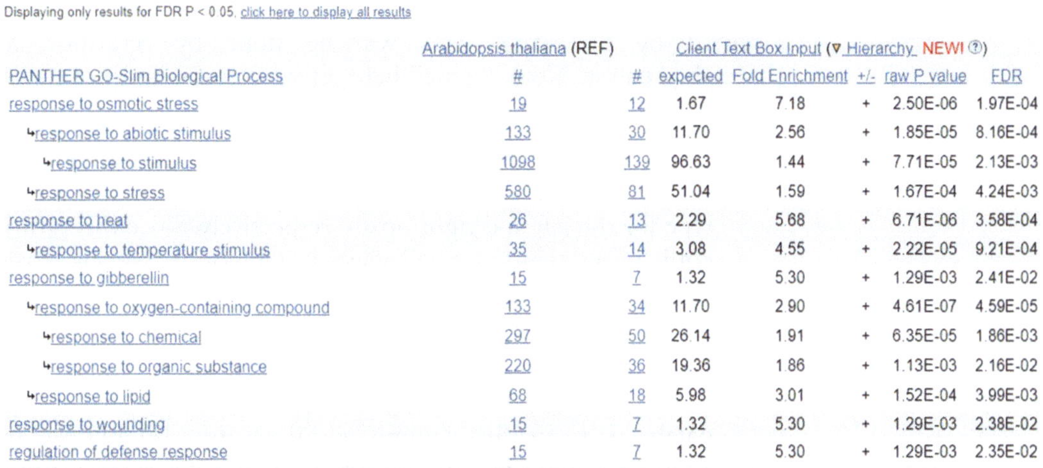

Displaying only results for FDR P < 0.05, click here to display all results

	Arabidopsis thaliana (REF)	Client Text Box Input (▽ Hierarchy NEW! ?)					
PANTHER GO-Slim Biological Process	#	#	expected	Fold Enrichment	+/-	raw P value	FDR
response to osmotic stress	19	12	1.67	7.18	+	2.50E-06	1.97E-04
↳response to abiotic stimulus	133	30	11.70	2.56	+	1.85E-05	8.16E-04
↳response to stimulus	1098	139	96.63	1.44	+	7.71E-05	2.13E-03
↳response to stress	580	81	51.04	1.59	+	1.67E-04	4.24E-03
response to heat	26	13	2.29	5.68	+	6.71E-06	3.58E-04
↳response to temperature stimulus	35	14	3.08	4.55	+	2.22E-05	9.21E-04
response to gibberellin	15	7	1.32	5.30	+	1.29E-03	2.41E-02
↳response to oxygen-containing compound	133	34	11.70	2.90	+	4.61E-07	4.59E-05
↳response to chemical	297	50	26.14	1.91	+	6.35E-05	1.86E-03
↳response to organic substance	220	36	19.36	1.86	+	1.13E-03	2.16E-02
↳response to lipid	68	18	5.98	3.01	+	1.52E-04	3.99E-03
response to wounding	15	7	1.32	5.30	+	1.29E-03	2.38E-02
regulation of defense response	15	7	1.32	5.30	+	1.29E-03	2.35E-02

Fig. 4 GO overrepresentation test with TAIR IDs using PANTHER GO-Slim

that more DEGs have membership in each GO term with p_{adj} (FDR here) < 0.05 than would be expected by chance. For instance, if osmotic stress genes weren't actually differentially expressed between the treatment factors (in this example, oilseed crop variety), the probability that 12 of the genes involved in osmotic stress are in the DEG list out of 19 total in the GO term is so low that the null hypothesis can be rejected, concluding that osmotic stress is altered between treatment factors.

3.2.12 Coexpression Analysis

Experiments dealing with many treatment factors (i.e., multiple accessions of an organism, samples from different geographical locations, etc.) may require the investigator to perform an analysis looking at trends of response variables with each other rather than each treatment factor's specific response to individual variables. A coexpression analysis is a powerful tool for determining how genes, whether specific genes or large modules of genes, are correlated to one another based on expression. Certain coexpression approaches generally require the entire transcriptome to be analyzed; however performing a network in a way that will find central hub genes (those centrally tied to treatment factor effects) can be effectively performed on just DEGs. Furthermore, biological processes can be correlated with the coexpression results via GO terms. An effective protocol for performing this using a DEG set of this type is available in Orlando Contreras-López et al. [26].

3.3 Multi-omics Integration

We now describe a routine integration of both transcriptomic and metabolomic data into a pathway analysis. This will assist in gaining biological context from our results, and we can use a database that contains pathways with genes and metabolites which are represented by our data. This will be done by returning to MetaboAnalyst. Note: "gene" here refers to transcript, not locus, in a typical transcriptomic dataset.

1. In the main module list (Fig. 1), select joint-pathway analysis. Before entering data, select the organism and metabolomics type for the left and right boxes, respectively. Since this protocol is using plants, we will select "Arabidopsis thaliana (thale cress)" for the organism gene names and "Targeted (compound list)" for metabolomics type.
2. Enter data as tab-separated text for both options. The ID type must be specified for data to be properly connected to the databases. In the first column, add the feature ID followed by a tab-separated value for log fold-change. Click submit. Once submitted, MetaboAnalyst will automatically tell which input features were properly correlated with the database. Be sure to check this for both genes and compounds prior to continuing.
3. There are now a few important considerations to be made for an accurate analysis. For "Pathway Database," select "Metabolic pathways (integrated)." This is appropriate when one is only interested in how both transcriptomic and metabolomic data work together; if also interested in gene-only pathways, one can select all pathways for this option.
4. The algorithm parameters then need to be specified under "Algorithm Selection" prior to running the analysis. As above, this can change depending on what the user is looking for. In a typical integrative analysis where hub features (genes

or metabolites) are being searched for that connect large branches of metabolic pathways, betweenness centrality is a good choice. If only interested in immediate relationships of pathway features, degree centrality is a quick choice that will return the hub features simply connected to the most other features directly as a first neighbor.

5. Finally, the integration method must be carefully considered. Targeted metabolomics datasets are generally many orders of magnitude smaller than transcriptomic datasets, as most RNA-Seq platforms simply collect all mRNA present while our metabolomics protocol contains pre-selected metabolites of interest. Therefore, combining queries in this case would prove statistically inappropriate because there may be thousands of significant genes, but no more than the maximum amount of measured metabolites as significant metabolites in targeted metabolomics are often only ~100. For getting around this, the software can run the metabolite data alone, simultaneously run the gene data alone, and then combine results by selecting one of three different assumptions about feature weights [4, 5]:
   - Universally consider the relevance of metabolites and genes to be the same regardless of the pathway analyzed or database used; do this by selecting "combine *p* values (unweighted)." This may be helpful if one is trying to do a quick screen of affected pathways making sure both datasets are considered no matter what. In this situation, 4 significant metabolites would hold the same weight as 25 significant transcripts in any given pathway. Thus, the effect of the significant differences on the overall pathway activity would equally be considered regardless of how many significant features there were of each type; one can think of this as getting results where half the effect on the pathway is forced to be from metabolites and half is considered to be from genes. One downside of this approach is that if one is working with databases or pathways that contain many more genes than measurable metabolites (which they often do), one could be artificially inflating the relevance of significant metabolites or artificially deflating the relevance of a significant genes since *p*-value weights are forced to be equal in both data types. This is common (among other reasons) due to the fact that most pathways/organisms often have multiple genes encoding for one reaction but generally only a reactant and product metabolite for said reaction.
   - Take into account the amount of metabolites and genes in the entire database when weighing results. Therefore, if a database contains 500 metabolites but 6000 genes, the fact that metabolites are less well-represented is accounted for in

the calculations. If one is looking more at global-scale effects on metabolism, where simply the number of pathways that are affected is of interest, this can be an effective option. The downside here is that one does not have much conclusion power with regard to specific pathways; although there is a 1–13 ratio of metabolites to genes in the database searched, the most significant pathways may be inaccurately represented, because that individual pathway could have a metabolite to gene ratio much lower, such as 1–100. In this situation, the user may erroneously conclude that the top pathway in the results was indeed the most affected; however the weights of features would be completely misapplied if trying to judge effects on that pathway versus others in the results.

- Take into account separately the amount of metabolites and genes for each pathway contained in the database. If one wants to make biological interpretations by how certain pathways are ranked in comparison to others, this is the most appropriate option.

Following submission, proceed to observe the results.

3.4 Interpretation of Results

An overview of the pathway analysis is displayed, showing a graph depicting various metabolic pathways and how significantly they were affected by the experimental treatment. This is based on two metrics: (a) a log-transformed *p*-value on the *x*-axis indicating the overrepresentation ("enrichment" in MetaboAnalyst) of the differentially expressed features and (b) a pathway impact value on the *y*-axis, which indicates how perturbed a specific pathway was based on the location of significantly different features *within that pathway*. Hence, if using betweenness centrality as mentioned in Subheading 3.3, having a significantly different gene/metabolite in a central location (e.g., hub) within a metabolic pathway would result in high pathway impact, which increases as it approaches 1.00. Staying consistent with the example stated in the introduction, consider that the results in Fig. 5 display the joint pathway analysis output when comparing embryos between the high-yielding oilseed crop variety "A" and the poorly yielding variety "B".

Notable KEGG pathways overrepresented ("enriched") by the significantly different features are the pentose phosphate pathway and carbon fixation in photosynthetic organisms, while the position of the significant features within the pathway was most influential in glycolysis and starch and sucrose metabolism [27, 28]. When clicking on an individual pathway, a detailed figure will be displayed (Fig. 6), showing the position of significant features within the pathway, and whether they were up- or downregulated. Now that the results have been obtained, multiple types of conclusions can be made.

1. As shown in the pathway impact figure, most of what is generally considered central carbon metabolism (glycolysis, pentose phosphate pathway, tricarboxylic acid cycle, carbon fixation) is significantly impacted and overrepresented. In the example stated and shown in Fig. 5, variety "A", with higher oil content in its seeds, has higher flux of overall carbon flow toward oil. Previous research on oilseed species pennycress found that all pathways involved in central carbon metabolism, when comparing a high- and low-oil accession of the species, were significantly impacted [2]. Furthermore, upon inspecting which features were up- and downregulated in each pathway, evidence was found that increased glycolytic flow, malate precursor accumulation, increased pentose phosphate pathway activity, and increased anaplerotic pathway activity were occurring in the high-oil accession. Such comparisons between a high-yielding oilseed crop variety "A" and the poorly yielding variety "B" have the potential to identify targets to genetically engineer this crop to improve its oil content. This kind of inference may be

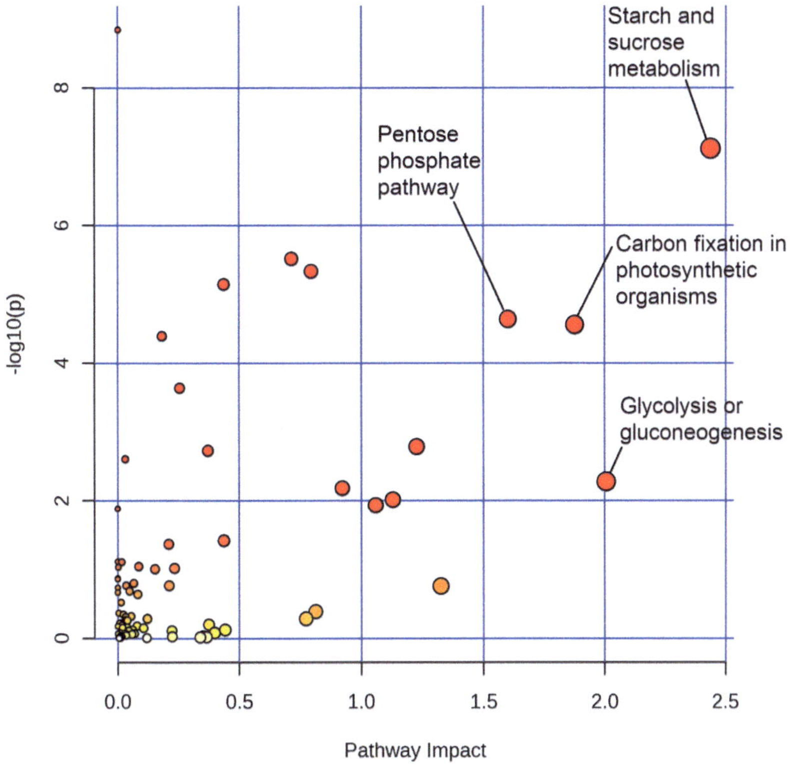

Fig. 5 Example results from joint pathway analysis in MetaboAnalyst, with select pathways highlighted. (Adapted using open metabolomic and transcriptomic data from Johnston et al. [2])

displayed using Fig. 6: *PGL1* and *PFK5* (green nodes) are highly downregulated in oilseed variety "A", while metabolite content shows consistent upregulation. With this information, one could investigate a further hypothesis that carbon is redirected toward pentose phosphate pathway intermediates, which would result in higher production of NADPH to support oil synthesis [2, 29]. Furthermore, it appears that the rate of converting intermediates is slowed down at least in part due to reduced expression of *PFK5* and/or *PGL1*.

2. Sometimes, more specific metabolic differences are observed within specialized parts of metabolism. Research on eggplant identified a significant contribution of glycine, serine, and

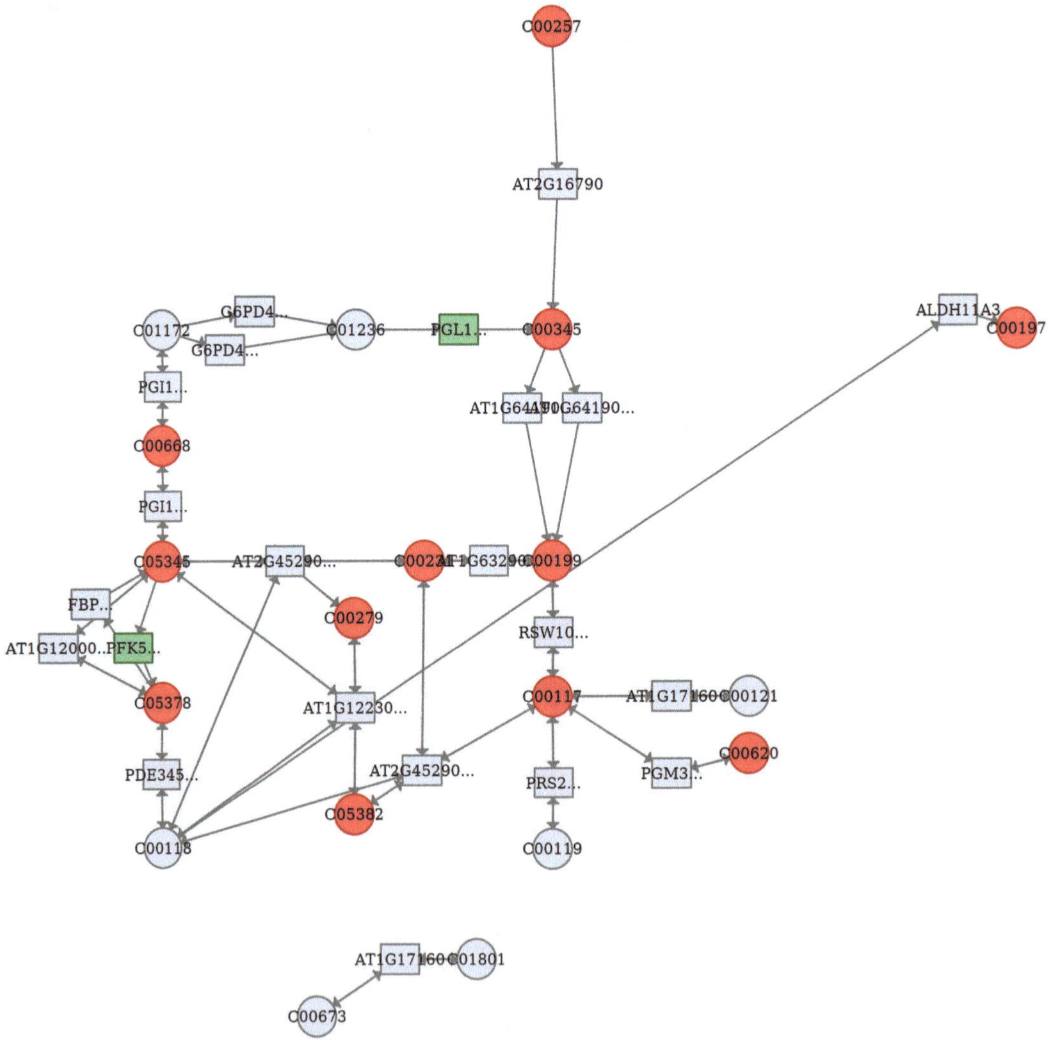

Fig. 6 Detailed view of pentose phosphate pathway from joint pathway analysis, including example data. (Adapted using open metabolomic and transcriptomic data from Johnston et al. [2])

threonine metabolism to nitrogen use efficiency, as serine was accumulated in nitrogen-use-inefficient genotypes of the species [30]. Furthermore, increased accumulation of starch was shown in the inefficient genotypes, which conceivably is related to carbon redirection toward storage.

3. The application of this methodology is not limited to species producing conventional crop products. One group used joint pathway analysis to identify orchid responses to biotic stress, showing that flavonoid biosynthesis in the plant was significantly impacted by fungal interaction [31]. This research identified multiple secondary metabolites that accumulated in response to the interaction and proposed that this interaction could be exploited to use the *Dactylorhiza* sp. orchids as an agronomic source of such compounds.
4. This methodology is equally applicable to animal systems provided proper considerations are made. Wooden breast myopathy, a trait negatively affecting chicken meat quality, was investigated by a research group using combined transcriptomics and metabolomics [32]. The pathway analysis directed researchers to find specific metabolic associations with progression of the disease, one of which was that downregulation of carnosine synthase expression. This was corroborated by the inclusion of metabolomics, which showed that the products of carnosine synthase, carnosine and anserine, were also reduced in diseased samples.

4 Notes

1. The guides and tutorials for the Analyst software are installed automatically with the software and are available from the Start menu. For more information on how to use the Analyst software, please check the software manual: https://sciex.com/content/dam/SCIEX/pdf/software/an_ref_d1000064246_en.pdf.
2. The use of the samples without dilution is discouraged, as we have observed considerable peak distortions in such cases.
3. For amino acids, switch between positive and negative polarities was used. Negative ionization was selected to specifically differentiate homoserine from threonine that has unique product ions and therefore transitions.
4. For leaf tissue, RNA concentration is generally high enough to obtain a good yield and quality from kits. The following extraction protocol applies to seeds and other tissues that may contain oils, lignin, high protein content, and other insoluble biomass components.

5. Commands using -p or -@ indicate how many cores are to be used. If performing on a supercomputer, you may want to increase this to save time and use all available computing power. These steps can take quite some time otherwise.
6. The flag "--max memory" tells Trinity how much memory to use per core. This example is thus assuming each core as 2 GB of memory.
7. A lot of warnings and notices may appear while AGAT is running. This is usually just logging information; just make sure the output is correct in these cases.
8. The `-maxmem` flag in this case is in MB, not GB.
9. This is assuming the biological sample reads are paired-end and strand-specific Illumina reads. If not, use the appropriate algorithm in the HISAT2 instructions above. Since we are no longer doing assembling, the "`--dta`" option is not necessary.
10. Generally, this predicted gene product will be the same as all of the gene's corresponding child transcript(s), but it is good practice to confirm.
11. Different instruments or sequencing data from different technologies always require adjustments when performing research. Even two groups using their own instrument of the model may find a big difference in signal intensity, reads per lane, sequencing errors, etc.
12. Programs are regularly updated, so options and menus may differ if using different versions.
13. What is considered statistically appropriate is always changing. Assumptions about one's data and what kind of analysis is to be performed may be considered void with the advent of new methods. We encourage regular literature reviews to stay current and obtain the maximum research impact possible.

Acknowledgments

- This work was supported by the United Soybean Board grant no. 2332-203-0102 to APA.
- This work was supported by the Department of Energy Office of Science, Office of Biological and Environmental Research (BER), grant # DE-SC0020325 to APA.
- This work is supported by Agricultural and Food Research Initiative (AFRI) grant no. 2021-67034-35137 to CRJ and 2021-67013-3777 to APA from the USDA National Institute of Food and Agriculture.

- Any opinions, findings, conclusions, or recommendations expressed in this publication are those of the author(s) and do not necessarily reflect the view of the US Department of Agriculture.
- The authors would like to acknowledge the BioAnalytical Facility (bdi.unt.edu/baf) and Jean-Christophe Cocuron at the University of North Texas.

References

1. Subramanian I, Verma S, Kumar S, Jere A, Anamika K (2020) Multi-omics data integration, interpretation, and its application. Bioinform Biol Insights 14:117793221989905. https://doi.org/10.1177/1177932219899051
2. Johnston C, García Navarrete LT, Ortiz E, Romsdahl TB, Guzha A, Chapman KD, Grotewold E, Alonso AP (2022) Effective mechanisms for improving seed oil production in pennycress (Thlaspi arvense L.) highlighted by integration of comparative metabolomics and transcriptomics. Front Plant Sci 13: 943585. https://doi.org/10.3389/FPLS.2022.943585/FULL
3. Cocuron JC, Alonso AP (2014) Liquid chromatography tandem mass spectrometry for measuring 13C-labeling in intermediates of the glycolysis and pentose phosphate pathway. Methods Mol Biol 1090:131–142. https://doi.org/10.1007/978-1-62703-688-7_9
4. Pang Z, Zhou G, Ewald J, Chang L, Hacariz O, Basu N, Xia J (2022) Using MetaboAnalyst 5.0 for LC–HRMS spectra processing, multi-omics integration and covariate adjustment of global metabolomics data. Nat Protoc 17(8):1735–1761. https://doi.org/10.1038/s41596-022-00710-w
5. Pang Z, Chong J, Zhou G, De Lima Morais DA, Chang L, Barrette M, Gauthier C, Jacques PÉ, Li S, Xia J (2021) MetaboAnalyst 5.0: narrowing the gap between raw spectra and functional insights. Nucleic Acids Res 49: W388–W396. https://doi.org/10.1093/NAR/GKAB382
6. Cocuron JC, Tsogtbaatar E, Alonso AP (2017) High-throughput quantification of the levels and labeling abundance of free amino acids by liquid chromatography tandem mass spectrometry. J Chromatogr A 1490:148–155. https://doi.org/10.1016/j.chroma.2017.02.028
7. Cocuron JC, Anderson B, Boyd A, Alonso AP (2014) Targeted metabolomics of Physaria fendleri, an industrial crop producing hydroxy fatty acids. Plant Cell Physiol 55:620–633. https://doi.org/10.1093/pcp/pcu011
8. Kim D, Langmead B, Salzberg SL (2015) HISAT: a fast spliced aligner with low memory requirements. Nat Methods 12:357–360. https://doi.org/10.1038/nmeth.3317
9. Li H, Handsaker B, Wysoker A, Fennell T, Ruan J, Homer N, Marth G, Abecasis G, Durbin R (2009) The sequence alignment/map format and SAMtools. Bioinformatics 25: 2078–2079. https://doi.org/10.1093/bioinformatics/btp352
10. Pertea M, Pertea GM, Antonescu CM, Chang TC, Mendell JT, Salzberg SL (2015) StringTie enables improved reconstruction of a transcriptome from RNA-seq reads. Nat Biotechnol 33(3):290–295. https://doi.org/10.1038/nbt.3122
11. Pertea M, Pertea G (2020) GFF utilities: GffRead and GffCompare. F1000Res 9:304. https://doi.org/10.12688/F1000RESEARCH.23297.1
12. Grabherr MG, Haas BJ, Yassour M, Levin JZ, Thompson DA, Amit I, Adiconis X, Fan L, Raychowdhury R, Zeng Q, Chen Z, Mauceli E, Hacohen N, Gnirke A, Rhind N, Di Palma F, Birren BW, Nusbaum C, Lindblad-Toh K, Friedman N, Regev A (2011) Full-length transcriptome assembly from RNA-seq data without a reference genome. Nat Biotechnol 29(7):644–652. https://doi.org/10.1038/nbt.1883
13. Robertson G, Schein J, Chiu R, Corbett R, Field M, Jackman SD, Mungall K, Lee S, Okada HM, Qian JQ, Griffith M, Raymond A, Thiessen N, Cezard T, Butterfield YS, Newsome R, Chan SK, She R, Varhol R, Kamoh B, Prabhu AL, Tam A, Zhao Y, Moore RA, Hirst M, Marra MA, Jones SJM, Hoodless PA, Birol I (2010) De novo assembly and analysis of RNA-seq data. Nat Methods 7(11): 909–912. https://doi.org/10.1038/nmeth.1517
14. Patro R, Duggal G, Love MI, Irizarry RA, Kingsford C (2017) Salmon provides fast and bias-aware quantification of transcript

expression. Nat Methods 14(4):417–419. https://doi.org/10.1038/nmeth.4197

15. Quinlan AR (2014) BEDTools: the Swiss-army tool for genome feature analysis. Curr Protoc Bioinformatics 47:11.12.1–11.12.34. https://doi.org/10.1002/0471250953.BI1112S47
16. Li H (2018) Minimap2: pairwise alignment for nucleotide sequences. Bioinformatics 34:3094–3100. https://doi.org/10.1093/BIOINFORMATICS/BTY191
17. Love MI, Huber W, Anders S (2014) Moderated estimation of fold change and dispersion for RNA-seq data with DESeq2. Genome Biol 15:550. https://doi.org/10.1186/s13059-014-0550-8
18. Soneson C, Love MI, Robinson MD (2015) Differential analyses for RNA-seq: transcript-level estimates improve gene-level inferences. F1000Res 4:1521. https://doi.org/10.12688/F1000RESEARCH.7563.2
19. Andrews S (2010) FastQC: a quality control tool for high throughput sequence data. https://www.bioinformatics.babraham.ac.uk/projects/fastqc/
20. Conesa A, Götz S, García-Gómez J, Terol J et al (2005) Blast2GO: a universal tool for annotation, visualization and analysis in functional genomics research. Bioinformatics 21(18):3674–3676. academic.oup.com
21. Mi H, Muruganujan A, Huang X, Ebert D, Mills C, Guo X, Thomas PD (2019) Protocol Update for large-scale genome and gene function analysis with the PANTHER classification system (v.14.0). Nat Protoc 14(3):703–721. https://doi.org/10.1038/s41596-019-0128-8
22. Thomas PD, Ebert D, Muruganujan A, Mushayahama T, Albou LP, Mi H (2022) PANTHER: making genome-scale phylogenetics accessible to all. Protein Sci 31:8–22. https://doi.org/10.1002/PRO.4218
23. Alonso AP, Piasecki RJ, Wang Y, LaClair RW, Shachar-Hill Y (2010) Quantifying the labeling and the levels of plant cell wall precursors using ion chromatography tandem mass spectrometry. Plant Physiol 153:915–924. https://doi.org/10.1104/pp.110.155713
24. Koubaa M, Cocuron JC, Thomasset B, Alonso AP (2013) Highlighting the tricarboxylic acid cycle: liquid and gas chromatography-mass spectrometry analyses of 13C-labeled organic acids. Anal Biochem 436:151–159. https://doi.org/10.1016/j.ab.2013.01.027
25. Horn PJ, Liu J, Cocuron JC, McGlew K, Thrower NA, Larson M, Lu C, Alonso AP, Ohlrogge J (2016) Identification of multiple lipid genes with modifications in expression and sequence associated with the evolution of hydroxy fatty acid accumulation in Physaria fendleri. Plant J 86:322–348. https://doi.org/10.1111/tpj.13163
26. Contreras-López O, Moyano TC, Soto DC, Gutiérrez RA (2018) Step-by-step construction of gene co-expression networks from high-throughput Arabidopsis RNA sequencing data. In: Root development. Methods in molecular biology. Humana Press Inc, pp 275–301
27. Kanehisa M, Goto S (2000) KEGG: Kyoto encyclopedia of genes and genomes. Nucleic Acids Res 28:27–30
28. Kanehisa M, Furumichi M, Sato Y, Ishiguro-Watanabe M, Tanabe M (2021) KEGG: integrating viruses and cellular organisms. Nucleic Acids Res 49:D545–D551. https://doi.org/10.1093/nar/gkaa970
29. Tsogtbaatar E, Cocuron J-C, Alonso AP, Lunn J (2020) Non-conventional pathways enable pennycress (Thlaspi arvense L.) embryos to achieve high efficiency of oil biosynthesis. J Exp Bot 71:3037–3051. https://doi.org/10.1093/jxb/eraa060
30. Mauceri A, Aci MM, Toppino L, Panda S, Meir S, Mercati F, Araniti F, Lupini A, Panuccio MR, Rotino GL, Aharoni A, Abenavoli MR, Sunseri F (2022) Uncovering pathways highly correlated to NUE through a combined metabolomics and transcriptomics approach in eggplant. Plan Theory 11:700. https://doi.org/10.3390/PLANTS11050700/S1
31. Hampejsová R, Berka M, Berková V, Jersáková J, Domkářová J, von Rundstedt F, Frary A, Saiz-Fernández I, Brzobohatý B, Černý M (2022) Interaction with fungi promotes the accumulation of specific defense molecules in orchid tubers and may increase the value of tubers for biotechnological and medicinal applications: the case study of interaction between Dactylorhiza sp. and Tulasnella calospora. Front Plant Sci 13:2295. https://doi.org/10.3389/FPLS.2022.757852/BIBTEX
32. Wang Z, Brannick E, Abasht B (2023) Integrative transcriptomic and metabolomic analysis reveals alterations in energy metabolism and mitochondrial functionality in broiler chickens with wooden breast. Sci Rep 13(1):1–11. https://doi.org/10.1038/s41598-023-31429-7

Chapter 5

RNA-Seq Data Analysis for Differential Gene Expression Using HISAT2–StringTie–Ballgown Pipeline

Vivek Thakur

Abstract

Comparison of transcriptome for candidate gene discovery has become an important tool for biologists. While such studies lack the degree of resolution one gets from well-designed forward or reverse genetic studies, nevertheless, this has been a method of choice for giving coarse insight into the underlying biological processes or mechanisms. This was further accelerated with the availability of sequencing technologies. While many pipelines are available for RNA-seq data analysis, the protocol discussed here will guide the first-time users for conducting routine RNA-seq analysis using whole genome sequence as reference.

Key words Transcript abundance, Differential gene expression, HISAT2, StringTie, RSEM, Ballgown, Mapping to whole genome reference, Plant dataset

1 Introduction

For understanding underlying mechanisms behind a trait or any phenotypic differences, one of the key methods has been to examine the genes which are differentially expressed between two or more conditions. Earlier, microarray had been one of the popular high-throughput approaches for comparison of transcriptomes, which has been gradually taken over by transcriptome sequencing (RNA-seq). This book chapter presents a basic RNA-seq protocol which allows them to compare a pair of conditions to find the transcripts which are differentially abundant.

Rajeev K. Azad (ed.), *Transcriptome Data Analysis*, Methods in Molecular Biology, vol. 2812,
https://doi.org/10.1007/978-1-0716-3886-6_5,

2 Materials

2.1 Obtaining Software and its Installation

The RNA-seq analysis will be performed using open-source software which can be compiled and run on Linux or Mac operating systems (OS). The text following every explanation are commands run from `Bash` terminal in *Ubuntu 22.04* operating system.

2.1.1 SRA Toolkit

The SRA Toolkit is a commonly used software for obtaining high-throughput sequencing data from NCBI's Short Read Archive (SRA) database and their processing. The SRA Toolkit download page can be accessed at the URL: https://github.com/ncbi/sra-tools/wiki/01.-Downloading-SRA-Toolkit.

Download current binaries for Ubuntu as below, followed by extraction of the contents of the compressed archieve file (`tar.gz`) (*see* **Note 1**):

```
wget --output-document sratoolkit.tar.gz https://ftp-trace.ncbi.nlm.nih.gov/sra/sdk/current/sratoolkit.current-ubuntu64.tar.gz

tar -vxzf sratoolkit.tar.gz
```

2.1.2 HISAT2, StringTie, and RSEM

Download and extract the binaries of software tools, namely, `HISAT2` [1] and `stringTie` [2], for transcriptome analysis.

```
wget --output-document hisat2-2.2.1_Linux_x86_64.zip https://cloud.biohpc.swmed.edu/index.php/s/oTtGWbWjaxsQ2Ho/download

unzip hisat2-2.2.1_Linux_x86_64.zip

wget http://ccb.jhu.edu/software/stringtie/dl/stringtie-2.2.1.Linux_x86_64.tar.gz

tar -xvzf stringtie-2.2.1.Linux_x86_64.tar.gz
```

RSEM [3] is an alternate RNA-seq analysis pipeline, which can be downloaded and installed as below:

```
wget https://github.com/deweylab/RSEM/archive/refs/tags/v1.3.3.tar.gz

tar -xvzf RSEM-1.3.3.tar.gz
```

To compile RSEM, simply enter the extracted `RSEM` folder and run `make`:

```
cd RSEM-1.3.3

make
```

2.1.3 Ballgown

Ballgown [4] software will be used for differential gene expression analysis. To install this package, start `R` software (version "4.3") and enter

```
> if (!require("BiocManager", quietly = TRUE))
install.packages("BiocManager")

> BiocManager::install("ballgown")
```

This will also install all dependencies of the `ballgown` package (*see* **Note 2**).

2.1.4 Samtools

Samtools software will be required for the purpose of manipulation of the alignment files. Follow the commands to download, extract, and install it:

```
wget https://github.com/samtools/samtools/releases/download/
1.20/samtools-1.20.tar.bz2
tar -xvjf samtools-1.20.tar.bz2
cd samtools-1.20
make
```

This will also install all dependencies of the `ballgown` package (*see* **Note 2**).

2.2 Obtaining Transcriptome Datasets

For demonstration of the current protocol (*see* **Note 3**), a study titled "Dissecting the Dynamic Changes of Methylome and Transcriptome During Rice Transformation" has been chosen [5]. The raw transcriptome sequence dataset is available at NCBI's GEO database with GEO accession number GSE202715 [6]. Although there are many samples involved in this study, but for the sake of simplicity, only four samples have been selected—two replicates of seedling samples and another two for callus:

(i) *O. sativa* Seedling RNA-seq rep1 (SRA Run ID: SRR19165926) [7].
(ii) *O. sativa* Seedling RNA-seq rep2 (SRA Run ID: SRR19165925) [8].
(iii) *O. sativa* Callus RNA-seq rep1 (SRA Run ID: SRR19165924) [9].
(iv) *O. sativa* Callus RNA-seq rep2 (SRA Run ID: SRR19165923) [10].

Using "prefetch <SRA-run-ID>" command in sratoolkit, download the raw transcriptome sequencing data of all four samples/runs mentioned above. For example, the command for downloading raw reads of SRA run SRR19165926 will be:

```
prefetch SRR19165926
```

This will create a folder in the name of the respective SRA run ID, and the sequence data will be downloaded in a file with .sra extension (*see* **Note 4**).

Extract the reads from the downloaded SRA files using sratoolkit's fastq-dump command. For example, the command for extracting reads of sample/run ID SRR19165926 will be:

```
fastq-dump SRR19165926/SRR19165926.sra
```

It will generate a text file with extension .fastq. The command for other three samples/runs should be executed by replacing the respective sample/run IDs.

2.3 Obtaining Reference Genome

Download cDNA and reference genome sequences from Rice Genome Annotation Project (RGAP 7) (files named all.cdna and all.con, respectively) [11, 12].

```
wget http://rice.uga.edu/pub/data/Eukaryotic_Projects/o_sativa/annotation_dbs/pseudomolecules/version_7.0/all.dir/all.con

wget http://rice.uga.edu/pub/data/Eukaryotic_Projects/o_sativa/annotation_dbs/pseudomolecules/version_7.0/all.dir/all.cdna
```

3 Methods

3.1 Preparing the Reference Genome

For aligning the reads to the reference genome, firstly an index of the reference genome is required. Creation of the index requires genome annotation (in .gtf format), splice sites (.ss), and exons (.exon). If annotation in .gtf format is unavailable, then it can be obtained from the .gff file available at rice genome annotation web page [12]. The following set of commands/programs need to be executed to prepare reference sequence (rice_trans_ref):

```
RSEM-1.3.3/rsem-gff3-to-gtf all.gff3 all.gtf #to be executed when .gtf file is unavailable

hisat2-2.2.1/extract_splice_sites.py all.gtf >rice.ss
```

```
hisat2-2.2.1/extract_exons.py all.gtf >rice.exons
```

```
hisat2-2.2.1/hisat2-build --ss rice.ss --exon rice.exons all.
con rice_trans_ref
```

3.2 Mapping Reads onto the Reference Genome

Align the reads of each of the four samples/runs (in `.fastq` format) against the indexed rice reference genome (`rice_-trans_ref`) using `hisat2` command to obtain the alignment files (in `.sam` format):

```
hisat2-2.2.1/hisat2 --dta -p 19 -x rice_trans_ref -U
SRR19165926.fastq -S SRR19165926.sam
```

```
hisat2-2.2.1/hisat2 --dta -p 19 -x rice_trans_ref -U
SRR19165925.fastq -S SRR19165925.sam
```

```
hisat2-2.2.1/hisat2 --dta -p 19 -x rice_trans_ref -U
SRR19165924.fastq -S SRR19165924.sam
```

```
hisat2-2.2.1/hisat2 --dta -p 19 -x rice_trans_ref -U
SRR19165923.fastq -S SRR19165923.sam
```

3.3 Processing of Alignment Files

Convert the format of alignment files from text to sorted binary (from `.sam` to `.bam` format) using "`samtools sort`" command [13]:

```
samtools-1.20/samtools sort -@ 4 -m 500M -o SRR19165926.srt.
bam SRR19165926.sam
```

```
samtools-1.20/samtools sort -@ 4 -m 500M -o SRR19165925.srt.
bam SRR19165925.sam
```

```
samtools-1.20/samtools sort -@ 4 -m 500M -o SRR19165924.srt.
bam SRR19165924.sam
```

```
samtools-1.20/samtools sort -@ 4 -m 500M -o SRR19165923.srt.
bam SRR19165923.sam
```

3.4 Updation of Genome-Wide Transcript Structure

Generate transcript information (or annotation) for each sample/run based on the above generated alignment of respective transcriptome reads data using "`stringtie`" command:

```
stringtie-2.2.1.Linux_x86_64/stringtie -p 4 -G all.gtf -l
SRR19165926 -o SRR19165926.gtf SRR19165926.srt.bam
```

```
stringtie-2.2.1.Linux_x86_64/stringtie -p 4 -G all.gtf -l
SRR19165925 -o SRR19165925.gft SRR19165925.srt.bam

stringtie-2.2.1.Linux_x86_64/stringtie -p 4 -G all.gtf -l
SRR19165924 -o SRR19165924.gtf SRR19165924.srt.bam

stringtie-2.2.1.Linux_x86_64/stringtie -p 4 -G all.gtf -l
SRR19165923 -o SRR19165923.gtf SRR19165923.srt.bam
```

Now merge the transcripts constructed from each sample/run in the above step into a single file using "`stringtie —merge`" command:

```
stringtie-2.2.1.Linux_x86_64/stringtie --merge -p 4 -G all.gtf
-o merged.gtf SRR19165926.gtf SRR19165925.gtf SRR19165924.gtf
SRR19165923.gtf
```

3.5 Estimation of Transcript Abundances

With the help of annotation inferred from this merged file (`merged.gtf`), the expression of all transcripts of each sample/run will be estimated using "`-e`" option of `stringtie`, which will only estimate the abundance of given reference transcripts:

```
stringtie-2.2.1.Linux_x86_64/stringtie -e -b SRR19165926 -p
4 -G merged.gtf -o SRR19165926.gtf SRR19165926.srt.bam

stringtie-2.2.1.Linux_x86_64/stringtie -e -b SRR19165925 -p
4 -G merged.gtf -o SRR19165925.gtf SRR19165925.srt.bam

stringtie-2.2.1.Linux_x86_64/stringtie -e -b SRR19165925 -p
4 -G merged.gtf -o SRR19165924.gtf SRR19165924.srt.bam

stringtie-2.2.1.Linux_x86_64/stringtie -e -b SRR19165924 -p
4 -G merged.gtf -o SRR19165923.gtf SRR19165923.srt.bam
```

The count tables will be created for each sample/run in newly created folders named after sample/run names. These files will serve as an input for differential gene expression analysis by `ballgown` tool (*see* **Note 5**).

3.6 Organizing the Gene Expression Data

For each sample/run, three of the output files obtained above, namely, `e_data.ctab`, `i_data.ctab`, and `t_data.ctab`, store expression measurements at exon, intron, and transcript level, respectively (Table 1). The remaining two output files `e2t.ctab` and `i2t.ctab` have information about mapping of exons and introns, respectively, to the transcripts (Table 2).

Table 1
The first five entries from "`t_data.ctab`" file having transcript-level expression measurements

t_id	chr	strand	start	end	t_name	num_exons	length	gene_id	gene_name	cov	FPKM
1	Chr1	+'	2903	10817	LOC_Os01g01010.1	12	3017	MSTRG.1	LOC_Os01g01010	0.000	0.000
2	Chr1	+'	2950	10817	MSTRG.1.2	12	2957	MSTRG.1	.	13.719	5.054
3	Chr1	+'	2984	10817	LOC_Os01g01010.2	12	2473	MSTRG.1	LOC_Os01g01010	0.216	0.080
4	Chr1	+'	11218	12435	LOC_Os01g01019.1	2	1127	LOC_Os01g01019	LOC_Os01g01019	0.606	0.223
5	Chr1	+'	12648	16291	LOC_Os01g01030.1	5	2840	MSTRG.2	LOC_Os01g01030	0.438	0.161

In these entries, the gene_id MSTRG.1 has three transcript isoforms, namely LOC_Os01g01010.1, MSTRG.1.2, and LOC_Os01g01010.2, and the second isoform (MSTRG.1.2) shows higher expression than the other two. Description of column headers: `t_id`, transcript id; `chr`, `strand`, `start`, `end`, genomic location of the transcript; `t_name`, Cufflinks-generated transcript id; `num_exons`, number of exons comprising the transcript; `length`, transcript length, including both exons and introns; `gene_id`, gene the transcript belongs to; `gene_name`, HUGO gene name for the transcript, if known; `cov`, per-base coverage for the transcript (available for each sample); `FPKM`, estimated expression level of the transcript (available for each sample) measured in terms of fragments per kilobase per million reads (FPKM)

Table 2
The first eight entries from "e2t.ctab" file having information of mapping of exons to transcripts

e_id	t_id
1	1
2	1
2	2
2	3
3	1
3	2
3	3

In these entries, the first exon ID (e_id: 1) is present uniquely in first transcript (t_id: 1), and the other two exons (e_id: 2 and 3) are common to all three transcripts (t_id: 1, 2, and 3)

For easy processing of this data in later steps (i.e., differential gene expression analysis), move all folders with expression data-related files to a new folder (`expressionData`):

```
mkdir expressionData #creates a new folder

mv SRR19165926/ SRR19165925/ SRR19165924/ SRR19165923/ expres-
sionData/
```

3.7 Differential Gene Expression Using `Ballgown` Tool

The ballgown library needs to be loaded first in R:

```
R

> library(ballgown)
```

The parent folder (`expressionData`), having folders of the expression data of all samples/runs, needs to be specified for creation of a `ballgown` object:

```
> data_directory=system.file('expressionData',package='ball-
gown')

> bg=ballgown(dataDir=data_directory, samplePattern='SRR',
meas='all')
 Sat Nov 4 13:52:51 2023
 Sat Nov 4 13:52:51 2023: Reading linking tables
 Error in file(file, "rt") : invalid 'description' argument
```

The creation of the `ballgown` object failed in the above step possibly due to the samplePattern argument which did not match with the expected pattern (i.e., "SRR01," "SRR02"). As an alternative, the same can be achieved by directly specifying samples as argument:

```
> bg=ballgown(samples=c('SRR19165923','SRR19165924',
'SRR19165925','SRR19165926'),meas='all')
```

3.8 Exploring the ballgown Object for Stored Expression Data

The data slots present in the above created `ballgown` object (`bg`) can be viewed by `str()` function:

```
> str(bg)
 Formal class 'ballgown' [package "ballgown"] with 7 slots
  ..@ expr :List of 3
 .. ..$ intron:'data.frame': 207477 obs. of 17 variables:
 .. ..$ exon :'data.frame': 286596 obs. of 33 variables:
  .. ..$ trans :'data.frame': 83432 obs. of 18 variables:
  ..@ indexes :List of 5
  .. ..$ e2t :'data.frame': 429519 obs. of 2 variables
  .. ..$ i2t :'data.frame': 346087 obs. of 2 variables
  .. ..$ t2g :'data.frame': 83432 obs. of 2 variables:
  .. ..$ bamfiles: NULL
  .. ..$ pData :'data.frame': 4 obs. of 2 variables:
  ..@ structure :List of 3
  .. ..$ intron:Formal class 'GRanges' ... with 7 slots
   .. ..$ exon :Formal class 'GRanges' ... with 7 slots
 .. ..$ trans :Formal class 'CompressedGRangesList'... with 5 ...
   ..@ dirs  :  Named chr  [1:4]  "SRR19165923"  "SRR19165924"
"SRR19165925" "SRR19165926"
  ..@ mergedDate: chr "Thu Nov 2 21:45:48 2023"
  ..@ meas : chr "all"
  ..@ RSEM : logi FALSE
```

The description of these slots is as follows:

expr	tables containing expression data for the three genomic features, namely, introns, exons, and transcripts
structure	genomic locations of features
indexes	tables connecting components of the assembly and providing other experimental information (e.g., phenotype data)
dirs	directories holding data created by Tablemaker
mergedDate	date of ballgown object creation
meas	expression measurement(s) the `ballgown` object contains in its data slot. Vector of one or more of "rcount," "ucount," "mrcount," "cov," "cov_sd," "mcov," "mcov_sd," or "FPKM," if Tablemaker output is used. Can also be "all" for all measurements
RSEM	*true* if object was made from RSEM output, *false* if object was made from Tablemaker/Cufflinks output

Among the abovementioned slots, data in the `expr` slot will be used for differential gene expression analysis. The expression data of top few transcripts can be viewed by printing top few entries in the `trans` data-frame of the `expr` slot of the `bg` object by either of the following ways:

```
> head(expr(bg)$trans)

> head(texpr(bg), 'FPKM'))
```

3.9 Pairwise Differential Expression Testing

For finding genes which are differentially expressed between two groups (in present case, seedling and callus), the group structure is first specified by updating `indexes` slot for phenotypic data (`pData`):

```
> sampleNames(bg)
 [1] "SRR19165923" "SRR19165924" "SRR19165925" "SRR19165926"

> pData(bg)=data.frame(id=sampleNames(bg), group=c(1,1,0,0))

> pData(bg)
 id group
1 SRR19165923 1
2 SRR19165924 1
3 SRR19165925 0
4 SRR19165926 0
```

Finally, the group-wise abundances of every transcript (or gene) are tested for any statistically significant differences using `stattest()` function:

```
> stat_results=stattest(bg, feature='transcript',
meas='FPKM', covariate='group',getFC=TRUE)

> head(stat_results)
 feature id fc pval qval
1 transcript 1 9.641415e-07 0.425456256 0.8829002
2 transcript 2 1.216899e+06 0.418782526 0.8829002
3 transcript 3 3.467012e-03 0.308720389 0.8829002
4 transcript 4 2.701933e+00 0.666701189 0.8884825
5 transcript 5 5.760627e+00 0.008440984 0.8829002
6 transcript 6 1.585459e+01 0.355136641 0.8829002

> stat_results2=stattest(bg, feature='gene', meas='FPKM',
covariate='group',getFC=TRUE)

> head(stat_results2)
 feature id fc pval qval
```

```
1 gene ChrSy.fgenesh.gene.1 0.9026398 0.7931116 0.9928428
2 gene ChrSy.fgenesh.gene.10 1.0000000 NaN NaN
3 gene ChrSy.fgenesh.gene.11 1.0000000 NaN NaN
4 gene ChrSy.fgenesh.gene.12 1.1261754 0.7382399 0.9928428
5 gene ChrSy.fgenesh.gene.13 1.1454396 0.7130775 0.9928428
6 gene ChrSy.fgenesh.gene.14 0.8340431 0.6101121 0.9928428
```

3.10 Statistically Significant Differentially Expressed Genes

In order to select the genes or transcripts which are significantly differentially expressed between two groups, both `p-valve` and `q-value` thresholds are considered depending upon the user's preferences. In the present case, the genes with `p-value` <=0.05 were first selected, which were about ~2000 in count:

```
> gene_res=subset(stat_results2, pval<=0.05)

> dim(gene_res)
 [1] 2060 5
```

Since this kind of analysis involves hypothesis testing of thousands of genes/transcripts, there is a high chance of false discovery even when `p-value` threshold is set to 0.01, so appropriate correction(s) for multiple testing (is) are applied, which is reported in terms of `q-value`. Sorting the above set of genes based on `q-value` showed that the lowest observed `q-value` was 0.503293, indicating that while a couple of thousands of genes had `p-value` below threshold (0.05), however, none of them qualified when the false discovery rate is considered (Table 3) (*see* **Note 6**).

```
> head(gene_res[order(gene_res$qval),])
```

Table 3
Top few gene entries, showing differential expression between seedlings and callus tissues, filtered for p-value<=0.05

S. no.	Feature	id	fc	pval	qval
44766	Gene	MSTRG.18981	6.75E+07	1.32E-05	0.5032934
36257	Gene	MSTRG.11321	5.79E+03	6.28E-05	0.5999482
51366	Gene	MSTRG.4653	4.00E-02	5.75E-05	0.5999482
53051	Gene	MSTRG.617	1.71E-01	6.01E-05	0.5999482
35544	Gene	MSTRG.10680	3.54E+02	1.28E-04	0.6124427
43896	Gene	MSTRG.18198	8.00E+00	1.16E-04	0.6124427

The genes have also been ordered in ascending order based on `q-value` to evaluate the original results of pairwise tests for multiple testing. Description of column headers: id, Gene ID; fc, fold change; pval, p-value; qval, q-value

4 Notes

1. Add path to the `sratoolkit` binaries folder to the PATH environment variable. Since the extracted `sratoolkit` folder name may vary, so use the one you got after extraction.

```
export PATH=$PATH:$PWD/sratoolkit.3.0.7-ubuntu64/bin/
```

2. This step will require internet connection and may take at least 10 min depending on the internet speed and number of dependencies.
3. A very detailed and extensive RNA-seq data analysis protocol involving HISAT2, StringTie, and Ballgown softwares is also available for further reference [14].
4. As an alternative to `prefetch` command, the data can also be downloaded from the following weblink by using the respective SRA run ID. For example, the URL for downloading raw sequences of SRA run `SRR19165926` is

   https://sra-pub-run-odp.s3.amazonaws.com/sra/SRR19165926/SRR19165926
5. Among the alternative RNA-seq analysis pipelines for read count estimation, RSEM [3] and Salmon [15] can be further explored. For differential expression testing, R-based packages like DESeq2 [16] and edgeR [17] can be used.
6. Readers are strongly encouraged to refer to the best practices for RNA-seq analysis before following the protocol discussed in this chapter [18]. It's also suggested to get a detailed conceptual understanding from reviews [19, 20].

References

1. Kim D, Paggi JM, Park C et al (2019) Graph-based genome alignment and genotyping with HISAT2 and HISAT-genotype. Nat Biotechnol 37:907–915. https://doi.org/10.1038/s41587-019-0201-4
2. Pertea M, Pertea GM, Antonescu CM, Chang TC, Mendell JT, Salzberg SL (2015) StringTie enables improved reconstruction of a transcriptome from RNA-seq reads. Nat Biotechnol 3(3):290–295. https://doi.org/10.1038/nbt.3122
3. Li B, Dewey CN (2011) RSEM: accurate transcript quantification from RNA-Seq data with or without a reference genome. BMC Bioinformatics 12:323. https://doi.org/10.1186/1471-2105-12-323
4. Frazee A, Pertea G, Jaffe A et al (2015) Ballgown bridges the gap between transcriptome assembly and expression analysis. Nat Biotechnol 33:243–246. https://doi.org/10.1038/nbt.3172
5. Hsieh J-WA, Chang P, Kuang L-Y, Hsing Y-IC, Chen P-Y (2023) Rice transformation treatments leave specific epigenome changes beyond tissue culture. Plant Physiol 193(2):1297–1312. https://doi.org/10.1093/plphys/kiad382
6. GEO database accession GSE202715. https://www.ncbi.nlm.nih.gov/geo/query/acc.cgi?acc=GSE202715
7. O. sativa Seedling RNA-seq rep1 SRA Run: SRR19165926. https://trace.ncbi.nlm.nih.gov/Traces/?view=run_browser&acc=SRR19165926
8. O. sativa Seedling RNA-seq rep2 SRA Run: SRR19165925. https://trace.ncbi.nlm.nih.gov/Traces/?view=run_browser&acc=SRR19165925

9. O. sativa Callus RNA-seq rep1 SRA Run: SRR19165924. https://trace.ncbi.nlm.nih.gov/Traces/?view=run_browser&acc=SRR19165924
10. O. sativa Callus RNA-seq rep2 SRA Run: SRR19165923. https://trace.ncbi.nlm.nih.gov/Traces/?view=run_browser&acc=SRR19165923
11. Kawahara Y, de la Bastide M, Hamilton JP et al (2013) Improvement of the Oryza sativa Nipponbare reference genome using next generation sequence and optical map data. Rice 6:4. https://doi.org/10.1186/1939-8433-6-4
12. RGAP 7 database. http://rice.uga.edu/pub/data/Eukaryotic_Projects
13. Danecek P, Bonfield JK, Liddle J, Marshall J et al (2021) Twelve years of SAMtools and BCFtools. GigaScience 10(2):giab008. https://doi.org/10.1093/gigascience/giab008
14. Pertea M, Kim D, Pertea G et al (2016) Transcript-level expression analysis of RNA-seq experiments with HISAT, StringTie and Ballgown. Nat Protoc 11:1650–1667. https://doi.org/10.1038/nprot.2016.095
15. Patro R, Duggal G, Love MI, Irizarry RA, Kingsford C (2017) Salmon provides fast and bias-aware quantification of transcript expression. Nat Methods 14(4):417–419. https://doi.org/10.1038/nmeth.4197
16. Love MI, Huber W, Anders S (2014) Moderated estimation of fold change and dispersion for RNA-seq data with DESeq2. Genome Biol 15:550. https://doi.org/10.1186/s13059-014-0550-8
17. Chen Y, Lun AAT, Smyth GK (2016) From reads to genes to pathways: differential expression analysis of RNA-Seq experiments using Rsubread and the edgeR quasi-likelihood pipeline. F1000Research 5:1438. https://doi.org/10.12688/f1000research.8987.2
18. Conesa A, Madrigal P, Tarazona S et al (2016) A survey of best practices for RNA-seq data analysis. Genome Biol 17:13. https://doi.org/10.1186/s13059-016-0881-8
19. Wang Z, Gerstein M, Snyder M (2009) RNA-Seq: a revolutionary tool for transcriptomics. Nat Rev Genet 10:57–63. https://doi.org/10.1038/nrg2484
20. Van-den Berge K, Hembach KM et al (2019) RNA sequencing data: Hitchhiker's guide to expression analysis. Ann Rev Biomed Data Sci 2:139–173. https://doi.org/10.1146/annurev-biodatasci-072018-021255

Chapter 6

RNA Sequencing Experimental Analysis Workflow Using *Caenorhabditis elegans*

Jose Robledo, Saifun Ripa Nahar, Manuel A. Ruiz, Raymond J. Hendricks, David J. Burks, Mary L. Ladage, Taegun Kwon, Rajeev K. Azad, and Pamela A. Padilla

Abstract

RNA sequencing is an approach to transcriptomic profiling that enables the detection of differentially expressed genes in response to genetic mutation or experimental treatment, among other uses. Here we describe a method for the use of a customizable, user-friendly bioinformatic pipeline to identify differentially expressed genes in RNA sequencing data obtained from *C. elegans*, with attention to the improvement in reproducibility and accuracy of results.

Key words RNA sequencing, Bioinformatic pipeline, Differentially expressed genes, Reproducibility

1 Introduction

RNA sequencing is a powerful tool for transcriptomic profiling across many species and platforms that has been widely used for more than 10 years. While the basic workflow of RNA sequencing remains largely unchanged, improvements in RNA collection techniques, sequencing platform accessibility, and computer hardware and software have streamlined the process [1]. As we detail here, the use of *C. elegans* can facilitate the investigation of many genotypes and treatments owing to its rapid generation time, the ability to precisely control development, and the many resources available from the community to obtain or generate genetic alleles [2, 3]. Experimental reproducibility depends on many factors including careful experimental planning and animal husbandry. Here, we describe a powerful and user-friendly bioinformatic pipeline for the identification of differentially expressed genes from raw RNA sequencing data (Fig. 1). This pipeline can be implemented

Rajeev K. Azad (ed.), *Transcriptome Data Analysis*, Methods in Molecular Biology, vol. 2812,
https://doi.org/10.1007/978-1-0716-3886-6_6,

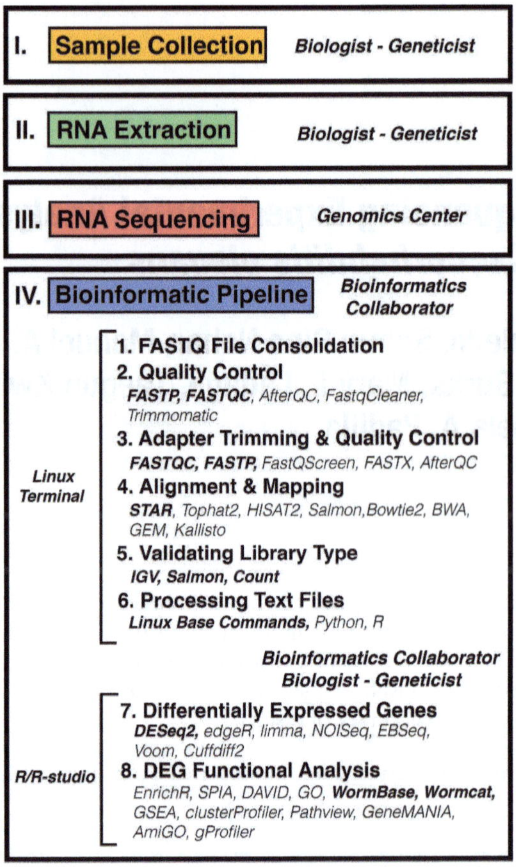

Fig. 1 Experimental workflow for designing RNA-seq experiments using *C. elegans*. We provide collaboration considerations for designing and analyzing an RNA-seq experiment. Programs that are used in this workflow are in bold; programs not in bold are included as alternatives to consider at each step of the bioinformatic pipeline

using open-source software and knowledge of only two programming languages. Further, we present options for customization based around needs and the experience of the user.

2 Materials

Prepare solutions using ultrapure deionized (DI) water and molecular-grade reagents.

1. *Caenorhabditis elegans* strains (*Caenorhabditis* Genetics Center), *E. coli* OP50 (*Caenorhabditis* Genetics Center), incubator (Tritech Cat No. DT2-MP-47L), and Petri dish plates (Sarstedt, 92 mm × 16 mm, Cat. No. 82.1473.001 and 60 mm × 15 mm, Cat. No. 82.1194.500) for worm maintenance.

2. Omega Bio-Tek E.Z.N.A. HP Total RNA Kit (R6812-01).
3. TruSeq Stranded Total RNA Library Prep Kit (Illumina, Cat. No. 20020598).
4. TruSeq RNA UD Indexes (Illumina, Cat. No. 20020591).
5. M9 solution: 3 g KH_2PO_4, 6 g Na_2HPO_4 (anhydrous) or 11 g $Na_2HPO_4 \cdot 7H_2O$, 5 g NaCl, 1 mL 1 M $MgSO_4$; bring up to 1 L with diH_2O, mix thoroughly, dispense into bottles, and autoclave for 25 min. $MgSO_4$ may form a precipitate during the autoclave cycle; if this occurs, sterile $MgSO_4$ can be added separately after autoclaving.
6. LB media plates: 10 g Bacto Tryptone, 5 g Bacto Yeast Extract, 10 g NaCl, 15 g agar. Bring up to 1 L with diH_2O and mix. Autoclave for 25 min, cool slightly, and pour into 25 mL Petri dishes. Store at 15 °C until needed.
7. LB liquid media: 10 g Bacto Tryptone, 5 g Bacto Yeast Extract, 10 g NaCl. Bring up to 1 L with diH_2O and mix well. Dispense into bottles and autoclave for 25 min. Store at room temperature in a low humidity area.
8. Nematode Growth Medium (NGM): 3 g NaCl, 20 g agar, 2.5 g peptone, 25 mL of 1 M PPB, 975 mL diH_2O. Cover with foil, for 25 minutes, and cool slightly to ~55 °C. Media can be kept stirring on a hot plate for a short time until dispensing into Petri dishes. Before dispensing media into Petri dishes, add 1 mL of 1 M $CaCl_2$ (filter sterilized), 1 mL of 2 mg/mL uracil (filter sterilized), 1 mL of 5 mg/mL cholesterol (filter sterilized), and 1 mL of 1 M $MgSO_4$ (filter sterilized). To avoid boiling over in the autoclave, use a flask at least three times the volume of media.
9. A computer with a Linux processor is required for running this bioinformatic analysis pipeline. A computer with more processing power (cores) will run the necessary programs more quickly. The STAR aligner program requires 16 GB of RAM to align the *C. elegans* genome. Ensure Linux and R/RStudio are compatible with the system being used (for additional computer system considerations, *see* **Notes 74** and **75**).
10. Internet access using Google Chrome is recommended to access links and files. Among other considerations, the location and browser differences can impact the ability to access or download large files from the Internet. Google Chrome has demonstrated consistency, speed, and availability across all platforms.

3 Methods

3.1 Sample Collection

Caenorhabditis elegans strains are raised at 20 °C on NGM plates seeded with *E. coli* OP50 (unless otherwise indicated by the strain or experiment) as described in [2]. At least three biological replicates are collected for each sample (*see* **Notes 1** and **2**).

1. Perform a T-streak of *E. coli* OP50 onto LB plates from a glycerol stock of OP50 following proper sterile technique. Incubate for approximately 16 h (overnight) at 37 °C.
2. Inoculate a bottle of LB liquid media with one OP50 colony and shake for 12–16 h at 37 °C (*see* **Note 3**). Seed NGM plates with OP50 and spread evenly to cover the whole plate; allow to dry for at least 24 h. Large Petri dishes (92 mm × 16 mm filled with 25 mL NGM) are seeded with 900 μL of OP50 and small Petri dishes (60 mm × 15 mm filled with 10 mL NGM) are seeded with 300 μL of OP50.
3. For each sample, allow approximately 200–300 day 1 adult hermaphrodites to lay eggs on 2–3 large plates (92 × 16 mm filled with 25 mL NGM; 100 animals per plate) for an hour to generate a synchronized population (*see* **Notes 4–6**).
4. For each sample, use a worm pick to transfer a total of ~600 embryos from the large plates onto small plates (60 mm × 15 mm filled with 10 mL NGM; 60 animals per plate), and culture to the desired developmental stage. Verify the developmental stage using germline morphology and hours post-egg lay (*see* **Notes 7–10**).
5. Wash the animals from the plates using M9 and collect in 15 mL conical tubes. Wash three times with M9 to remove excess bacteria, allowing the animals to settle by gravity after each wash. Remove supernatant and collect animals into 1.5 mL Eppendorf tubes with a final volume of approximately 50 μL of worm sample.
6. Immediately snap freeze in liquid nitrogen and store at −80 °C (*see* **Note 11**).

3.2 RNA Extraction and Isolation

1. Preheat water bath to 65 °C and keep samples in a container with crushed ice before starting the extraction and isolation. The RNA extraction and isolation procedure can be completed at room temperature. When possible, keep samples on ice. Read through protocol prior to starting RNA extraction and isolation procedure (*see* **Notes 12–18**).
2. Use the Omega Bio-Tek E.Z.N.A. HP Total RNA Kit to perform the RNA extraction and isolation. Allow sample to thaw for 2–3 min before adding 350 μL of GTC lysis buffer to the sample. The final volume should be approximately 400 μL (*see* **Note 13**).

3. Place sample in water bath (65 °C) for 2 min.
4. Insert RNA homogenizer mini column into a 1.5 mL Eppendorf tube (provided in kit). Transfer 400 μL sample lysate into the RNA homogenizer mini column and centrifuge at 13,000 ×*g* for 2 min. Discard the RNA homogenizer mini column and keep the filtrate.
5. Transfer 350 μL filtrate to a new 1.5 mL Eppendorf tube (not provided by kit), and add 1 volume (350 μL) of 70% ethanol to sample, and vortex to mix thoroughly (no centrifugation step necessary).
6. Transfer ~700 μL sample into a HiBind RNA Mini column inserted into a new 1.5 mL Eppendorf collection tube and centrifuge at 10,000 ×*g* for 1 min. Discard filtrate and centrifuge HiBind RNA Mini column at 10,000 ×*g* for 1 min again.
7. Add 500 μL of RNA Wash Buffer to the HiBind RNA Mini column and centrifuge at 10,000 ×*g* for 30 s. Discard the filtrate and reuse the 1.5 mL Eppendorf collection tube.
8. Add 500 μL of RNA Wash Buffer II to the HiBind RNA Mini column and centrifuge at 10,000 ×*g* for 1 min. Discard the filtrate and reuse the 1.5 mL Eppendorf collection tube. Repeat.
9. Centrifuge HiBind RNA Mini column with 1.5 mL Eppendorf tube at 13,000 ×*g* for 2 min. Allow the HiBind RNA Mini column with 1.5 mL Eppendorf collection tube to dry for 1 min before final elution step to ensure residual ethanol evaporation.
10. Discard the 1.5 mL Eppendorf tube. Insert HiBind RNA Mini column into a new 1.5 mL Eppendorf tube. Add 70 μL of nuclease-free water to the HiBind RNA Mini column. Centrifuge 1.5 mL Eppendorf tube with HiBind RNA Mini column at 13,000 ×*g* for 2 min.
11. From the eluate, aliquot two 6 μL–10 μL samples into sterile 1.5 mL Eppendorf tubes for determination of concentration and to test for degradation and contamination. Reserve the remainder of the eluate for RNA sequencing. Quantify the concentration of RNA in each sample using a NanoDrop spectrophotometer (in our hands, sample concentrations range between 32 ng/μL and 121 ng/μL).
12. Check for degradation and contamination of the RNA by using a Qubit, NanoDrop spectrophotometer, and/or bleached gel protocol [4] before RNA sequencing (*see* **Note 18**).
13. Store the remaining 50–60 μL RNA sample at −80 °C and send samples to the genomics center for sequencing.

3.3 RNA Sequencing

1. After collection, RNA samples are sent to a genomics center for sequencing. For this method, samples were processed at the University of North Texas Genomics Center—similar facilities are accessible in most areas (*see* **Notes 19–21**).
2. Using 1 μg of total RNA as the starting material, prepare RNA-seq libraries using TruSeq Stranded Total RNA Library Prep Kits and following the manufacturer's instructions.
3. Deplete ribosomal RNA using RiboZero Gold provided in the kit.
4. Fragment resulting RNA samples and convert to double-stranded DNA for sequencing. Uniquely index individual libraries using TruSeq RNA UD Indexes and pool in equimolar ratio.
5. Sequence the pooled libraries on an Illumina NextSeq 500 sequencing system for 75 bp paired-end reads. Sequences are then made available for analysis.
6. After publication, the raw sequence reads are deposited in the Sequence Read Archive (SRA) with reference ID "SRA##########" in GenBank.

3.4 Bioinformatic Pipeline: Linux Command Line (Terminal)

Part 1 of the bioinformatic pipeline is written for the use of a MacBook Pro using the Linux bash command line (e.g., terminal application). Create a folder where all bioinformatic analysis files are located for part 1 of the analysis (*see* **Notes 22–27**).

3.4.1 Installing Required Programs

1. Use the Anaconda (Conda) repository environment management system to install all free open-source packages for Windows, macOS, or Linux. Anaconda (Conda) installation can be downloaded for free using the Google Chrome Internet browser at https://www.anaconda.com/. Follow the directions to install the Conda environment on Linux terminal provided by the user guide at https://docs.conda.io/projects/conda/en/latest/user-guide/index.html (*see* **Note 28**).
2. Search for the desired program on the Conda page at https://www.anaconda.com/ or the Anaconda Navigator application. Copy and paste a Conda installer into the Linux terminal to run installation (*see* **Note 29**).
3. Install the following packages using a Conda environment:
   (a) FASTQC—https://anaconda.org/bioconda/fastqc
   (b) FASTP—https://anaconda.org/bioconda/fastp [5].
   (c) Salmon—https://anaconda.org/bioconda/salmon
   (d) STAR—https://anaconda.org/bioconda/star [6].
   (e) SAMTOOLS—https://anaconda.org/bioconda/samtools [7, 8].

3.4.2 FASTQ Files

1. The RNA Illumina sequencing system for paired-end reads exports a total of eight FASTQ files per sample. Plus and minus strands (R1 and R2) were sequenced for each lane (resulting in four lanes total) for each sample provided (*see* **Note 30**). To consolidate the FASTQ files for each sample, concatenate (cat command in Linux) all lanes (001–004) for each strand (R1 and R2). This will result in two FASTQ files (R1 and R2) for each sample (*see* **Note 31**).

3.4.3 Quality Control

1. Use the FASTQC package to perform quality control analysis by following the documentation directions at https://www.bioinformatics.babraham.ac.uk/projects/fastqc/. Quality control analysis of each sample FASTQ file was completed. Example reports to aid quality report interpretation are provided in the Example Reports section of the FASTQC documentation (*see* **Note 32**).

```
Sample Linux Code: ~% cat /file/location/sample1_L001_R1.fastq
/file/location/sample1_L002_R1.fastq
/file/location/sample1_L003_R1.fastq
/file/location/sample1_L004_R1.fastq >
/file/location/sample1_R1.fastq
Sample Linux Code: ~% cat /file/location/sample1_L001_R2.fastq
/file/location/sample1_L002_R2.fastq
/file/location/sample1_L003_R2.fastq
/file/location/sample1_L004_R2.fastq >
/file/location/sample2_R2.fastq
```

3.4.4 Adapter Trimming

1. Perform additional quality control and adapter trimming using the FASTP package following the documentation directions at https://github.com/OpenGene/fastp#fastp (*see* **Note 33**). Reports are generated to assess the quality of the data and can be interpreted using the documentation at https://github.com/OpenGene/fastp#fastp or by using FASTQC report examples.

```
Sample Linux Code: ~% fastp -w 4 -i
/file/location/sample1_R1.fastq -o
/file/location/fastp/sample1_R1.fastq -I
/file/location/sample1_R2.fastq -O
/file/location/fastp/sample1_R2.fastq -h
/file/location//fastp/ sample1_fastq.html
```

2. Quality control analysis is performed on both pre- and post-adapter trimmed FASTQ files to check for adapter contamination. FASTP performs the adapter trimming step and generates new FASTQ files that are used for the final analysis.

3.4.5 Alignment/Mapping

1. Using Google Chrome, download the *Caenorhabditis elegans* genome assembly FASTA file (Caenorhabditis_elegans. WBcel235.106.**dna.toplevel.fa.gz** file extension) from the Ensembl website [https://useast.ensembl.org/ Caenorhabditis_elegans/Info/Index] (*see* **Note 34**).
2. Using Google Chrome, download the *Caenorhabditis elegans* genome annotation file (Caenorhabditis_elegans. WBcel235.106.**gtf.gz** file extension) from the Ensembl website [https://useast.ensembl.org/Caenorhabditis_elegans/ Info/Index] (*see* **Note 34**).
3. Create a "Bioinformatic analysis" folder for the experimental data set being analyzed. Within, create a folder named "cel235_ensembl." Transfer the FASTA and annotation files from Ensembl into the "cel235_ensembl" folder.
4. Generate a genome index for *C. elegans* using the STAR genome. Generate command and follow the package documentation directions at https://github.com/alexdobin/ STAR/blob/master/doc/STARmanual.pdf. For the small genome of *C. elegans*, set the –genomeSAindexNbases to 12. To reduce RAM consumption, set the –genomeChrBinNbits to 12 (*see* **Note 35**).

```
Sample Linux Code: ~% STAR --runMode genomeGenerate --run-ThreadN 4 --genomeDir /folder/location/cel235_ensembl --genomeFastaFiles /file/location/cel235_ensembl/Caenorhabditis_elegans.WBcel235.dna.toplevel.fa --genomeSAindexNbases 12 --genomeChrBinNbits 12 --outFileNamePrefix /folder/location/cel235_ensembl/cel235_ensembl --sjdbGTFfile /file/location/cel235_ensembl/Caenorhabditis_elegans.WBcel235.106.gtf --sjdbOverhang 100
```

5. For each sample, run a STAR alignment using the default settings and follow the package documentation directions at https://github.com/alexdobin/STAR/blob/master/doc/ STARmanual.pdf. Additional commands to include are the export of unmapped reads files, GeneCounts option (ReadsPerGene.out.tab files), BAM coordinated/uncoordinated files, and the adjustments described in **step 4** (*see* **Notes 34–37**).

```
Sample Linux Code: ~% STAR --runThreadN 4 --genomeDir /folder/location/cel235_ensembl --readFilesIn /file/location/fastp/sample1_trial1_R1.fastq /file/location/fastp/sample1_-trial1_R2.fastq --quantMode GeneCounts --outSAMtype BAM Unsorted SortedByCoordinate --outFileNamePrefix /file/location/star_rawcounts/sample1_trial1 --outReadsUnmapped Fastx
```

6. Perform alignment on both pre- and post-trimmed data to examine adapter contamination. However, only use post-trimmed data for final analysis (*see* **Notes 34–37**).

3.4.6 Validating Library Strandedness (IGV, Salmon, and STAR Gene count) (See Note 38)

1. To differentiate between a stranded and unstranded sample library, first install the Integrative Genomics Viewer (IGV) following the package documentation directions at https://software.broadinstitute.org/software/igv/.
2. Index all the STAR exported BAM coordinated files using SAMTOOLS following package documentation directions for indexing BAM files at http://www.htslib.org/doc/samtools.html.

```
Sample Linux Code: ~% samtools index /file/location/sample1_-trial1_Aligned.sortedByCoord.out.bam /file/location/sample1_-trial1_Aligned.sortedByCoord.out.bam.bai
```

3. Load the indexed BAM coordinated files onto IGV, select the *C. elegans* WS235 genome, and determine the strandedness of the library by following the instructions at https://darencard.net/blog/2019-09-13-determining-stranded-rnaseq/ (*see* **Note 39**). The number of counts associated with each gene can also be explored using the genome browser.
4. Next, to ensure that the library is properly analyzed, Salmon is used to detect the strand specificity of sequenced reads. Download the *C. elegans* genome FASTA file (Caenorhabditis_elegans.WBcel235.109.dna.toplevele.fa.gz file extension) again using the Google Chrome Ensembl files found in https://useast.ensembl.org/Caenorhabditis_elegans/Info/Index, and place into a folder called "wormgenome_salmon" (*see* **Note 34**).
5. Index the *C. elegans* genome FASTA file following the Salmon package documentation directions at https://salmon.readthedocs.io/en/latest/ (*see* **Note 40**).

```
Sample Linux Code: ~% salmon index -t /file/location/salmon/wormgenome_salmon/c_elegans.PRJNA13758.WS285.genomic.fa -i /file/location/salmon/wormgenome_salmon --gencode
```

6. Using one of the previously generated, post-trim sample FASTQ files, quantify gene counts in mapping-based mode following the Salmon package documentation directions at https://salmon.readthedocs.io/en/latest/. Determine the strandedness of the library by interpreting the acronym output of the library using the package documentation (*see* **Note 41**).

```
Sample Linux Code: ~% salmon quant -i /file/location/salmon/wormgenome_salmon -l A -1 /file/location/fastp/sample1_-trial1_R1.fastq -2 /file/location/fastp/WT_trial1_R2.fastq --validateMappings -o transcript_quant
```

7. To further validate the strandedness of the samples, the STAR Gene count option from Subheading 3.4.5, **step 5** exports a text file (extension "ReadsPerGene.out.tab") with four columns corresponding to strandedness possibilities: gene ID, counts for unstranded RNA-seq, counts for first read strand aligned with RNA, and counts for second read strand aligned with RNA. Use a grep/awk script command in the Linux terminal to determine the number of reads in each column for each sample in ReadsPerGene.out.tab text file. The count method here can be referenced in https://biocorecrg.github.io/RNAseq_course_2019/alnpractical.html, and strandedness can be determined by the column with the most reads (*see* **Notes 42** and **43**).

```
Sample Linux Code: ~% grep -v "N_" /file/location/star_raw-
counts/Sample1_trial1ReadsPerGene.out.tab | awk '{unst+=$2;
forw+=$3;rev+=$4}END{print unst,forw,rev}'
```

3.4.7 STAR Gene count Processed Files

1. Once library strandedness is determined, follow the STAR package documentation to extract the appropriate columns from the ReadsPerGene.out.tab text files for all samples exported in Subheading 3.4.5, **step 5**. Column 1 reports the gene list and columns 2–4 report gene counts corresponding to library strandedness. Perform the processing of ReadsPerGene.out.tab text files for all samples using a cut and grep command to extract column 1 and the appropriate second column onto a new text file; this file will be used for all further analysis (*see* **Notes 43** and **44**).

```
Sample Linux Code: ~% cut -f1,4 /file/location/star_rawcounts/
Sample1_trial1ReadsPerGene.out.tab | grep -v "_" > Sample1_-
trial1_counts.txt
```

3.5 Bioinformatic Pipeline: R and RStudio

Part 2 of the bioinformatic pipeline utilizes R and RStudio on a macOS/Linux platform. Multiple free R packages are available to analyze RNA sequencing data. The R language can perform statistical analysis and graphing analysis (*see* **Note 22**).

1. Create a folder where all R/RStudio analysis will be performed for the project.
2. Inside the R project folder, create a folder labeled "star_rawcounts" and transfer all the processed STAR files from Subheading 3.4.7 (*see* **Note 44**).

3. Create a "sample_description.txt" file with a description of the samples, genotypes, sample names, file names (e.g., from STAR raw counts file folder), trial numbers, etc. (*see* **Note 45**).
4. Using a previously downloaded *Caenorhabditis elegans* genome annotation file (Caenorhabditis_elegans. WBcel235.106.gtf.gz file extension) from the Ensembl [https://useast.ensembl.org/Caenorhabditis_elegans/Info/Index], create a "wormgenes.txt" file that includes all *C. elegans* genes with associated WormBaseIDs. Extracting wormgenes and WormBaseIDs using the *C. elegans* annotation file from Ensembl can be done by using a cat and awk command in the Linux terminal application (*see* **Note 46**).

```
Sample Linux Code: ~% cat Caenorhabditis_elegans.WBcel235.106.
gtf | awk -F "\t" 'BEGIN{OFS="\t"}{if($3=="transcript"){split
($9, a, "\""); print a[2],a[6]}}' > wormgenes.txt
```

5. Download a "functional_description.txt" file from the WormBase website [https://downloads.wormbase.org/releases/current-production-release/] with file name "*c_elegans. PRJNA13758.WS285.functional_descriptions.txt.gz.*"
6. The functional_description.txt file will be used to extract the following descriptions: WBgenes, gene class description, concise description, and automated description. Use multiple awk commands to create four functional gene description text files using WBgenes and a different WormBase gene description column (i.e., class description, concise description, or automated description) (*see* **Note 47**).

```
Sample Linux Code:
gene list: ~% grep "WBGene" c_elegans.PRJNA13758.WS285.func-
tional_descriptions.txt > WB285_WBgenes.txt
gene class description: ~% awk '/Gene class description:/'
c_elegans.PRJNA13758.WS285.functional_descriptions.txt >
WB285_geneclass.txt
concise description: ~% awk ' /Concise description:/ { flag=1;
pfx="" } /Automated description/ { flag=0; print "" } flag {
printf "%s%s",pfx,$0; pfx=" " } ' c_elegans.PRJNA13758.WS285.
functional_descriptions.txt > WB285_concise.txt
automated description: ~% awk ' /Automated description:/ {
flag=1; pfx="" } /Gene class description:/ { flag=0; print ""
} flag { printf "%s%s",pfx,$0; pfx=" " } ' c_elegans.
PRJNA13758.WS285.functional_descriptions.txt > WB285_auto-
mated.txt
```

7. Download and install R using the website directions at https://cran.r-project.org/. Download and install RStudio Desktop using the website directions at https://posit.co/download/rstudio-desktop/. Open RStudio Desktop, create a new project, and assign the folder created in Subheading 3.5, **step 1** as the directory. All further analyses to identify DEGs use RStudio Desktop or R.
8. Install all the following library packages required to identify differentially expressed genes (DEGs) by following package documentation directions (*see* **Note 48**):
   (a) devtools—https://www.r-project.org/nosvn/pandoc/devtools.html
   (b) DESeq2—https://bioconductor.org/packages/release/bioc/html/DESeq2.html
   (c) ashr—https://github.com/stephens999/ashr
   (d) dplyr—https://cran.r-project.org/web/packages/dplyr/dplyr.pdf
   (e) writexl—https://cran.r-project.org/web/packages/writexl/index.html
   (f) ggplot2—https://ggplot2.tidyverse.org/index.html
   (g) ggrepel—https://cran.r-project.org/web/packages/ggrepel/index.html
   (h) pheatmap—https://www.r-bloggers.com/2022/10/the-pheatmap-function-in-r/
   (i) pcaExplorer—https://bioconductor.org/packages/release/bioc/html/pcaExplorer.html
9. Load all library packages required for the desired analysis to identify DEGs.

```
Sample R Code: > library(devtools) #install.packages("dev-
tools")
> library(DESeq2) #BiocManager::install("DESeq2")
> library(ashr) #install_github("stephens999/ashr")
> library(dplyr) #install.packages("dplyr")
> library(writexl) #install.packages("writexl")
> library(ggplot2) #install.packages("ggplot2")
> install.packages("tidyverse")
> library(ggrepel) #install.packages("ggrepel")
> library(pheatmap) #install.packages("pheatmap")
> library(pcaExplorer) #BiocManager::install("pcaExplorer")
```

10. Import the sample_description.txt file using the read.table function in RStudio (*see* **Note 49**).

```
Sample R Code: > sample_description <- read.table("sample_de-
scription.txt", header=T, sep="\t")
```

11. Import the wormgenes.txt file using the read.table function in RStudio (*see* **Note 49**).

```
Sample R Code: > wormgenes <- read.table("wormgenes.en-
sembl.106.txt", header=FALSE, sep="\t")
```

12. Import WormBase functional gene description text files from Subheading 3.5, **step 6** using read.table function in RStudio (*see* **Note 49**).

```
Sample R Code: > worm1 <- read.table("WB285_WBgenes.txt",
header=F, sep="\t", rownames(F))
> worm2 <- read.table("WB285_geneclass.txt", header=F,
sep="\t", rownames('Gene class description:'))
> worm3 <- read.table("WB285_concise.txt", header=F,
sep="\n", rownames('Concise description'))
> worm4 <- read.table("WB285_automated.txt", header=F,
sep="\t", rownames('Automated description:'))
```

13. Combine all WormBase functional gene descriptions onto a new data frame using the cbind function (four columns total). Delete the extra column names found on each gene description so that only the gene description text for each column remains (*see* **Notes 50** and **51**).

```
Sample R Code: > wormgene_desc <- cbind(worm1,worm2,worm3,
worm4)
> Colnames(wormgene_desc)
> Colnames(wormgene_desc) <- c("WormBaseID", "GeneName",
"Transcript_Name", "Gene_class_description", "Concise_de-
scription", "Automated_description")
> wormgene_desc$Gene_class_description <- gsub("Gene class
description: ","",wormgene_desc$Gene_class_description)
> wormgene_desc$Concise_description <- gsub("Concise descrip-
tion: ","",wormgene_desc$Concise_description)
> wormgene_desc$Automated_description <- gsub("Automated de-
scription: ","",wormgene_desc$Automated_description)
```

14. Export this data as an Excel file using the write_xlsx function. This is the final WormBase functional gene description file (*see* **Note 52**).

```
Sample R Code: > write_xlsx(wormgene_desc, "wormbase.WS285_ge-
nedescriptions.xlsx")
```

15. Construct a DESeq2 data set object using the DESeqDataSet-FromHTSeqCount function. Load in the sample_description data frame, directory for processed STAR files (Subheading 3.5, **step 2**), and design of the experiment using column(s) in the sample_description data frame (*see* **Notes 53** and **54**).

```
Sample R Code: > experiment1 <- DESeqDataSetFromHTSeqCount
(sampleTable = sample_description, directory = "star_raw-
counts", design = ~ Genotype)
```

16. Filter the resulting data by removing genes with less than ten summed raw counts [9] (*see* **Note 55**).

```
Sample R Code: > experiment1 <- experiment1[rowSums(counts
(experiment1)) > 10, ]
```

17. Fit statistical model of DESeq2 using the DESeq function (*see* **Note 56**).

```
Sample R Code: > stat1 <- DESeq(experiment1)
```

18. Create a norm_counts data frame for normalized counts using the log2(counts) function of the statistical model variable. Add WormBaseIDs and gene names to the norm_counts data frame using the merge(unique) function (*see* **Notes 57** and **58**).

```
Sample R Code: > norm_counts <- log2(counts(se_star2, normal-
ized = TRUE)+1)
> norm_counts_genes <- merge(unique(wormgenes[,1:2]), data.
frame(ID=rownames(norm_counts), norm_counts), by=1, all=F)
```

19. Using the norm_counts data frame, create new data frame (s) for each sample to include the following columns: Worm-BaseIDs, gene name, and sample 1 normalized counts for each trial replicate using base R functions (*see* **Notes 59** and **60**). Use the rowMeans function to add a column of average normalized counts to each sample data frame (*see* **Note 61**).

```
Sample R Code: > sample1$sample1_avg_ncounts <- rowMeans(sam-
ple1[,3:5])
> sample2$sample2_avg_ncounts <- rowMeans(sample2[,3:5])
> Sample_1 <- sample1[,c(1,2,6)]
> Sample_2 <- sample2[,c(1,2,6)]
> avg_norm_counts <- merge(sample_1, sample_2)
```

20. Export the average normalized counts as a text file using the write.table function and as an Excel table using the write_xlsx function (*see* **Note 62**).

```
Sample R Code: > write.table(norm_counts_genes, "experiment1_-
normcounts.txt", quote=F, col.names=T, row.names=F, sep="\t")
> write_xlsx(norm_counts_genes, "experiment1_normcounts.
xlsx")
> write.table(avg_norm_counts, "experiment1_avg_normcounts.
txt", quote=F, col.names=T, row.names=F, sep="\t")
> write_xlsx(avg_norm_counts, "experiment1_avg_normcounts.
xlsx")
```

3.5.1 Visualization: Principal Component Analysis and Sample Distance Heatmap

1. To visualize the variability between samples, perform a variance stabilizing transformation (vst function) using the statistical model of DESeq2. Calculate between sample distances matrix using the vst transformed variable and create a heatmap (*see* **Notes 63** and **64**).

```
Sample R Code: > vsd <- vst(stat1)
> sampleDistMatrix <- as.matrix(dist(t(assay(vsd))))
> png("sample_distance_heatmap.png")
> pheatmap(sampleDistMatrix)
```

2. Create a principal component analysis (PCA) plot (plotPCA function) using the transformed variable and a sample column description from the sample_description data frame (Subheading 3.5, **step 3**) (*see* **Notes 65–67**).

```
Sample R Code: > png("PCA1.png")
> plotPCA(object = vsd, intgroup = "Genotype_ortreatment")
```

3.5.2 Differentially Expressed Genes

1. To identify DEGs between two groups, use the results function, and include the object statistical model of DESeq2, using a contrast argument to include the two comparison names (*see* **Notes 68** and **69**).

```
Sample R Code: > deg1 <- results(object = stat1, contrast=c
("Genotype_ortreatment","sample1_treatment1","sample1_treat-
ment2"))
```

2. Use the lfcShrink function and include the object statistical model of DESeq2 and a contrast argument to include the two comparison names (*see* **Notes 68** and **69**).

```
Sample R Code: > deg1_shrink <- lfcShrink(dds = stat1,
contrast=c("Genotype_ortreatment","sample1_treatment2","sam-
ple1_treatment1"), type="ashr")
```

3. Merge the wormgene description data frame (Subheading 3.5, **steps 1–4**) with the lfcShrink variable using gene names as a common column in both data frames (*see* **Note 70**).

```
Sample R Code: > deg1_genes <- merge(unique(wormgene_desc[1:
6]), data.frame(ID=row.names(deg1_shrink), deg1_shrink),
by=1, all=F)
```

4. Merge the average norm counts (Subheading 3.5, **steps 19–20**) for the samples to the lfcShrink variable (*see* **Note 71**). Organize columns in the order desired (*see* **Note 72**).

```
Sample R Code: > deg1_genes <- merge(deg1_genes, sample1_-
treatment1)
> deg1_genes<- merge(deg1_genes, sample1_treatment2)
> colnames(deg1_genes)
> deg1_genes <- deg1_genes[c("WormBaseID", "Transcript_Name",
"GeneName", "baseMean", "sample1_treatment1_avg_ncounts",
"sample1_treatment2_avg_ncounts", "log2FoldChange", "lfcSE",
"pvalue", "padj", "Gene_class_description", "Concise_descrip-
tion", "Automated_description")]
```

5. Export the DEGs of each comparison to a text file using the write.table function and as an Excel file using the write_xlsx function (*see* **Note 73**).

```
Sample R Code: > write.table(deg1_genes, "deseq2_sample1treat-
ment2_vs_sample1treatment1.txt", quote=F, col.names=T, row.
names=F, sep="\t")
> write_xlsx(deg1_genes, "deseq2_sample1treatment2_vs_sam-
ple1treatment1.xlsx")
```

Repeat **steps 1–5** from Subheading 3.5.2 for each comparison made for all samples. Each file should include the following columns: WormBaseIDs, gene name, Sample1 average norm counts, Sample2 average norm counts, log2FoldChange, lfcSE, stat, pvalue, padj, gene class description, gene concise description, and gene automated description.

4 Notes

1. Consistency in animal husbandry methods is important to maintain reproducibility for each biological replicate in an experiment. This will ensure reproducibility of the RNA-seq data and ultimately result in more accurate interpretation of the data. Differences in experimental protocols between sample replicates could have an impact on gene expression,

complicating interpretation of RNA-seq results. Always use the same sample collection procedure and track the timing of steps wherever possible. This ensures the results are reproducibly reporting differences that are due to experimental conditions and not subtle changes in the collection protocol.

2. Maintain animals in proper culture conditions. Provide sufficient food and space for at least seven generations prior to using the animals for experiments to reduce intergenerational or transgenerational impacts from starvation or crowding [10, 11].
3. Measuring the optical density of *E. coli* OP50 prior to seeding NGM plates ensures the animals are feeding on the same amount of bacteria; differences in OD can impact metabolic rate [12].
4. Determine the egg-laying rate of animals of different genotypes and treatment groups before the experiment. Preliminary experiments establishing the egg-laying rate can help ensure enough animals can be collected for RNA sequencing.
5. The 1–2-h egg lay method is used to obtain a synchronized population. Hypochlorite treatment is an alternative method to synchronize a population in which adult animals are dissolved with hypochlorite solution and eggs are collected and hatched overnight in M9 solution. Animals arrest in the L1 stage when hatched in the absence of food and resume development in the presence of food. However, this method to synchronize a population should be avoided if possible, to minimize stress responses due to starvation or reactive oxygen species induction by hypochlorite [13].
6. Day 1–2 adult animals are used for the egg lay method to synchronize animals. Older animals may retain embryos longer in the uterus prior to laying them, resulting in a more asynchronous population.
7. Carefully control the number of animals that are on each plate to avoid stress responses induced by crowding.
8. Consider the total number of animals that are collected for RNA extraction and isolation. A smaller population will have a different yield on the amount of RNA that is collected and sequenced compared to a larger population. Using a larger number of animals per sample will facilitate higher genome coverage and the detection of genes with low expression. However, certain mutations can have low egg-laying rates which can pose a challenge to collecting sufficient animals for RNA extraction and isolation.

9. A mutation may have an impact on developmental progression, and therefore care should be taken to understand developmental timing when compared to other genotypes to ensure animals can be collected at the same time.
10. Developmental progression is monitored by keeping track of the hours post-egg lay and germline morphology. For example, the mid-L4 stage is identified by the location of the distal tip using a compound microscope [14]. It is vital to perform baseline experiments to determine the anticipated time for animals to grow from embryo to the desired developmental stage for RNA collection. This is to minimize the number of times the animals must be removed from the incubator to monitor developmental progression, which could result in transcriptional changes unrelated to the experimental conditions [15].
11. Keep track of the time it takes to collect the animals starting with the plate rinse, in between washes, and through snap freezing the Eppendorf tube. The time it takes to collect the animals should be consistent across all samples. This is to prevent variation resulting from subtle protocol changes, which could impact gene expression and the interpretation of the RNA-seq results.
12. Keep track of timing during the extraction protocol and keep consistency across all samples to ensure reproducibility of the experiment. Wear gloves and use RNAase-free pipette tips and tubes. To become familiar with the RNA extraction and isolation protocol, it is helpful to practice going through the protocol multiple times before the actual experiment.
13. We found that the addition of a 65 °C incubation step, additional homogenization time, and additional membrane drying time resulted in increased RNA yield. These steps are noted in the protocol described here, which differs slightly from the protocol given by the manufacturer. To become familiar with the RNA extraction and isolation protocol, it is helpful to practice going through the protocol multiple times before the actual experiment.
14. Keep samples in crushed ice to avoid RNA degradation.
15. The RNA extraction and isolation protocol can be performed at room temperature. However, when working with more than one sample, keep samples in crushed ice.
16. The centrifugation step is done at room temperature but can be performed at 4 °C if RNA degradation occurs.
17. Ensure that the Eppendorf tube and the column are closed to prevent spilling of sample. During heat bath use, ensure clips are locked or use screw cap tubes to prevent sample from spilling or evaporation.

18. Prior to sending samples to the genomics center, ensure degradation of RNA is not occurring in any samples. There are multiple ways to detect this degradation. For practice samples use the bleach protocol and/or a NanoDrop spectrophotometer. Final degradation check before sequencing is completed using a Qubit. This will ensure high-quality RNA is sequenced.
19. RNA sequencing is performed by the University of North Texas Genomics Center using an Illumina sequencer.
20. Determine sequencing method in collaboration with the genomics center to ensure mRNA targets are sequenced properly. This method may change depending on the biological question.
21. Determine library choice for sequencing method to ensure mRNA sequencing method answers biological question.
22. Programs are invoked from shell or within R (through RStudio). Source code and additional text files are provided at https://github.com/robledoj13/Celegans_RNASeq2023. The programs used in this pipeline are publicly available, open-source, and typically available for modern Linux distributions and macOS. This pipeline does not assume that the end user has any programming experience. While the R language has many built-in statistical methods and analysis options, we will be using a collection of additional publicly available libraries for R to handle some of our downstream analysis. Review additional conceptual program considerations to ensure computer system is built for your analysis [https://biocorecrg.github.io/RNAseq_course_2019/] [16, 17].
23. Please refer to the documentation manuals for each program used in this pipeline. Depending on the data being analyzed, there may be options and configurations better suited to each use-case scenario. While many aspects of this pipeline use the default parameters for the programs used, understanding the assumptions of each program's default parameters and configurations is an invaluable step toward optimizing your data workflow. Troubleshooting issues can be solved by reading through the documentation manuals for each program. The following online communities can also help if questions arise:
    (a) github.com
    (b) biostars.org
    (c) support.bioconductor.org
    (d) r-bloggers.com
24. Prior knowledge of using a Linux shell environment, such as bash, as well as knowing some R programing syntax is helpful for this analysis. A review of common syntax used to select

directories, activate programs, and edit text files will help in processing RNA Sequencing data.

25. Use simple naming conventions for all files and folders to reduce typing errors when running code. Avoid spaces to simplify navigation through the command line interface. Organize the bioinformatic analysis files by using multiple folders and directories if needed. Keep thorough notes for troubleshooting errors.
26. Ensure that any compressed files are decompressed before program access. Gzip format is popular in the field of bioinformatics, and files can be compressed and decompressed with the gzip and gunzip commands, respectively.
27. In all programs, the thread option is adjusted depending on computer specifications.
28. To ease installation of programs, we recommend using a universal package manager such as Conda. Bioconda is a repository specifically focused on biomedical software and should be able to provide all of the packages that we will be using in this guide.
29. In some cases, there may be more than one package or version available for programs in Bioconda. We recommend installing the most up-to-date, compatible version for the system being used.
30. In our test data, each sample produces eight sequencing files (. fastq extension). Each file includes sample name, lane number (001, 002, 003, 004), and strandedness (plus (R1) and minus (R2)) though this format is ultimately determined by the source of the sequencing data.
31. Use the *cat* command in Linux to reduce the number of input files for our downstream steps. This is also a good opportunity to rename files to include pertinent information about the samples (e.g., genotype, treatment) to keep things organized.
32. To gain experience in interpretation of Quality Control Reports, users should become familiar with examples of high- and low-quality data. The online manual for FASTQC includes many examples of both good- and bad-quality sequencing data [https://www.bioinformatics.babraham.ac.uk/projects/fastqc/Help/].
33. The FASTP package will perform the adapter trimming step and quality control. FASTP will generate new FASTQ files provided into a new folder. Review quality reports and ensure high-quality data for analysis. We will be using the default parameters, but please note that custom adapters or the need for less/more stringent trimming can be invoked as needed.

34. Use a FASTA file and an annotation file from the same database. Ensembl and WormBase sometimes have different formats for files. Individual FASTA files separated by chromosome location can be used to lower system requirements, but more processing of files is required. Files that have the whole genome and gtf annotation files are recommended if the system has enough resources to do so.
35. Adjust for the small genome of *C. elegans* and RAM consumption according to computer system considerations in **Note 34**.
36. Ensure that the output FASTQ files from FASTP are used for alignment. BAM files are an alignment output format that can be used for additional exploratory analysis such as through the Integrative Genomics Viewer (IGV) or used as input for variant calling, differential expression, etc. The GeneCounts option is a STAR-specific parameter that exports a text file of raw gene counts as {SAMPLE_NAME}.ReadsPerGene.out.tab.
37. Unmapped reads can be analyzed with tools such as BLAST to determine their source. The alignment/mapping step can be performed with a genome and transcript-annotation file, or with a transcriptome index. For STAR, we first build an index from the genome with an annotated transcript file, but there are many other alignment tools that may use a different approach.
38. Library strandedness should be provided by the genomics center and bioinformatic validation performed using IGV, Salmon, and count methods previously described in Subheading 3.4.6. This crucial step will help the user explore each sample file and help understand RNA-seq method more in depth. The exploration of each sample in this step can help bioinformaticians/biologists have an insight about the experiment.
39. When indexing BAM coordinated files, both .bam and .bam.bai files need to be in the same folder for IGV.
40. Index genome using Salmon to automatically determine library method.
41. If necessary, Salmon can be run in mapping-based mode to automatically determine library method. The validateMappings option will automatically detect the most likely library type.
42. Another option is to count the number of reads using the STAR Gene count to determine library method. If the library is stranded, this should be fairly obvious to decipher based on which columns have significant counts across the transcriptome.

43. Extract the appropriate columns using STAR documentation manual. Column 1 includes gene list, while columns 2–4 contain gene counts corresponding to library strandedness.
44. Further process the STAR Gene count files to extract the appropriate columns. Ensure in multiple ways that the appropriate columns are extracted for the experiment. The goal is to have a text file with gene names and raw counts.
45. The sample_description.txt file has a column for each raw count file and other columns used to describe each sample such as genotype, treatment, trial number, etc. Each column is separated by a space for ease of import into R. Notice that some of the columns will be used for the experimental design to identify DEGs. Consider the comparisons required by using the different columns for the sample_description.txt file. Multiple questions can be addressed by assessing the different variables in samples collected.
46. The wormgenes.txt file includes WormBaseIDs commonly used for genes on wormbase.org. WormBase also has additional files that can be used to describe gene families (e.g., functional gene descriptions). The goal is to extract only two columns from the annotation file which includes gene names and WormBaseIDs. Other commands can also be used to get the same result.
47. wormbase.org provides additional *C. elegans* gene list files to further describe WormBase genes.
48. There are many more library packages available for further classification of genes that can help in understanding significant DEGs. The packages used in here are an example of common analyses to understand significant DEGs.
49. We can use the built-in read.table function of R to import simple text files as a data table. Ensure each column is separated by a similar character (e.g., space, tab, comma, semicolon, etc.), and adjust the separator parameter appropriately.
50. Validate the same number of genes in each file using the nrow function to ensure each file is imported correctly.
51. Edit column names accordingly to ensure each file is formatted correctly.
52. Export gene descriptions using the write_xlsx function. Ensure all columns are transferred correctly.
53. Create a DESeq2 data object if samples differ by genotype. Additional variables can be included in the experimental design by following DESeq2 documentation directions.
54. Additional biological questions can be addressed from the different samples examined. A column can be organized to

include both genotype and treatment differences or to make different comparisons. Experiments should be designed carefully to prepare the sample_description file.

55. A filtering step will help increase the statistical power of the DEG analysis.
56. Model DESeq2 to the experiment to identify significant DEGs.
57. Normalize counts using the log2 function. +1 will result in no errors for any zero present in the data.
58. Add WormBaseIDs to the normalized count data frame.
59. Add average normalized counts to DESeq2 model object (stat1). The numbers can be changed to represent the desired columns.
60. Ensure the correct columns are selected for each sample. Validate using base R functions. The head() function will preview the top six lines of the data frame.
61. Calculate mean norm counts to an average normalized counts object. Ensure all mean counts are calculated correctly.
62. Export file count files. Ensure all columns are transferred correctly.
63. Transformation and sample distance matrix calculations are used to create a heatmap and principal component analysis (PCA). This process includes calculations used to produce log2 scale data already normalized by DESeq2 according to library size.
64. Sample distance heatmap will help visually show the differences between samples. Samples clustering together ensure samples are reproducible, and sample groups spread apart suggest that the experimental design resulted in unique expression patterns.
65. PCA plot and sample distance heatmap are techniques used for reducing the dimensionality of large data sets. This process helps to interpret the variability of the samples of an experiment.
66. PCA plot will help to visually show the differences between samples.
67. Set the intgroup grouping column variable for the samples (e.g., genotype, treatment, trial number, etc.). Grouping columns is one that was created from sample_description data frame **step 3** in Subheading 3.5.
68. DEG analysis (treatment 2 = sample2 variable in (Subheading 3.5, **steps 19–20**); review syntax accordingly). Use "ashr" for LFC shrinkage. If used in published research, please cite Stephens, M. (2016) False discovery rates: a new deal. Biostatistics, 18:2. https://doi.org/10.1093/biostatistics/kxw04.

69. Be careful with the order when interpreting significantly DEGs. For example, gene A is significantly different in treatment 2 when compared to treatment 1. Therefore, the order for this example will be "sample1_treatment2", "sample1_treatment1".
70. Add average wormgene_descriptions to DEG file (Subheading 3.5, **steps 2–5**). Ensure the right columns are transferred using BaseR functions.
71. Add average norm counts of sample ((Subheading 3.5, **steps 19–20**). Ensure the right columns are transferred using BaseR functions.
72. Reorder column names for interpretation of significant differences for each gene as a row. Ensure the right columns are transferred using BaseR functions.
73. Write differential expression analysis results to a text file and an Excel file. Ensure the right columns are transferred to both files. Sometimes gene names may be interpreted as dates by Excel.
74. Cross-platform Bioinformatic Analysis Considerations.

    Initial setup of computer hardware and software to perform bioinformatic analyses on RNA sequencing data can be challenging due to the software complexity, version incompatibilities, and software dependencies involved. Having a thorough understanding of the software installed and the system that it is installed on can avoid potential problems for downstream analysis. Executing a bioinformatic pipeline requires multiple software packages to be installed that may vary due to the computer system's hardware and software specifications. Below are alternative considerations to keep in mind when using different computer systems:

    74.1 macOS,

    Apple manufactures computers with 64-bit Intel processors (AMD64 architecture) and Apple-designed Apple silicon processors (Apple ARM64 architecture). The Anaconda platform and R/RStudio support Apple silicon systems. R/RStudio applications are also supported although some packages may run via the Rosetta2 AMD64 emulation layer on macOS. If packages are not available on ARM64, it may be advisable to install the Anaconda AMD64 installer.

    74.2 Linux (Ubuntu).

    Linux offers additional flexibility and functionality for running bioinformatic analyses. This includes the ability to use larger amounts of memory, the flexibility to use a variety of different software versions, and greater customizability while retaining software. The Anaconda platform is compatible with Ubuntu Linux. Official R packages are available via the CRAN repository website [https://cran.r-project.org/bin/linux/ubuntu/fullREADME.html], and R/RStudio packages are available via the RStudio website [https://posit.co/download/rstudio-desktop/].

74.3 Microsoft Windows 10/11.

Running bioinformatic analyses is also possible on Windows 10/11 via Windows Subsystem for Linux (WSL). Instructions to install WSL [https://learn.microsoft.com/en-us/windows/wsl/install] and Ubuntu releases are both available from Microsoft. WSL allows for running a complete Ubuntu Linux system from within Windows at near-native speeds and is an option if not using macOS/Linux systems.

74.4 Hardware Considerations.

Hardware requirements may differ depending on programs used for RNA-seq analysis. For the RNA-seq pipeline reported here, the STAR aligner was used to index the *C. elegans* genome and the alignment step. In using the *C. elegans* genome, 16 gigabytes of RAM, 100 gigabytes of storage, and a modern 8-core processor are sufficient for most bioinformatic analysis. The use of external hard drives may also help in storing large files.

75. Reproducibility of Bioinformatic Pipelines.

75.1 Containerization.

Biologists may be interested in reproducing bioinformatic pipelines using similar system requirements as referenced in published work. An approach to configure software performed by a published bioinformatic pipeline is to use a containerization tool such as Docker. Docker allows bioinformatic pipeline programs to run on different computers by creating a "container" that encapsulates all the software requirements needed for a bioinformatic pipeline. Containerization of bioinformatic pipelines using Docker or Singularity can be performed using BioContainers [https://biocontainers-edu.readthedocs.io/en/master/what_is_biocontainers.html] providing several prebuilt bioinformatics packages. These tools allow one to package all the necessary software and dependencies into a self-contained environment that can run across different systems either natively such as on macOS or on Windows Subsystem for Linux. This ensures that the entire analysis can be reproduced the same way each time it is run, regardless of the operating system or hardware used. However, containerization software can be complex and cumbersome to set up for most workflows. Although the use of containerization software is practical to run large-scale workflows, it is generally more practical to use a package manager to run command line bioinformatic software using desired program version installations.

75.2 Package Management and Environment Specifications.

Another approach is to use package management software such as the Anaconda platform [https://www.anaconda.com] with the Bioconda repository [https://bioconda.github.io] or mamba [https://github.com/mamba-org/mamba] package manager which is a reimplementation of the Anaconda platform package manager in C++. Package managers such as Anaconda platform can automatically manage dependencies and ensure that the required software is installed in a consistent and reproducible way. When defined versions of software are intended to be used as an alternative to containerization, software versions can be specified in a YAML markup file (*environment.yml*) to share an Anaconda platform environment, or a *requierments.txt* file to share a Python virtual environment. These files can be shared to reproduce the environment with the specified software packages and versions in the files.

75.3 Workflow Management Software.

If reproducibility is desired in more complex workflows, workflow management software including Nextflow and Snakemake can be used. This software is designed to capture the entire workflow of data-driven computational pipelines and define the bioinformatic pipeline in a series of steps or processes that are executed in a specific order.

75.4 Pipeline Software Design and Documentation Considerations.

To ensure reproducibility, it is important to provide documentation of the entire analysis pipeline, including all software packages and versions of software packages used. Reproducible bioinformatic scripting can be completed using tools such as Jupyter notebooks, which allow the documentation of analysis and code in a reproducible and interactive way. Overall, setting up a bioinformatics platform for sequence analysis can be challenging, but with the right tools and approaches, it's possible to ensure reproducibility and ease of setup.

References

1. Stark R, Grzelak M, Hadfield J (2019) RNA sequencing: the teenage years. Nat Rev Genet 20(11):631–656. https://doi.org/10.1038/s41576-019-0150-2
2. Brenner S (1974) The genetics of *Caenorhabditis elegans*. Genetics 77(1):71–94. https://doi.org/10.1093/genetics/77.1.71
3. Davis P, Zarowiecki M, Arnaboldi V et al (2022) WormBase in 2022-data, processes, and tools for analyzing *Caenorhabditis elegans*. Genetics 220(4). https://doi.org/10.1093/genetics/iyac003
4. Aranda PS, LaJoie DM, Jorcyk CL (2012) Bleach gel: a simple agarose gel for analyzing RNA quality. Electrophoresis 33(2):366–369. https://doi.org/10.1002/elps.201100335
5. Chen S, Zhou Y, Chen Y, Gu J (2018) fastp: an ultra-fast all-in-one FASTQ preprocessor. Bioinformatics 34(17):i884–i890. https://doi.org/10.1093/bioinformatics/bty560
6. Dobin A, Davis CA, Schlesinger F et al (2013) STAR: ultrafast universal RNA-seq aligner. Bioinformatics 29(1):15–21. https://doi.org/10.1093/bioinformatics/bts635
7. Li H, Handsaker B, Wysoker A et al (2009) The sequence alignment/map format and SAMtools. Bioinformatics 25(16): 2078–2079. https://doi.org/10.1093/bioinformatics/btp352
8. Danecek P, Bonfield JK, Liddle J et al (2021) Twelve years of SAMtools and BCFtools. Gigascience 10(2). https://doi.org/10.1093/gigascience/giab008
9. Chen Y, Lun AT, Smyth GK (2016) From reads to genes to pathways: differential expression analysis of RNA-Seq experiments using Rsubread and the edgeR quasi-likelihood pipeline. F1000Res 5:1438. https://doi.org/10.12688/f1000research.8987.2
10. Klosin A, Casas E, Hidalgo-Carcedo C et al (2017) Transgenerational transmission of environmental information in *C. elegans*. Science 356(6335):320–323. https://doi.org/10.1126/science.aah6412
11. Miersch C, Doring F (2012) Paternal dietary restriction affects progeny fat content in *Caenorhabditis elegans*. IUBMB Life 64(7): 644–648. https://doi.org/10.1002/iub.1042
12. Preez GD, Fourie H, Daneel M et al (2020) Oxygen consumption rate of *Caenorhabditis elegans* as a high-throughput endpoint of toxicity testing using the seahorse XF(e)96 extracellular flux analyzer. Sci Rep 10(1):4239. https://doi.org/10.1038/s41598-020-61054-7
13. Whiteman M, Spencer JP, Szeto HH et al (2008) Do mitochondriotropic antioxidants prevent chlorinative stress-induced mitochondrial and cellular injury? Antioxid Redox Signal 10(3):641–650. https://doi.org/10.1089/ars.2007.1879
14. Antebi A, Norris CR, Hedgecock EM et al (1997) Cell and growth cone migrations. In: Riddle DL, Blumenthal T, Meyer BJ, Priess JR (eds) C. elegans II, 2nd edn. Cold Spring Harbor, New York
15. Melnick M, Gonzales P, Cabral J et al (2019) Heat shock in *C. elegans* induces downstream of gene transcription and accumulation of double-stranded RNA. PLoS One 14(4): e0206715. https://doi.org/10.1371/journal.pone.0206715
16. Gimenez G, Stockwell PA, Rodger EJ et al (2023) Strategy for RNA-Seq experimental design and data analysis. Methods Mol Biol 2588:249–278. https://doi.org/10.1007/978-1-0716-2780-8_16
17. Chatterjee A, Ahn A, Rodger EJ et al (2018) A guide for designing and analyzing RNA-Seq data. Methods Mol Biol 1783:35–80. https://doi.org/10.1007/978-1-4939-7834-2_3

Chapter 7

Inferring Novel Cells in Single-Cell RNA-Sequencing Data

Ziyi Li and Peng Yang

Abstract

Single-cell RNA-sequencing (scRNA-seq) is a powerful technology that allows researchers to study gene expression heterogeneity within a tissue or cell population. One of the major advantages of scRNA-seq is that it allows researchers to identify and characterize novel cell types or subpopulations within a tissue that may be missed by traditional bulk RNA-sequencing methods. Although many existing methods have been developed to recognize known cell types, inferring novel cells may still be challenging in routine scRNA-seq analysis. Here we describe three lines of methods for inferring novel cells: unsupervised and outlier-detection-based methods, supervised and semi-supervised methods, and copy number variation (CNV)-based methods, as well as the corresponding situations that each method applies. We also provide implementation code and example usages to illustrate the available methods.

Key words Copy number variation, Novel cell detection, Single-cell RNA-seq, Supervised annotation, Unsupervised clustering

1 Introduction

The analysis of single-cell RNA sequencing (scRNA-seq) data consists of several procedures including preprocessing, dimension reduction, visualization, and downstream analysis [1]. Cell type annotation, usually performed as the first step of downstream analysis, characterizes different cell populations in the samples and provides crucial information for additional analysis methods [2–4]. The accuracy of the obtained annotated cell labels can impact the findings and conclusions greatly.

The cell annotation methods can generally be classified into two groups, the unsupervised clustering methods and the supervised annotation methods. The unsupervised clustering methods adopt algorithms such as community detection, consensus-based clustering, and statistical model-based clustering to divide the cell population into a few clusters. The popular methods include Seurat [5], SHARP [6], SC3 [7], and TSCAN [8]. However, labeling the obtained clusters can be a subjective procedure that requires

Rajeev K. Azad (ed.), *Transcriptome Data Analysis*, Methods in Molecular Biology, vol. 2812,
https://doi.org/10.1007/978-1-0716-3886-6_7,

biological knowledge and lacks reproducibility [9]. As a result, several supervised methods have been proposed to use existing annotated single-cell data as reference and annotate the new data with machine learning or statistical approaches. Examples of methods in this category include scmap [3], CHETAH [10], singleR [11], and CellAssign [12].

Although many existing methods are available to cluster and annotate the cells, inferring novel cells, especially the cells that are subject-specific or disease-associated, from scRNA-seq data can still be a challenging task. In the unsupervised analysis setting, the novel cells are the rare cells that existing studies have little knowledge about and few biological markers to define them [13]. In the supervised annotation setting, the novel cells are the ones that were not present in the reference data but present in the data of interest [14]. Such novel cells can be a distinct and unique subpopulation of a major cell lineage, e.g., a specific subpopulation of T cells [15], or a brand new cell population, e.g., a new tumor clone [16]. It is important to identify these novel cells as they have been frequently reported to associate with disease progression and patients' outcomes. For example, a unique microglia subpopulation has been found to play a key role in restricting development of Alzheimer's disease [17].

Here we present the use of bioinformatics methods to infer novel cells in three settings. In the first two settings, we consider the general scRNA-seq data from any conditions or diseases. We illustrate the unsupervised clustering and outlier-detection-based methods to identify rare cell types. When there is existing annotated scRNA-seq data that can serve as reference, we describe novel cell identification in the supervised annotation setting to leverage the reference information. Lastly, we focus on novel cell identification in cancer studies, where the neoplastic cells usually bear high level of copy number variations (CNV). We describe how to use the existing CNV-based method to distinguish normal and tumor cells.

2 Novel Cell Identification by Unsupervised Clustering and Outlier Detection-Based Methods

After obtaining and preprocessing the scRNA-seq data, a typical analysis pipeline involves unsupervised clustering to explore potential cell types in the sample. It is important to remove low-quality cells (e.g., those with excessively high levels of mitochondrial gene expression) and doublets during the preprocessing steps. Assuming the expression count matrix is denoted by `expX`, we illustrate how to use Seurat to cluster the cells and visualize cluster-specific markers in the R code below.

```
library(Seurat)

# create seurat object
thisdata <- CreateSeuratObject(counts = expX, meta.
data = meta)
thisdata <- NormalizeData(thisdata)
thisdata <- FindVariableFeatures(thisdata, selec-
tion.method = "vst", nfeatures = 2000)
all.genes <- rownames(thisdata)
thisdata <- ScaleData(thisdata, features = all.
genes)
thisdata <- RunPCA(thisdata, features = Variable-
Features(object = thisdata))
thisdata <- FindNeighbors(thisdata, dims = 1:20)

# cluster cells and perform dimension reduction
thisdata <- FindClusters(thisdata, resolution = 0.5)
thisdata <- RunUMAP(thisdata, dims = 1:20)

# visualize the clusters on a UMAP plot
DimPlot(thisdata, reduction = "umap", label = TRUE)

# find cell type-specific markers
thisdata.markers <- FindAllMarkers(thisdata, only.
pos = TRUE)
thisdata.markers %>% group_by(cluster) %>% top_n(n = 4,
wt = avg_log2FC) -> top10

# visualize the top markers via heatmap
DoHeatmap(thisdata, features = top10$gene, size = 3)
```

In the above code, we use Seurat to perform preliminary clustering with a community detection algorithm and visualize the data using Uniform Manifold Approximation and Projection (UMAP). There are several parameters in these functions that users can tune, and more details can be found in the Seurat tutorial (https://satijalab.org/seurat/articles/pbmc3k_tutorial.html). This is done with or without concern for the presence of novel cells. However, if novel cells are present, the observation of an outlier cluster and unconventional cluster-specific markers from these preliminary explorations can be the first sign. For example, in Fig. 1, cluster 10 appears to be far from the other clusters, and cluster 11 does not express any of the known markers listed, indicating that they may not belong to the major cell type lineages and can potentially be novel cell types. Calling a cluster a novel cell type in this setting can be a subjective decision and relies heavily on prior biological knowledge.

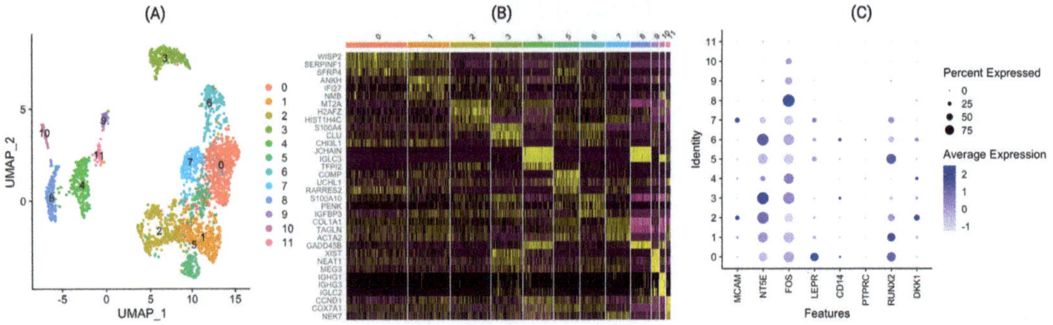

Fig. 1 An example output from Seurat for exploring potential clusters in the sample includes UMAP plot (**a**), expression heatmap (**b**), and dot plot for selected markers (**c**). The clusters that locate far from other known cell types or contain novel cell type-specific markers can potentially be novel cell types

In addition to subjectivity, another concern is that the clustering algorithm tends to form major clusters. When there is a rare cell type with a limited cell number, the clustering algorithm may force the cells to be included in a major cluster rather than having a separate cluster. This can prevent the identification of rare cell types. Several bioinformatics methods have been developed to address this. For example, GiniClust overcomes the challenge of rare cell type detection by using the Gini index, a metric originally developed to evaluate social inequality [13]. RaceID is another popular rare cell type identification method that uses a combination of unsupervised clustering and parametric modeling to distinguish common cell types from rare ones [18]. Below, we demonstrate how to implement RaceID to detect rare cell populations.

```
library(RaceID)

sc <- SCseq(expX)
sc <- filterdata(sc,mintotal=2000)
sc <- compdist(sc,metric="pearson")
sc <- clustexp(sc)

# plot the within-cluster dispersion
# If this decrease becomes constant, saturation is
reached
plotsaturation(sc,disp=FALSE)
plotjaccard(sc)
sc <- clustexp(sc,cln=6,sat=FALSE)

# identify outliers with an internally computed
background model
sc <- findoutliers(sc)

# plot the number of outliers as a function of the p-
value
```

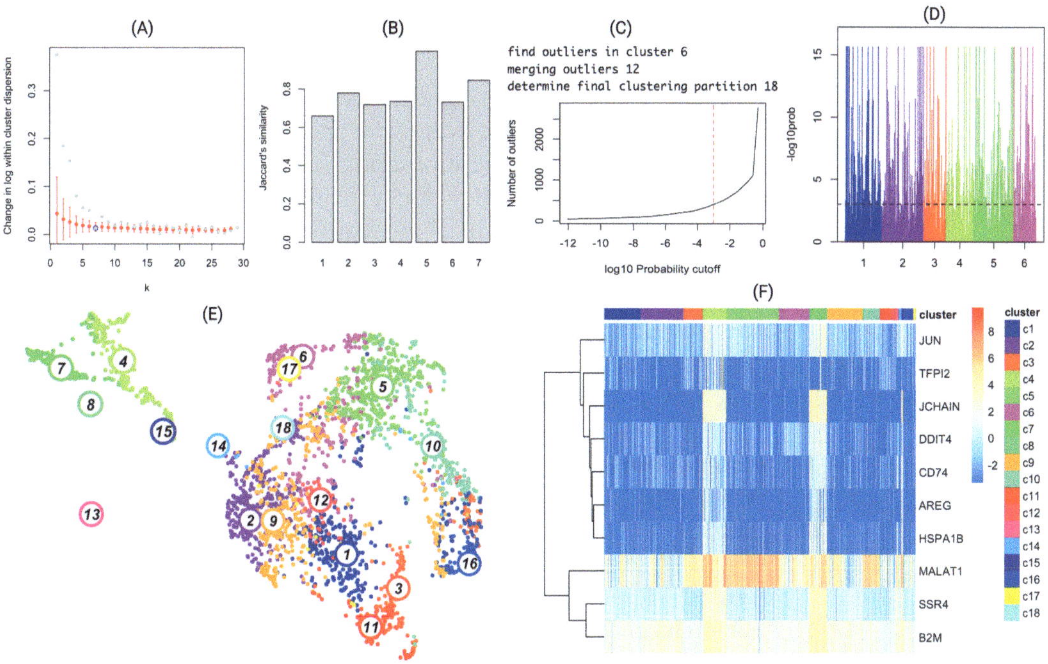

Fig. 2 Example outputs from RaceID explore cell clusters and identify rare cell types, including dispersion saturation plot (**a**), Jaccard's similarity plot (**b**), the number of outliers plot (**c**), barplot of the p values of all cells in each cluster (**d**), the UMAP and the marker expression plot (**e** and **f**)

```
plotsensitivity(sc)

# barplot of p-values across all cells in each
cluster:
plotoutlierprobs(sc)

# draw UMAP plot
sc <- compumap(sc)
plotmap(sc,um=TRUE)

# identify cluster specific differential genes
d <- clustdiffgenes(sc,4,pvalue=.01)
dg <- d$dg
genes <- head(rownames(dg)[dg$fc>1],10)
plotmarkergenes(sc,genes)
```

Figure 2 shows example outputs from applying the CRAN package `RaceID`. Some functions offered by RaceID overlap with the functions from Seurat, but RaceID provides specific functions to compute an internal background model for rare cell type identification (panels C and D). As indicated by the tutorial (https://cran.r-project.org/web/packages/RaceID/vignettes/RaceID.html), the final confirmation of the rare cell types still relies on the known marker genes provided by the investigators (panel F).

3 Novel Cell Identification in Supervised Annotation Framework

With the accumulation of annotated scRNA-seq databases, many supervised methods have been developed to provide more efficient and reproducible cell type annotation. The idea behind these methods is quite similar: They use existing scRNA-seq data as the training data and train a machine-learning-based classifier or statistical model. Under this supervised annotation framework, it is quite challenging to identify novel cell types because their information is not included in the training data. These methods usually use a cut-off threshold to separate a few cells as "unassigned" cells, which are potentially novel cells that require further confirmation.

For example, scmap uses a combination of three similarity metrics computed between the cells in the current data and the cells in the training data to assign cell labels. If the maximum similarity value across all three similarity metrics is below a similarity threshold (default value 0.7), then the cells are marked as "unassigned" [3]. CHETAH (CHaracterization of cEll Types Aided by Hierarchical classification) is another method based on hierarchical structure that explicitly allows the assignment of cells to an intermediate or unassigned state [10]. The assignment is performed based on a confidence score that measures the magnitude of the evidence of assigning a cell to a certain branch. The cells with confidence scores smaller than a threshold (default 0.1) are assigned as unassigned. We illustrate the usage of CHETAH below.

Following the notation from Subheading 2, we use `expX` as the scRNA-seq data to be labeled. We denote `refX` as the count matrix and `labs` as the annotated cell type labels of existing and annotated scRNA-seq data.

```
# create single cell experiment object for the
current data
sce <- SingleCellExperiment(list(counts=expX))
# add log normalized counts
sce <- logNormCounts(sce)

# create sce object for the reference data and
labels
# (obtained from existing publication)
sceB <- SingleCellExperiment(list(counts=refX),
colData=DataFrame(celltype=labs))
sceB <- logNormCounts(sceB)

# apply CHETAH to assign the labels
sce <- CHETAHclassifier(input = sce, ref_cells = sceB,
                        ref_ct = "celltype")
```

```
# this part is only needed if there is no dimen-
sional reduction available
library(Rtsne)
x=logcounts(sce)
res = Rtsne(t(x))
rownames(res$Y) <- colnames(sce)
colnames(res$Y) <- c("tSNE_1", "tSNE_2")
reducedDims(sce) <- SimpleList(TSNE = res$Y)

# visualize the CHETAH results
PlotCHETAH(sce)
```

As shown in Fig. 3, CHETAH assigns a number of cells to intermediate node ("Node 1/2/3") and unassigned group. The intermediate nodes (e.g., node 2) represent the cells that do not have enough confidence to be assigned to the terminal nodes (e.g., delta and beta) but are similar to these two cell types. They could be transition cells or proliferating cells that will ultimately become these mature cell types. The unassigned cells can be novel cells that resemble none of these listed cell types.

In addition to methods like scmap and CHETAH that label novel cells as unassigned, there is some new method developed to specifically identify novel cells in the supervised version. CAMLU (Cell Annotation using Machine-Learning-based method for the presence of Unknown cells) uses autoencoder to compute the distance between the reference and the current data and identify novel cells if their transcriptome profiles are significantly different from the reference cells as shown in Fig. 4 [14].

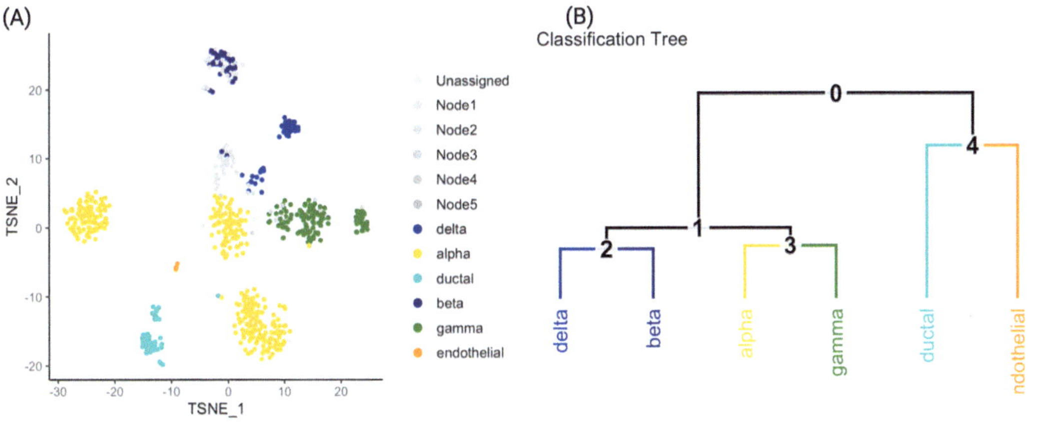

(C)

alpha	beta	delta	ductal	endothelial	gamma	Node1	Node2	Node3	Unassigned
323	46	54	64	9	120	6	72	7	1

Fig. 3 Example outputs of CHETAH for annotating cells: TSNE plot (**a**), classification tree of the cell types (**b**), and the numbers of cells assigned to each cell type (**c**). CHETAH allows assignment of cells intermediate node or unassigned category, and these "Node1/2/3" and unassigned cells are potential novel cells in the data

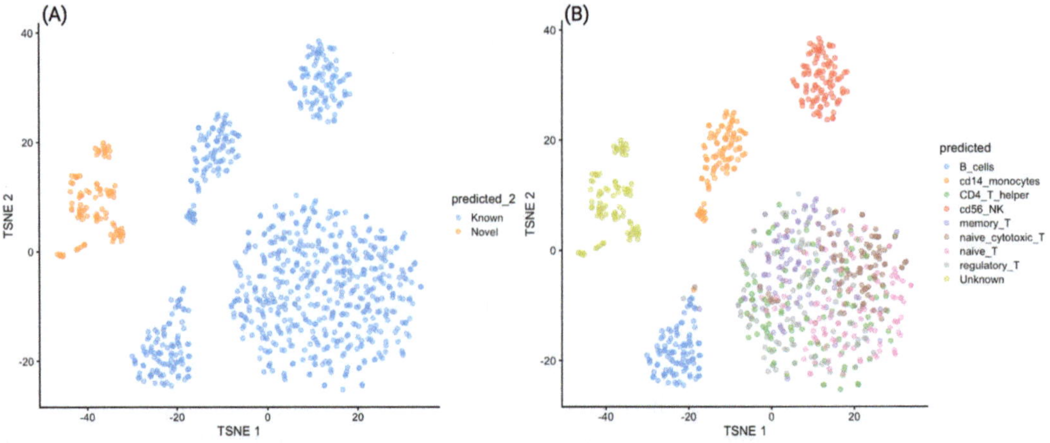

Fig. 4 Visualization of the results from CAMLU. The labels can be obtained for classifying known/novel (**a**) or for full spectrum of cell types (**b**)

```
library(CAMLU)

# classify the labels using CAMLU
label_all <- CAMLU(x_train = refX,
                   x_test = expX,
                   y_train = labs,
                   full_annotation = TRUE,
                   ngene=3000,
                   lognormalize=TRUE)

     # get the dimension reduction of the data
     bin_labels <- factor(label_all$label_known_un-
known,
                          levels = c(0,1),
                          labels = c("Known", "Novel"))
     sce <- SingleCellExperiment(list(counts=expX),
                             colData=DataFrame(celltype=as.
factor(labs),
                                                    predicted =
label_all$label_full,
                                                    predicted_2
= bin_labels))
sce <- logNormCounts(sce) ## add log normalized
counts
sce <- runTSNE(sce, perplexity=10)

# visualize the predicted labels
par(mfrow = c(1,2))
plotTSNE(sce, colour_by = "predicted_2")
plotTSNE(sce, colour_by = "predicted")
```

4 Novel Cell Identification in Cancer Research

In cancer studies, tumor cells continually evolve by acquiring genomic alterations and often deviate from diploid states. This motivates the development of methodologies to distinguish DNA-level copy number variations (CNVs) directly from scRNA-seq data. Large-scale chromosomal CNVs in transcriptomic data can be inferred by the increased or reduced average gene expression compared to an approximate reference profile from diploid cells, which represents amplifications or deletions, respectively. InferCNV [19] is a method for estimating the residual expression of genes across genomic positions and revealing evidence of supporting CNVs. We introduce the application of InferCNV below. The required input data includes a raw count matrix `expX` as denoted in Subheading 2, a sample annotation file `annX` that provides potential cell type labels or clustering results for each cell, and a gene ordering file `ordG` to denote the chromosomal location of each gene, which can be obtained from https://data.broadinstitute.org/Trinity/CTAT/cnv/.

```
# create the infercnv object
infercnv_obj = CreateInfercnvObject(raw_counts_ma-
trix = expX,
                                annotations_file = annX,
                                  delim = "\t",
                                 gene_order_file = ordG,
                                    ref_group_names = c
("cluster1", "cluster2"))

# perform infercnv operations to reveal cnv signal
infercnv_obj = infercnv::run(infercnv_obj,
                          # use 1 for smart-seq, 0.1 for
10x-genomics
                           cutoff = 0.1,
                             # dir is auto-created for
storing outputs
                           out_dir = "output_dir",
                           # cluster shows up by groups
                           cluster_by_groups = T,
                           denoise = T,
                           HMM=T)
```

In the above code, we input clustering results from Seurat to denote the cell groups and choose "cluster1" and "cluster2" as reference cell groups due to some prior knowledge about the cell types. For instance, clusters with strong immune signaling and evidence of epithelial cell absence are usually recommended

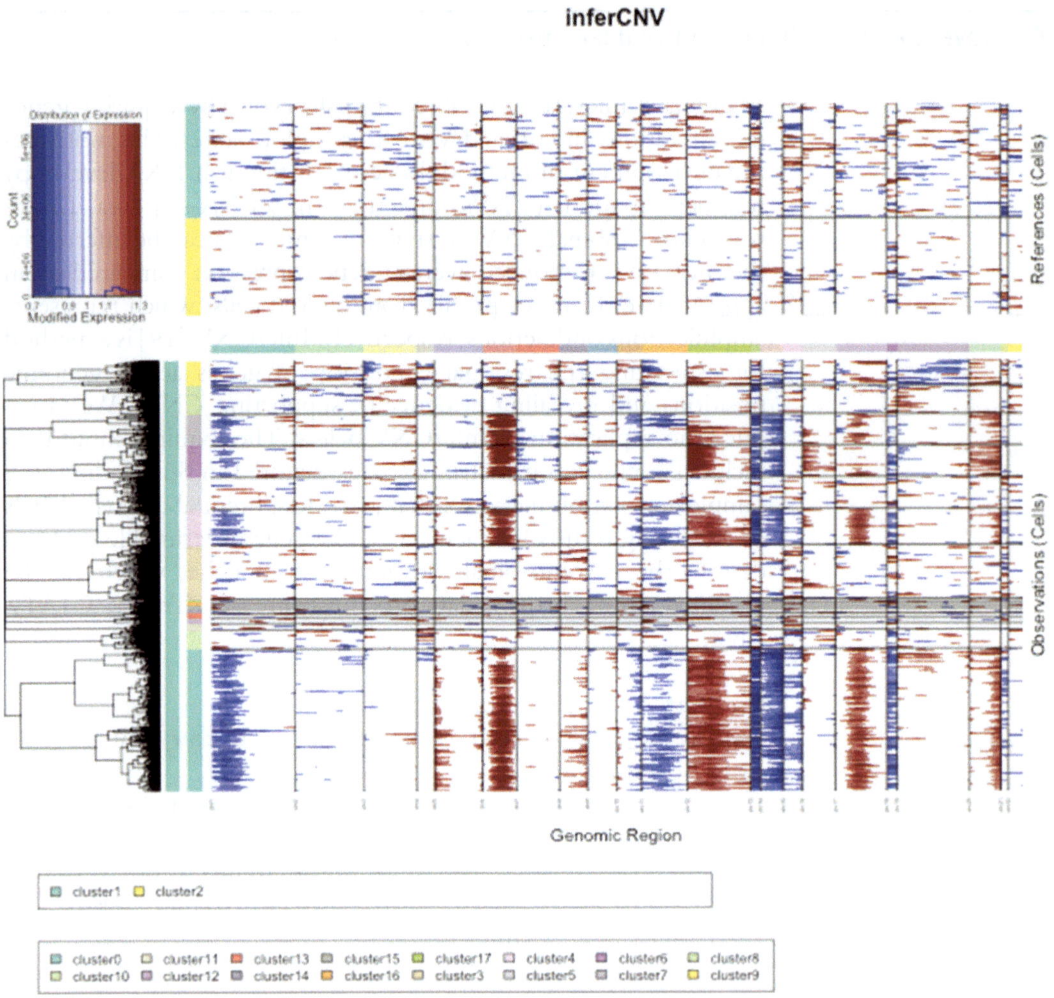

Fig. 5 Heatmap of the results from inferCNV. Each row of the heatmap denotes a cell, and each column corresponds to a gene ordered from left to right across the chromosomes. The CNV profiles from reference cells are located on the top, and CNV profiles for the observation cells are shown below

candidates for reference. Figure 5 depicts the heatmap of CNV profiles from both reference diploid cells (top) and observation cells (bottom). The positive and negative residual expression shows up as red and blue blocks, respectively, highlighting likely amplified or deleted segments of chromosomes. In general, the deviations of gene expression appearing in the reference diploid cells should be minimal, with the exception of certain outlier transcripts. Researchers can determine clusters from observation cells with systemic amplification or deletion detected in certain chromosomal segments as tumor cells.

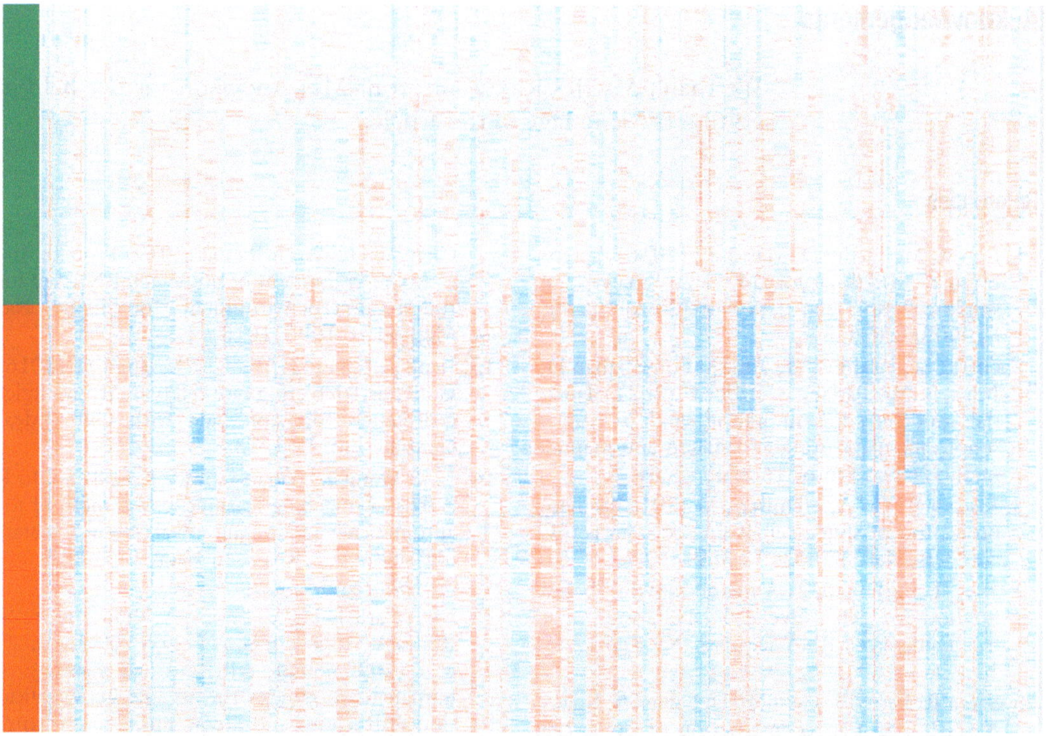

Fig. 6 Heatmap of the results of CNVs from copyKAT. Each row of the heatmap denotes a cell, and each column corresponds to a gene ordered from left to right across the chromosomes. The left bar represents the predicted aneuploid (red) and diploid cells (green)

In addition to taking reference diploid cells as prior knowledge, copyKAT [20] (copy number karyotyping of aneuploid tumors) applies the Gaussian mixture model (GMM) to detect a subset of diploid cells with high confidence as the baseline value of reference cells. And then predict the tumor cells by the normal cell enrichment and GMM distribution tests. Subsequent analysis, i.e., hierarchical clustering of the CNVs profile from aneuploid cells, is applied for subclonal genotype identification as shown in Fig. 6.

```
## Run copyKAT
library(copykat)
copykat.test <- copykat(rawmat=expX, id.type="S",
        ngene.chr = 5, win.size = 25, KS.cut = 0.1,
        sam.name = "test", distance = "euclidean",
       norm.cell.names = "", output.seg = "FLASE",
         plot.genes = "TRUE", genome = "hg20", n.
cores = 1)
```

Acknowledgements

We thank Ms. Jessica Swann from MD Anderson for her help in editing the language of the article.

References

1. Lun A, McCarthy D, Marioni J (2016) A step-by-step workflow for low-level analysis of single-cell RNA-seq data with Bioconductor. F1000Research. 5
2. Clarke Z, Andrews T, Atif J, Pouyabahar D, Innes B, MacParland S, Bader G (2021) Tutorial: guidelines for annotating single-cell transcriptomic maps using automated and manual methods. Nat Protoc 16:2749–2764
3. Kiselev V, Yiu A, Hemberg M (2018) scmap: projection of single-cell RNA-seq data across data sets. Nat Methods 15:359–362
4. Li Z, Feng H (2022) A neural network-based method for exhaustive cell label assignment using single cell RNA-seq data. Sci Rep 12:1–12
5. Satija R, Farrell J, Gennert D, Schier A, Regev A (2015) Spatial reconstruction of single-cell gene expression data. Nat Biotechnol 33:495–502
6. Wan S, Kim J, Won K (2020) SHARP: hyperfast and accurate processing of single-cell RNA-seq data via ensemble random projection. Genome Res 30:205–213
7. Kiselev V, Kirschner K, Schaub M, Andrews T, Yiu A, Chandra T, Natarajan K, Reik W, Barahona M, Green A et al (2017) SC3: consensus clustering of single-cell RNA-seq data. Nat Methods 14:483–486
8. Ji Z, Ji H (2016) TSCAN: pseudo-time reconstruction and evaluation in single-cell RNA-seq analysis. Nucleic Acids Res 44:e117–e117
9. Sun X, Lin X, Li Z, Wu H (2022) A comprehensive comparison of supervised and unsupervised methods for cell type identification in single-cell RNA-seq. Briefings Bioinform 23: bbab567
10. De Kanter J, Lijnzaad P, Candelli T, Margaritis T, Holstege F (2019) CHETAH: a selective, hierarchical cell type identification method for single-cell RNA sequencing. Nucleic Acids Res 47:e95–e95
11. Aran D, Looney A, Liu L, Wu E, Fong V, Hsu A, Chak S, Naikawadi R, Wolters P, Abate A et al (2019) Reference-based analysis of lung single-cell sequencing reveals a transitional profibrotic macrophage. Nat Immunol 20:163–172
12. Zhang A, O'Flanagan C, Chavez E, Lim J, Ceglia N, McPherson A, Wiens M, Walters P, Chan T, Hewitson B et al (2019) Probabilistic cell-type assignment of single-cell RNA-seq for tumor microenvironment profiling. Nat Method 16:1007–1015
13. Jiang L, Chen H, Pinello L, Yuan G (2016) GiniClust: detecting rare cell types from single-cell gene expression data with Gini index. Genome Biol 17:1–13
14. Li Z, Wang Y, Ganan-Gomez I, Colla S, Do K (2022) A machine learning-based method for automatically identifying novel cells in annotating single-cell RNA-seq data. Bioinformatics 38:4885–4892
15. Le J, Park J, Ha V, Luong A, Branciamore S, Rodin A, Gogoshin G, Li F, Loh Y, Camacho V et al (2020) Single-cell RNA-Seq mapping of human thymopoiesis reveals lineage specification trajectories and a commitment spectrum in T cell development. Immunity 52:1105–1118
16. Tirosh I, Izar B, Prakadan S, Wadsworth M, Treacy D, Trombetta J, Rotem A, Rodman C, Lian C, Murphy G et al (2016) Dissecting the multicellular ecosystem of metastatic melanoma by single-cell RNA-seq. Science 352: 189–196
17. Keren-Shaul H, Spinrad A, Weiner A, Matcovitch-Natan O, Dvir-Szternfeld R, Ulland T, David E, Baruch K, Lara-Astaiso D, Toth B et al (2017) A unique microglia type associated with restricting development of Alzheimer's disease. Cell 169:1276–1290
18. Grün D, Lyubimova A, Kester L, Wiebrands K, Basak O, Sasaki N, Clevers H, Van Oudenaarden A (2015) Single-cell messenger RNA sequencing reveals rare intestinal cell types. Nature 525:251–255
19. Patel A, Tirosh I, Trombetta J, Shalek A, Gillespie S, Wakimoto H, Cahill D, Nahed B, Curry W, Martuza R et al (2014) Single-cell RNA-seq highlights intratumoral heterogeneity in primary glioblastoma. Science 344:1396–1401
20. Gao R, Bai S, Henderson Y, Lin Y, Schalck A, Yan Y, Kumar T, Hu M, Sei E, Davis A et al (2021) Delineating copy number and clonal substructure in human tumors from single-cell transcriptomes. Nat Biotechnol 39:599–608

Chapter 8

Unsupervised Single-Cell Clustering with Asymmetric Within-Sample Transformation and Per-Cluster Supervised Features Selection

Stefano Maria Pagnotta

Abstract

This chapter shows applying the Asymmetric Within-Sample Transformation to single-cell RNA-Seq data matched with a previous dropout imputation. The asymmetric transformation is a special winsorization that flattens low-expressed intensities and preserves highly expressed gene levels. Before a standard hierarchical clustering algorithm, an intermediate step removes noninformative genes according to a threshold applied to a per-gene entropy estimate. Following the clustering, a time-intensive algorithm is shown to uncover the molecular features associated with each cluster. This step implements a resampling algorithm to generate a random baseline to measure up/downregulated significant genes. To this aim, we adopt a GLM model as implemented in DESeq2 package. We render the results in graphical mode. While the tools are standard heat maps, we introduce some data scaling to clarify the results' reliability.

Key words Single-cell RNA-Seq, Single-cell clustering, Asymmetric within-sample transformation, Generalized linear models, Resampling, Human breast epithelial tissue

1 Introduction

Clustering addresses the question of aggregating items according to some metric of similarity. In bioinformatics, a general task is to cluster bulk samples or single cells, in both cases described by a large set of omic features such as transcriptomic expression levels. This chapter shows how to implement the AWST protocol [1] to single cells and get a clustering. The core of the protocol is an asymmetric data transformation that down-weights noisy features so that their contributions to the analysis vanish. In general, any clustering procedure can follow AWST, but as long as data dimensionality allows, we prefer hierarchical algorithms matched with Ward's linkage applied to an Euclidean distance matrix. Consistently with these choices, the criterion for exploring the number of clusters is the Calinski–Harabasz [2] curve. These statistical

Rajeev K. Azad (ed.), *Transcriptome Data Analysis*, Methods in Molecular Biology, vol. 2812,
https://doi.org/10.1007/978-1-0716-3886-6_8,

methodologies share the key concept of maximizing variability between groups while minimizing the variability within.

Once the cell partition has been computed, a downstream procedure to characterize every cluster is definite. Essentially, the procedure performs a gene-wise ANOVA analysis with the help of the generalized linear models (GLMs) implemented in the DESeq2 package [3]. The large amount of variability induced in the data processed by the GLM assures robust results.

Although we present the two analyses in sequence (clustering + per-cluster features selection), they can be separated and included in other pipelines. Any clustering procedure can feed the group signature mining step. Any signature mining step can follow the proposed clustering procedure.

A few modifications also allow researchers to apply our proposal to bulk RNA-Seq data.

The chapter has been arranged as follows: Subheading 1 shows the results generated by the pipeline, with some molecular remark; Subheading 2 shows materials and software; the last section describes the methodology step by step. A few notes close the chapter.

The software can be retrieved from [4].

2 Result

We consider the collection of 6964 single cells from human breast epithelial tissue of donor [5] analyzed in [6]. Raw counts have been first treated for dropout imputation with [7] methodology, and then the AWST pipeline [1] has been applied. We performed an unsupervised hierarchical clustering with Ward's linkage and Euclidean distance using 2141 features out of about 20,000 gene expression levels. The Calinski–Harabasz [2] method suggests the number of clusters and proposes to investigate first a partition of three clusters, then a finer one of 5 groups. The partition *clust3* segregates the cells into three groups: Basal-like (c31 and c32) and Luminal (c33). Graphical results are in Fig. 1.

A sampling algorithm has been designed to detect the genes associated with each cluster. A synthetic control group was assembled by sampling the same number of items from each cluster; then, a gene-wise ANOVA was run to compare the synthetic control versus the clusters c31, c32, and c33, seen as treatment groups.

Figure 2a shows the statistical evidence of genes (top 200 features) uniquely associated with each cluster. Red means that the aggregated gene level in the cluster is above average (synthetic control); green indicates that the level is below average. The coloring follows the log fold-change of each group to the synthetic control. With the help of signatures from [8], c32 is a population

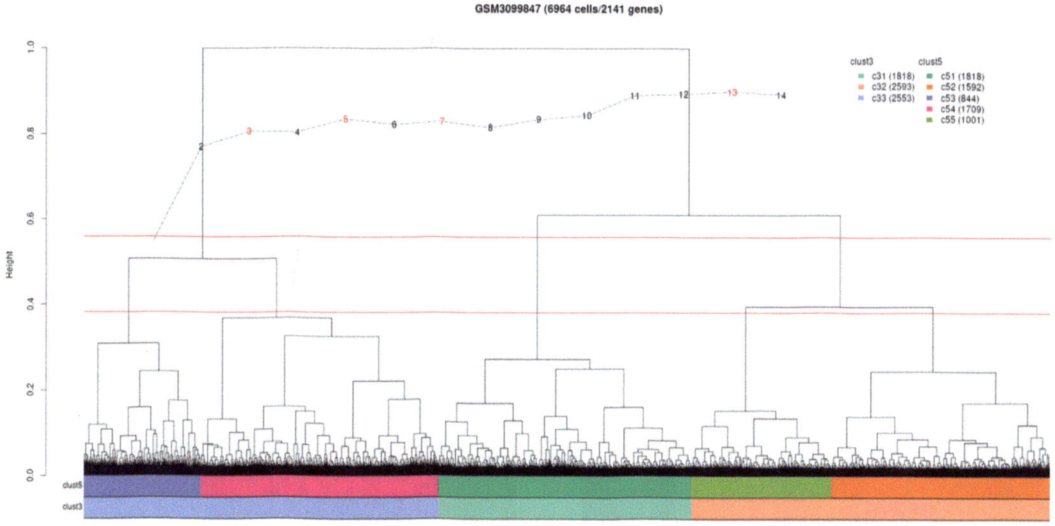

Fig. 1 Hierarchical clustering with Ward's linkage and Euclidean distance of 6964 single-cell profiles and 2141 features selected with AWST. The first annotation bar is the partition of cells into five groups (clust5); the second bar traces a partition of three clusters (clust3)

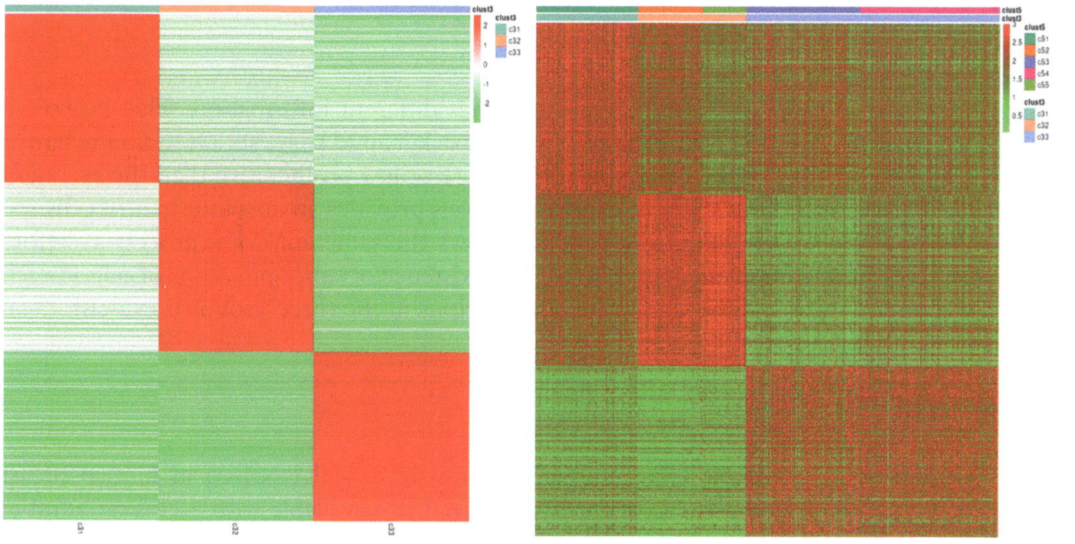

Fig. 2 Study of clust3. Heat map (left) of each cluster's top 200 standardized log fold-changes in clust3 to a synthetic background. Green means under-regulation to the background level; red means upregulation to the background. On the right is a heat map of the top 200 gene expression levels associated with each cluster in clust3. Green means no expression level; red means high expression level

of basal-like/stem cells, and c33 is a luminal group. From [9], c31 is a proliferative set of cells.

Figure 2b shows the same genes as Fig. 2a from a different point of view. This time, the normalized gene levels have been rendered. Green means no expression. As the color moves to red, the expression level increases.

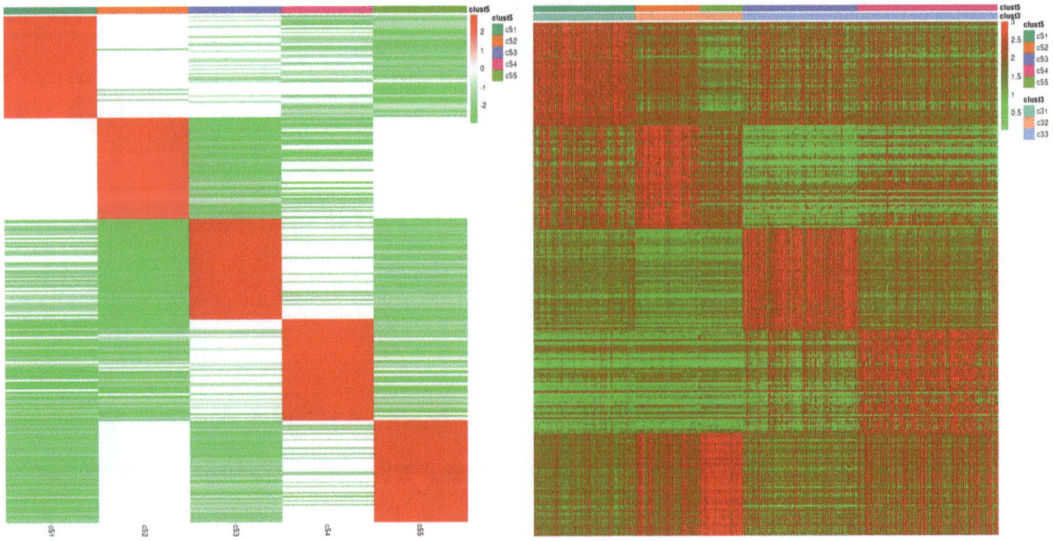

Fig. 3 Study of clust5. Heat map (left) of each cluster's top 50 standardized log fold-changes in clust5 to a synthetic background. Green means under-regulation to the background level; red means upregulation to the background. On the right is a heat map of the top 50 gene expression levels associated with each cluster in clust5. Green means no expression level; red means high expression level

Figure 2a and b carries different information. The analysis of log fold-change to a joint background suggests a trending behavior of the cell population (more or less proliferative than the average); the study of expression levels provides an absolute trait of cells.

Figure 3a and b follows the same analytical process as before but analyzes clust5. The existence of subpopulations of cells in each of the significant aggregations in the three groups is evident.

3 Material

Data The example data are single-cell RNAseq of human breast epithelial cells analyzed in [6] and available with accession number GSM3099847 [5] at NCBI.

Hardware Superdome server (224 CPU cores Intel(R) Xeon (R) Platinum 8280 CPU @ 2.70GHz, 1.5 TB RAM) matched with `docker.io/rocker/tidyverse:4.2.1`

Computational Software The analysis has been implemented in R version 4.2.3 (2023-03-15)—"Shortstop Beagle" (x86_64). The program includes many libraries specified chunk by chunk. The software is available at [4] to facilitate reproducibility.

4 Methods and Implementation

Here, we show the essential steps. Tedious or routine steps, such as data downloading, are not presented. They are in the script available from [4].

Chunk n.1: downloading data
Retrieve the data (stored in `ddata` as a matrix) from the NCBI repository and save it in a local file `GSM3099847.RData`.

Chunk n.2: preprocessing data
Apply the SAVER methodology to raw RNAseq data to recover gene expression [7, 10].

```
rm(list = ls())
load("GSM3099847.RData")
library(SAVER)
ddata <- saver(as.data.frame(ddata), estimates.only = TRUE, ncores = 60)
```

Then, a quantile normalization [11, 12] is applied to ddata.

```
library(EDASeq)
ddata <- betweenLaneNormalization(ddata, which = "full", round = FALSE)
save(ddata, file = "GSM3099847_SAVER_EDASeq.RData")
```

This version of `ddata` is stored in `GSM3099847_SAVER_EdaSeq.RData`. This version of data is the basis for any downstream analysis.

Chunk n.3: preparing data for clustering
Here, the Asymmetric Within-Sample Transformation [1, 13] is applied to data, together with the gene filtering. Then, the usual Euclidean distance matrix is computed, followed by the application of the hierarchical clustering with Ward's linkage method.

```
rm(list = ls())
load("GSM3099847_SAVER_EDASeq.RData")
library(awst)
exprData <- awst(ddata, poscount = TRUE, full_quantile = TRUE)
dim(exprData <- gene_filter(exprData)) #[1] 2141 6964
exprData <- t(exprData)
nrow_exprData <- nrow(exprData)
ncol_exprData <- ncol(exprData)
ddist <- dist(exprData)
hhc <- hclust(ddist, method = "ward.D2")
```

From the hierarchical data structure, the Calinski–Harabasz index is computed.

```
# full path in the .RMD file
source("https:// ... /calinski_20210214.R")
aCalinski <- calinski(hhc)
save(hhc, nrow_exprData, ncol_exprData, aCalinski,
     file = "GSM3099847_SAVER_EDASeq_AWST_hclust.RData")
```

Figure 1: Unsupervised hierarchical clustering.
This chuck generates the dendrogram in Fig. 1 and two data structures for later use. `clustering.df` is a data frame that collects the clusterings; `clust.colorCode` is instead a named list storing the association of each cluster with the corresponding color code. The variable `no_of_clusters` controls the cut of the dendrogram so that the required number of groups is gotten. From the setup of this variable, automatic cluster association, naming, and color coding follow. Here, we show the essential steps.

(continued)

The full code is in [4].

```
clustering.prefix <- "clust"; short.prefix <- "c"
clustering.df <- data.frame(row.names = hhc$labels, barcode = hhc$labels)
clust.colorCode <- NULL
###
mmain  <-  paste0("GSM3099847 (", nrow_exprData, " cells/",
           ncol_exprData, " genes)")
hhc$height <- hhc$height/max(hhc$height)
plot(hhc, hang = -1, labels = FALSE, xlab = "", sub = "", main = mmain)
plot(aCalinski, add = TRUE, from = 500, to = 5000, shift = 0.55, height = 0.35,
max_height = 1)###
# this clustering is for later use
clustering.df$reduced <- as.factor(cutree(hhc, k = 500))
###
no_of_clusters <- 3
clustering_name <- paste0(clustering.prefix, no_of_clusters)
clustering_name.col <- paste0(clustering_name, ".col")

hh <- (hhc$height[length(hhc$height)-no_of_clusters+2] +
      hhc$height[length(hhc$height)-no_of_clusters+1])/2
segments(1, hh, nrow_exprData, hh, col = "red") # cut the tree

tmp <- as.factor(cutree(hhc, k = no_of_clusters)) # get the clusters
levels(tmp) <- paste0(short.prefix, no_of_clusters, c(paste(1:9), letters))
[1:no_of_clusters]assign(clustering_name, tmp)

levels(tmp) <- brewer.pal(n = no_of_clusters, name = "Set2")
assign(clustering_name.col, tmp)

clustering.df$tmp <- get(clustering_name)
clustering.df$tmp.col <- get(clustering_name.col)

tt <- table(clustering.df$tmp, clustering.df$tmp.col)
colorCode <- colnames(tt)[apply(tt, 1, which.max)]
names(colorCode) <- rownames(tt)
clust.colorCode <- c(clust.colorCode, colorCode)

tmp <- grep("tmp", colnames(clustering.df))
colnames(clustering.df)[tmp] <- gsub("tmp", clustering_name,
colnames(clustering.df)[tmp])
```

After the running of this chunk, the content of `clustering.df` is

```
> head(clustering.df)
                                        barcode reduced clust3 clust3.col clust5 clust5.col
Ind5_AAACCTGAGAAACCTA-1 Ind5_AAACCTGAGAAACCTA-1       1    c31    #66C2A5    c51    #1B9E77
Ind5_AAACCTGAGACAGAGA-1 Ind5_AAACCTGAGACAGAGA-1       2    c31    #66C2A5    c51    #1B9E77
Ind5_AAACCTGAGCCAGAAC-1 Ind5_AAACCTGAGCCAGAAC-1       3    c32    #FC8D62    c52    #D95F02
Ind5_AAACCTGAGCCTATGT-1 Ind5_AAACCTGAGCCTATGT-1       4    c33    #8DA0CB    c53    #7570B3
Ind5_AAACCTGAGGAGTTTA-1 Ind5_AAACCTGAGGAGTTTA-1       5    c32    #FC8D62    c52    #D95F02
Ind5_AAACCTGAGTATTGGA-1 Ind5_AAACCTGAGTATTGGA-1       6    c32    #FC8D62    c52    #D95F02
```

and the content of `clust.colorCode` is

```
> head(clust.colorCode)
      c31       c32       c33       c51       c52       c53
"#66C2A5" "#FC8D62" "#8DA0CB" "#1B9E77" "#D95F02" "#7570B3"
```

Chunk 4: preparing data for Fig. 2a and b.
Figure 2 shows the result of a simulation experiment to discover the features associated with each cluster in the hierarchical clustering in Fig. 1. In this chunk, we prepare data for the graphical display of heat maps.

The numerical results come from $B = 50$ Monte Carlo experiments where, each time, a synthetic background has been assembled. If we consider the clustering clust3, the synthetic control `c00` is a group of randomly subsampled (without replacement) cells from each of the other clusters c31, c32, and c33 in equal numbers. This way, a new clustering clust3_b rises, with one more cluster than the original partition, $b = 1, 2, \ldots, 50$. clust3_b still classifies the $n = 6964$ initial cells. This is an example of a random clust3_b (`clust3` is the original partition, and `design` corresponds to clust3_b).

```
> addmargins(table(annotation.tmp$design, annotation.tmp$clust3))

      c31  c32  c33  Sum
c00   627  627  627 1881
c31  1255    0    0 1255
c32     0 2014    0 2014
c33     0    0 1814 1814
Sum  1882 2641 2441 6964
```

From each of the clusters in clust3_b, a sample with replacement (like bootstrap) of dimension m is sampled. m is set to one-third of the minimum size of the clusters in clust3_b. clust3_b^* is the collection of the "*bootstrap*" clusters. This is an example of the bootstrap partition.

```
> addmargins(table(annotation.tmp$design, annotation.tmp$clust3))

      c31  c32  c33  Sum
c00   135  138  145  418
c31   418    0    0  418
c32     0  418    0  418
c33     0    0  418  418
Sum   553  556  563 1672
```

clust3_b^*, transformed in a model matrix **X**, feeds the DESeq2 [3, 14] procedure to estimate a generalized linear model. In each row of **X**, all elements are zero, but one is set to 1. Given a gene j,

$$\log q_{ij} = \beta_{j0} + \beta_{1j} x_{i1} + \beta_{2j} x_{i2} + \beta_{j3} x_{i3},$$

where x_{i1}, x_{i2}, and x_{i3} are elements of the ith row of **X**, and q_{ij} is proportional to the true expected counts of the jth in the ith samples [3, 15]. The intercept estimate $\hat{\beta}_0$ is the average of the jth gene levels in `c00`. Given the setup, the estimate of β_{kj}, $k > 0$, is the difference between the average of levels in the kth cluster and the average in `c00`. Essentially, $\hat{\beta}_{kj}$ is a fold-change estimate between the gene level in the kth cluster and the reference cluster `c00`.

From each b run, we retain Wald's statistics associated with each β's. Wald's statistic is a standardized $\log_2$ fold-change with an asymptotic Gaussian distribution.

(continued)

Chunk 4: preparing data for Fig. 2a and b.(continued)

The construction of the partitions clust3_b and clust3_b^*, the setup, and the running of the linear model with DESeq2 are all enclosed in the function `single_shot()`. The value of such a function is the output of `DESeq()` function.

```
single_shot <- function() {
  annotation.tmp <- clustering.df
  mmin <- min(tt <- table(annotation.tmp$design))
  nooc <- length(tt) # no of clusters

  # assemble clust3$_b$
  tmp <- c()
  for(this_cluster in unique(annotation.tmp$design)) tmp <- c(tmp,
      sample(annotation.tmp$barcode[annotation.tmp$design == this_cluster],
      mmin/nooc, replace = FALSE))
  annotation.tmp$design[annotation.tmp$barcode %in% tmp] <- "000" # synthetic background

  # assemble clust3$_b^*$
  m <- min(table(annotation.tmp$design))/3
  tmp <- c()
  for(this_cluster in unique(annotation.tmp$design))
  tmp <- c(tmp,
     sample(annotation.tmp$barcode[annotation.tmp$design == this_cluster], m, replace = TRUE))
  annotation.tmp <- annotation.tmp[tmp,]

  # rename the items
  for(i in 1:nrow(annotation.tmp))
     annotation.tmp$id[i] <- paste(sample(c(letters, LETTERS, 0:9), 15, replace = TRUE),
  collapse = "")

  # setup the data-structure for DESeq2
  data.tmp <- ddata[, annotation.tmp$barcode]
  colnames(data.tmp) <- annotation.tmp$id
  rownames(annotation.tmp) <- annotation.tmp$id
  annotation.tmp$design <- factor(annotation.tmp$design)
  dds <- DESeqDataSetFromMatrix(countData = data.tmp, colData = annotation.tmp, design = ~ design)

  # in DESeq2, suggested correction for single cell data
  tmp <- scran::computeSumFactors(dds)
  dds@colData@listData[["sizeFactor"]] <- tmp@colData@listData$sizeFactor

  ans_DESeq <- DESeq(dds, test = "Wald", useT = T,
     minmu = 1e-6, fitType='local' , minReplicatesForReplace = Inf)
  ans_DESeq <- mcols(ans_DESeq, use.names=TRUE)
  ans_DESeq <- data.frame(ans_DESeq)
  invisible(ans_DESeq)
}
```

$B = 50$ times the `single_shot()` is called; Wald's statistics are extracted from the output and averaged across the runs. The strong law of large numbers assures that the average values are good approximations of the true values of the $\log_2$ fold-change (Wald's statistics) of each gene for each cluster to a generic background.

(continued)

Chunk 4: preparing data for Fig. 2a and b.(continued)

The result of this chunk is stored for the visualization task.

```
rm(list = ls())
load("GSM3099847_SAVER_EDASeq.RData")
# makes expression values similar to counts per millions (CPM)
ddata <- apply(ddata, 2, function(x) x/sum(x)*1e6)
mode(ddata) <- "integer" # transforms in integer with an implicit truncation

load("GSM3099847_SAVER_EDASeq_AWST_hclust_clustering.RData")
clustering.df$design <- as.character(clustering.df$clust3)
clustering.df$id <- NA

library(DESeq2)
library(scran)

ans <- single_shot()
WaldStatistic_design <- grep("WaldStatistic_design", colnames(ans))
tmp <- ans[, WaldStatistic_design]

runs <- 50
k <- 2
while(k <= runs) {
  system.time(ans <- single_shot())
  tmp <- tmp + ans[, WaldStatistic_design]
  k <- k + 1
}
ans <- tmp/runs
ans <- ans[-which(is.na(ans[, 1])),]
colnames(ans) <- unique(clustering.df$design)
save(ans, file = "GSM3099847_SAVER_EDASeq_ansDESeq2_clust3.RData")
```

Figure 2a: Heat map of each cluster's top 200 standardized log fold-changes in clust3 to a synthetic background.

Figure 2 shows the result steps of the simulation experiment to discover the features associated with each cluster in the hierarchical clustering in Fig. 1.

This chunk analyzes preparatory data from the Monte Carlo experiment and adapts to be displayed with the `pheatmap` [16] function. Here, we comment just some relevant and "obscure" step.

```
rm(list = ls())
load("GSM3099847_SAVER_EDASeq_AWST_hclust_clustering.RData")
load("GSM3099847_SAVER_EDASeq_ansDESeq2_clust3.RData")
colnames(ans) <- c("c31", "c32", "c33")

> head(ans)
                      c31          c32          c33
RP11-34P13.7  -0.003577552  -0.005607752  0.009281667
F0538757.2    -1.098342453   0.045402996  0.974026886
AP006222.2    -7.801465666   4.339955544  3.276145375
RP4-669L17.10  0.002787339  -0.005572933  0.002078891
RP11-206L10.9 -5.772635713 -11.026534351 11.822748057
LINC00115     -9.846021111  -1.623214357  8.424081508
##########################################################################
```

(continued)

ans is the data frame where each column is the Wald's test statistics associated with each cluster and for each gene (row-wise). We get the p-values from the test value and adjust for multiple tests, according to Benjamini–Hochberg's correction.

```
p.values <- 2*(1-pnorm(abs(as.matrix(ans))))
p.values <- apply(p.values, 2, p.adjust, method = "BH")
```

Given a level of significance alpha, we set to zero every nonsignificant entry of ans; then, we restrict the data frame to rows with only one component greater than zero.

```
alpha <- 0.01
ans <- ans * (p.values < alpha)
ans <- ans[rowSums(ans > 0) == 1,]
```

A statement like ans[rowSums(ans > 0) == 2,] would have selected the rows having two elements greater than zero. Then, the selected genes are upregulated in two clusters. Having many clusters, the condition allows the selection of genes upregulated in any combination of groups.

```
max_tops <- 200
ttable <- table(colnames(ans)[apply(ans, 1, which.max)])
K <- sapply(ttable, function(x) min(max_tops, x))

j <- 1
tops <- head(rownames(ans[order(ans[, j], decreasing = TRUE),]), K[j])
for(j in 2:ncol(ans))
  tops <- c(tops, head(rownames(ans[order(ans[, j], decreasing = TRUE),]), K[j]))
```

The following line implements a scaling of the w_{ij} Wald's test values to remove differences in variability associated with clusters:

$$w'_{ij} = w_{ij}/\tau_j,$$

where jth is associated with clusters and ith with genes. The w_{ij}'s are stored in the ans matrix. The scaling factor τ_j is a dispersion index from zero

$$\tau_j = \sqrt{\frac{1}{n}\sum_i (w_{ij} - 0)^2},$$

where n is the number of genes in ans. This way, the original zero value of data is preserved since there is no fold-change of the level for a cluster to the baseline (synthetic cluster c00).

```
ans <- apply(ans[tops,], 2, function(x) x/sqrt(mean(x^2)))
```

(continued)

The next step controls the contrast in the heat map and restricts the values in the range from -3 to $+3$. Values greater than 2τ from zero are full red; symmetrically, values lower than 2τ from zero are full green.

```
ans <- 6 * (pnorm(ans) - 0.5)

annotation_col <- data.frame(clust3 = colnames(ans), row.names = colnames(ans))
ann_colors = list(clust3 = clust.colorCode[grep("c3", names(clust.colorCode))])

suppressPackageStartupMessages(library(pheatmap))
pheatmap(ans, cluster_rows=FALSE, cluster_cols=FALSE, show_rownames = FALSE,
         annotation_col=annotation_col,
         annotation_colors = ann_colors,
         color = colorRampPalette(c("green2", "white", "red2"))(51))
```

Finally, in the heat map of Fig. 2, the levels of the standardized fold-change (Wald's test value) are associated with each cluster and to the synthetic control group c00.

Figure 2b: Heat map of the top 200 gene expression levels associated with each cluster in clust3.
This chunk is in charge of displaying the heat map of the expression level of the genes for each cell. As the number of cells is above the thousands, we aggregate the cells in 500 small clusters. Such an aggregation comes from chunk "Fig. 2," where we set

```
clustering.df$reduced<-as.factor(cutree(hhc,k=500))
```

The level associated with each cluster is summarized as the average of the levels participating in the group. `pheatmap` allows a dimensionality reduction but follows algorithms that can corrupt the external ordering coming from the hierarchical clustering.

```
load("GSM3099847_SAVER_EDASeq.RData")
###
rdata <- as.data.frame(t(ddata[tops,]))
rdata <- aggregate(rdata, list(reduced = clustering.df$reduced), mean)[, -1]
```

In the following two steps, the same transformation as in the chunk "Fig. 2" has been applied to data before feeding `pheatmap`. The meaning of the color is now different from Fig. 2. The gene level's zero value (in green) means no expression (negative intensities are not allowed). The gene levels are in red in the case of upregulation.

(continued)

```
rdata <- apply(rdata, 2, function(x) x/sqrt(mean(x^2)))
rdata <- t(rdata)

colnames(rdata) <- paste0("R", 1:ncol(rdata))
rdata <- 6*(pnorm(rdata) - 0.5)

tt <- table(clustering.df$reduced, clustering.df$clust3)
tt <- colnames(tt)[apply(tt, 1, which.max)]
annotation_col <- data.frame(row.names = colnames(rdata), clust3 = tt)
tt <- table(clustering.df$reduced, clustering.df$clust8)
tt <- colnames(tt)[apply(tt, 1, which.max)]
annotation_col$clust8 <- tt
annotation_col <- annotation_col[order(annotation_col$clust3, annotation_col$clust8),]
rdata <- rdata[, rownames(annotation_col)]

ann_colors <- list(clust3 = clust.colorCode[grep("c3", names(clust.colorCode))],
                   clust8 = clust.colorCode[grep("c8", names(clust.colorCode))])

pheatmap(rdata, cluster_rows=FALSE, cluster_cols=FALSE,
         show_colnames = FALSE, show_rownames = FALSE,
         annotation_col = annotation_col,
         annotation_colors = ann_colors,
         color = colorRampPalette(c("green", "red2"))(50))
```

References

1. Risso D, Pagnotta SM (2021) Per-sample standardization and asymmetric winsorization lead to accurate clustering of RNA-seq expression profiles. Bioinformatics 37(16):2356–2364. https://doi.org/10.1093/bioinformatics/btab091
2. Calinski T, Harabasz J (1974) A dendrite method for cluster analysis. Commun Stat 3(1):1–27. https://doi.org/10.1080/03610927408827101
3. Love MI, Huber W, Anders S (2014) Moderated estimation of fold change and dispersion for RNA-seq data with DESeq2. Genome Biol 15(12):550. https://doi.org/10.1186/s13059-014-0550-8
4. Pagnotta SM (2023) Unsupervised clustering of single-cells with asymmetric within-sample transformation, and per cluster supervised features selection. (supporting code). https://github.com/stefanoMP/MiMB
5. Kessenbrock K (2018) Sample GSM3099847 of adult human breast epithelium. https://www.ncbi.nlm.nih.gov/geo/query/acc.cgi?acc=GSM3099847
6. Nguyen QH, Pervolarakis N, Blake K, Ma D, Davis RT, James N, Phung AT, Willey E, Kumar R, Jabart E, Driver I, Rock J, Goga A, Khan SA, Lawson DA, Werb Z, Kessenbrock K (2018) Profiling human breast epithelial cells using single cell RNA sequencing identifies cell diversity. Nat Commun 9(1):2028. https://doi.org/10.1038/s41467-018-04334-1
7. Huang M, Wang J, Torre E, Dueck H, Shaffer S, Bonasio R, Murray JI, Raj A, Li M, Zhang NR (2018) Saver: gene expression recovery for single-cell RNA sequencing. Nat Methods 15(7):539–542. https://doi.org/10.1038/s41592-018-0033-z
8. Lawson DA, Bhakta NR, Kessenbrock K, Prummel KD, Yu Y, Takai K, Zhou A, Eyob H, Balakrishnan S, Wang CY, Yaswen P, Goga A, Werb Z (2015) Single-cell analysis reveals a stem-cell program in human metastatic breast cancer cells. Nature 526(7571): 131–135. https://doi.org/10.1038/nature15260
9. Alexandrou S, George SM, Ormandy CJ, Lim E, Oakes SR, Caldon CE (2019) The proliferative and apoptotic landscape of basal-like breast cancer. Int J Mol Sci 20(3). https://doi.org/10.3390/ijms20030667
10. Huang M (2006) Saver: Single-cell RNA-seq gene expression recovery. https://CRAN.R-project.org/package=SAVER
11. Risso D, Dudoit S, Geistlinger L (2011) EDA-Seq: exploratory data analysis and

normalization for RNA-seq. https://doi.org/doi:10.18129/B9.bioc.EDASeq
12. Risso D, Schwartz K, Sherlock G, Dudoit S (2011) GC-content normalization for RNA-seq data. BMC Bioinform 12(1):480. https://doi.org/10.1186/1471-2105-12-480
13. Risso D, Pagnotta SM (2021) Awst: asymmetric within-sample transformation. https://doi.org/doi:10.18129/B9.bioc.awst
14. Love M, Ahlmann-Eltze C, Forbes K, Anders S, Huber W (2014) DESeq2: differential gene expression analysis based on the negative binomial distribution. https://doi.org/doi:10.18129/B9.bioc.DESeq2
15. Anders S, Huber W (2010) Differential expression analysis for sequence count data. Genome Biol 11(10):R106. https://doi.org/10.1186/gb-2010-11-10-r106
16. Kolde R (2019) pheatmap: Pretty heatmaps. https://CRAN.R-project.org/package=pheatmap

Chapter 9

Inferring Tree-Shaped Single-Cell Trajectories with Totem

António G. G. Sousa, Johannes Smolander, Sini Junttila, and Laura L. Elo

Abstract

Single-cell transcriptomics allows unbiased characterization of cell heterogeneity in a sample by profiling gene expression at single-cell level. These profiles capture snapshots of transient or steady states in dynamic processes, such as cell cycle, activation, or differentiation, which can be computationally ordered into a "flip-book" of cell development using trajectory inference methods. However, prediction of more complex topology structures, such as multifurcations or trees, remains challenging. In this chapter, we present two user-friendly protocols for inferring tree-shaped single-cell trajectories and pseudotime from single-cell transcriptomics data with *Totem*. *Totem* is a trajectory inference method that offers flexibility in inferring both nonlinear and linear trajectories and usability by avoiding the cumbersome fine-tuning of parameters. The QuickStart protocol provides a simple and practical example, whereas the GuidedStart protocol details the analysis step-by-step. Both protocols are demonstrated using a case study of human bone marrow CD34+ cells, allowing the study of the branching of three lineages: erythroid, lymphoid, and myeloid. All the analyses can be fully reproduced in Linux, macOS, and Windows operating systems (amd64 architecture) with >8 Gb of RAM using the provided docker image distributed with notebooks, scripts, and data in Docker Hub (*elolab/repro-totem-ti*). These materials are shared online under open-source license at https://elolab.github.io/Totem-protocol.

Key words Single-cell RNA-seq, Trajectory inference, Tree-shaped topology, Pseudotime, Cell connectivity, Totem, Data analysis, Bioinformatics

1 Introduction

Advancements in single-cell technologies allow the measurement of hundreds to thousands of molecules across thousands of individual cells in a single sample. The sample may represent a mixture of cell populations collected at a given time, such as bone marrow CD34+ cells during hematopoiesis [1], or a collection of timepoints during a developmental process such as embryogenesis [2]. In either case, the heterogeneity intrinsic to cell populations is likely to represent related cell stages in dynamic processes such as cell cycle, activation, differentiation, or transition states driven by gene expression and/or chromatin accessibility [3]. To create a chronological "flip-book" of cell development, computational prediction of the

Rajeev K. Azad (ed.), *Transcriptome Data Analysis*, Methods in Molecular Biology, vol. 2812,
https://doi.org/10.1007/978-1-0716-3886-6_9,

pseudotime order of these cell snapshots can be used for trajectory inference. Such method has successfully unraveled, for instance, the developmental bifurcation between circulating and germinal center human follicular regulatory T cells from regulatory T cells [4].

Single-cell trajectory inference (TI) methods predict the time-dependent relatedness between cells by modeling molecular data that functionally characterize cell populations, such as gene or protein expression. These methods can be broadly divided into three main categories depending on whether they are applied on the (1) full gene space, (2) reduced gene space, or (3) RNA velocity [5]. Full gene-space methods rely on probabilistic models of gene expression, whereas dimensionality reduction methods reduce the gene space before projecting a spanning tree or graph across cells [5, 6]. RNA velocity uses the ratio of spliced and unspliced transcripts to derive cell state transitions [7]. This chapter focuses on dimensionality reduction methods (see Ding et al.'s (2022) [5] review for other methods).

Dimensionality reduction-based methods have two sequential steps: (1) dimensionality reduction and (2) trajectory modeling [6]. The first step involves reducing the gene and/or cell space while preserving most of the biological information and eliminating technical noise. This can be accomplished by low-dimensional reduction techniques such as principal component analysis (PCA), multidimensional scaling (MDS), t-distributed stochastic neighbor embedding (t-SNE) [8], uniform manifold approximation and projection (UMAP) [9], and clustering, which lead to identification of cell centroids (milestones) [5, 6]. The second step involves drawing a spanning tree or graph across the cells or milestones, for example, with a minimum spanning tree (MST) and principal curves algorithms [6]. The trajectories inferred can assume several types of topologies, such as linear, bifurcation, multifurcation, tree, cycle, connected, or disconnected, which can be fixed or free, depending on the method used.

According to Saelens et al. (2019) [10], the tree-shaped *Slingshot* TI method was ranked as the overall best method among the 45 TI methods benchmarked on more than 300 data sets. However, a comparison between *Slingshot* and *Totem*, a new tree-shaped TI method developed recently by Smolander et al. (2023) [11], showed that *Totem* outperformed *Slingshot* for nonlinear trajectories. The more complex topology structures, such as multifurcations or trees, are still challenging to predict with current methods. Therefore, the focus of this chapter is on the tree-shaped *Totem*'s TI method, which promises to address these challenges.

Totem is a user-friendly tree-shaped TI method implemented as an R software package by Smolander et al. (2023) [11]. It clusters cells L times using latent dimensions of the reduced gene space with the CLARA k-medoids algorithm [12]. For each clustering result, an MST [13] is built and smoothed using the principal curves algorithm from *Slingshot* [14]. *Totem* introduces a novel concept of cell connectivity, which is the ratio of a cell cluster's connections (edges) by the number of clusters (vertices) and is calculated for every clustering and respective MST result. These cell connectivity values are scaled and averaged across the MSTs to identify branching points and select the best clustering results. In addition, *Totem* was designed to be user-friendly by avoiding the cumbersome fine-tuning of parameters during trajectory inference, making it a suitable tool for beginners.

This chapter provides guidelines to perform tree-shaped single-cell trajectory and pseudotime inference analysis from processed single-cell transcriptomics data using *Totem*. We showcase the functionality, flexibility, and user-friendliness of *Totem* using two analysis protocols: QuickStart, a simple and practical protocol, and GuidedStart, a detailed step-by-step protocol. Both protocols are applied to a real single-cell RNA-sequencing (scRNA-seq) data set on hematopoiesis. All the analyses can be reproduced in Linux, macOS, and Windows operating systems running under an amd64 architecture with >8 Gb of RAM by using a docker image distributed with notebooks, scripts, and data in Docker Hub (*elolab/repro-totem-ti*). An online resource sharing the content materials described herein is available under open-source license at https://elolab.github.io/Totem-protocol.

2 Materials

2.1 Data Analysis Environment

The protocols presented in this chapter are distributed as R markdown (Rmd) notebooks which can run in the RStudio software. To allow flexibility, the notebooks were converted to R scripts. The notebooks and the respective scripts are provided in a containerized docker image publicly available at Docker Hub (*elolab/repro-totem-ti*), containing Ubuntu 20.04.5 LTS (Focal), R (v.4.2.1), RStudio Server (2022.07.2 Build 576), *Totem* (v.0.99.2) R package, and all the system and R libraries, as well as the data required to reproduce the protocols described herein. Instructions to run the container are provided in this protocol and in the GitHub repository https://github.com/elolab/Totem-protocol. The analyses were tested on a server using Slurm scheduler (v.22.05.5) and Singularity (Apptainer version 1.1.5-2.el7) container (*elolab/repro-totem-ti*) with 8Gb of RAM and four CPUs. All the code used in this protocol was made open-source under the General Public License (GPL) 3.0.

2.2 Case Study: Hematopoiesis scRNA-seq Data Set

To showcase the trajectory inference with Totem in practice, we used the scRNA-seq data set of human CD34+ bone marrow cells (replicate 1) from Setty et al. (2019) [15], which involves a known tree-shaped trajectory and is publicly available at the Human Cell Atlas portal (link: https://data.humancellatlas.

org/explore/projects/091cf39b-01bc-42e5-9437-f419a66c8a45). The *anndata* object "human_cd34_bm_rep1.h5ad" comprises counts, processed data (e.g., the dimensionality reduction t-SNE and transformed log-normalized counts), and cell metadata (e.g., cluster identities and Palantir's pseudotime estimates). The *anndata* object was downloaded from the Human Cell Atlas portal and imported to R as a *SingleCellExperiment* (SCE) class object using the function *readH5AD()* from *zellkonverter* (v.1.8.0 [16]). Counts were log1p normalized and cell-type annotations were retrieved from Palantir's GitHub issue: https://github.com/dpeerlab/Palantir/issues/40. The parsed SCE object was exported as an RDS file "human_cd34_bm_rep1.rds" and it is distributed in the docker image "repro-totem-ti". The object is also available in the Zenodo repository at https://doi.org/10.5281/zenodo.7845709. All these steps, with the exception of updating the download URL due to time limit use, can be reproduced with the script "download_h5ad_to_SCE_rds_script.R" (available in the GitHub repository https://github.com/elolab/Totem- protocol).

3 Methods

This section describes two *Totem* TI protocols: QuickStart and GuidedStart. QuickStart provides a simpler approach to TI, while GuidedStart offers a more detailed description of the parameter functions and intermediate results through intuitive visualizations, to obtain a refined trajectory (see Fig. 1). The QuickStart protocol is recommended for situations where one wants to preserve the dimensionality reductions (PCA, t-SNE, UMAP) obtained during a scRNA-seq integration and/or clustering analysis for trajectory inference. This is particularly useful for case studies where one obtains an integrated embedding after a batch-correction analysis, such as in *Seurat* [17] or *Scanorama* [18], overcoming potentially erroneous trajectory predictions that could be driven by batch effects. The GuidedStart protocol, on the other hand, is recommended for situations where one wants to perform a distinct dimensionality reduction method (Landmark MDS, diffusion maps [19]). To this end, the GuidedStart protocol provides step-by-step guidance on how to perform TI with *Totem*, starting from log-normalized counts till topology projection. Both protocols are demonstrated using the same data set of the human CD34+ bone marrow cells described in the Materials section to explore hematopoiesis, particularly erythroid, lymphoid, and myeloid lineages.

3.1 Software Installation

A docker image is available on Docker Hub (*elolab/repro-totem-ti*), containing the data, notebooks, scripts, and all software packages required for the analysis. This provides cross-OS, distributable, and reproducible analysis across Linux, macOS, and Windows

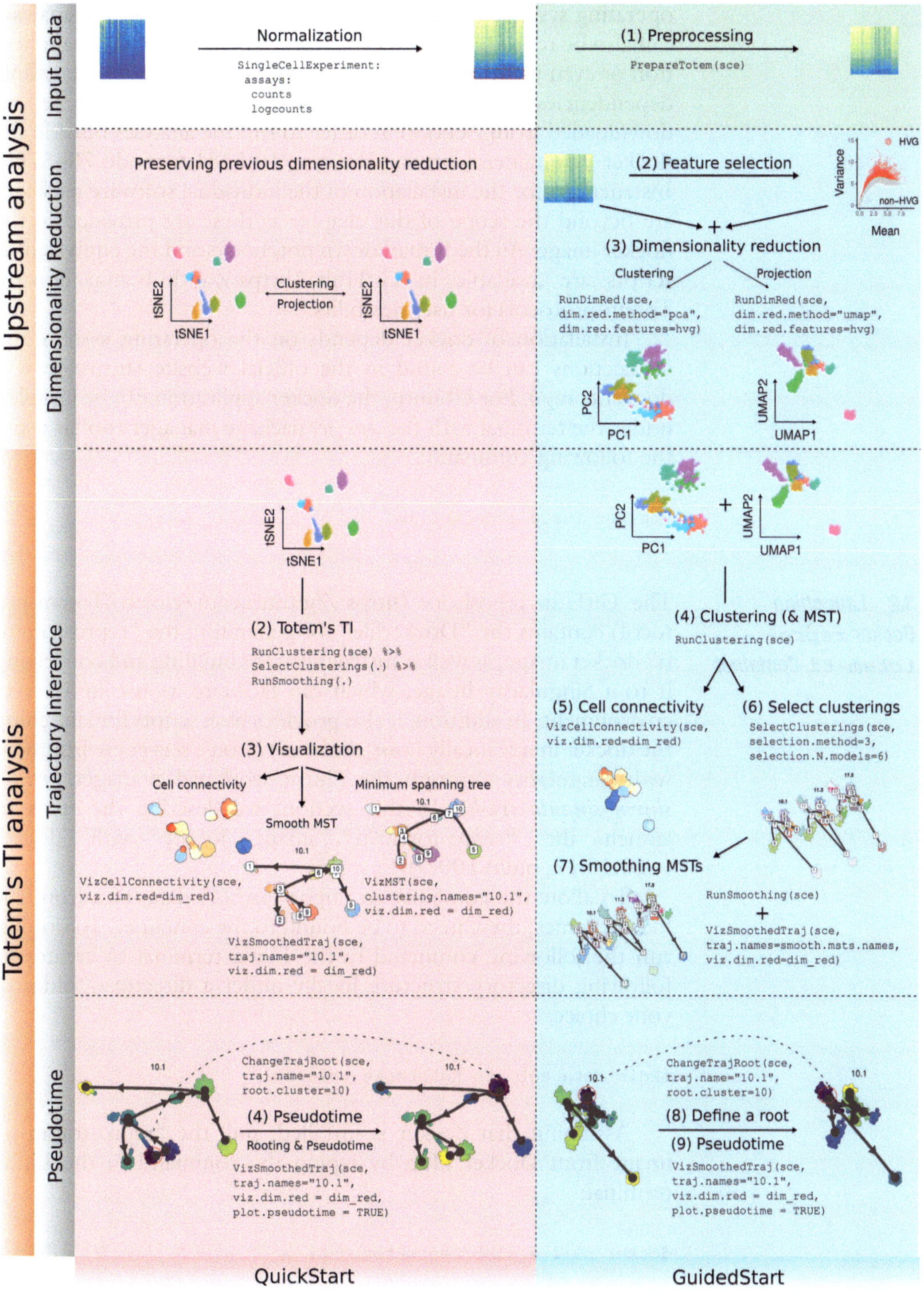

Fig. 1 Summary of the QuickStart and GuidedStart data analysis protocols for the inference of cell trajectory with Totem. The main protocol steps were numbered and the respective Totem's functions highlighted

operating systems (amd64 architecture). However, all the analyses can also be reproduced using the RStudio Desktop/server application or even just the R console, as long as *Totem* R package and its dependencies are installed. In this case, the data needs to be first downloaded from Zenodo in order to run the analyses outside the docker container: https://doi.org/10.5281/zenodo.7845709. Instructions for the installation of the individual software packages are beyond the scope of this chapter as these are provided in the docker image. All the R markdown notebooks and the equivalent R scripts are available in GitHub (https://github.com/elolab/Totem-protocol) for user flexibility.

Installation of docker depends on the operating system and instructions can be found in the official website (https://www.docker.com/). For Ubuntu, the docker application can be installed under the terminal with the *apt-get* package manager toolkit using the following command:

```
apt-get install docker.io
```

3.2 Launching Docker `repro-totem-ti` Container

The GitHub repository (https://github.com/elolab/Totem-protocol) contains the "Dockerfile" for generating the "repro-totem-ti" docker image, as well as instructions for building and converting it to a Singularity image, which can facilitate its use in a server environment. In addition, it also provides bash scripts for launching the docker image locally (*run_docker.sh*) or on a server environment with Singularity through the Slurm workload manager (*run_slurm_singularity.sh*). In this section, we describe the steps to launch the "repro-totem-ti" image locally with docker (v.20.10.17, build 100c701).

To allow the container to import/export data/files from the local volumes, these need to be bound to the container. To do this, run the following command in the Linux terminal to create the following directory structure locally under a directory/folder of your choice:

```
mkdir data scripts notebooks results
```

Assuming that docker is installed, pull the "repro-totem-ti" image from Docker Hub by typing the command in the Linux terminal:

```
docker pull elolab/repro-totem-ti
```

Then, launch the "repro-totem-ti" container in the Linux terminal by binding the "`results`" folder created above to the container's folder with the command:

```
docker run --rm -ti -e PASSWORD=Totem -p 8787:8787 \
-v $PWD/results:/home/rstudio/results \
repro-totem-ti
```

Finally, go to the browser of your choice (e.g., Chrome, Firefox) and type the following link:

```
http://localhost:8787
```

Use the following credentials to log into RStudio Server:

```
Username: rstudio
Password: Totem
```

The Files, Plots, Packages, Help, Viewer, Presentations RStudio pane layout displays the directory structure under the Files menu, including data, notebooks, results, and scripts. The data folder contains `human_cd34_bm_rep1.rds`, which stores the *SingleCellExperiment* class object of the human CD34+ bone marrow scRNA-seq data set used for all the analyses. The notebooks folder contains R markdown (Rmd) notebooks to reproduce analyses, including the QuickStart protocol (`QuickStart_Totem.Rmd`) and the GuidedStart protocol (`GuidedStart_Totem.Rmd`), as well as the bibliography used in the Rmd notebooks (`references.bib`). The results folder is empty and intended to save results, while the scripts folder contains R scripts equivalent to Rmd notebooks, including the QuickStart (`QuickStart_Totem.R`) and the GuidedStart (`GuidedStart_Totem.R`) protocols, and the R script used to generate the `human_cd34_bm_rep1.rds` file (`download_h5ad_to_SCE_rds_script.R`).

3.3 QuickStart Protocol

To access the "QuickStart_Totem.Rmd" notebook in RStudio, navigate to the Files menu and open the notebooks folder. This notebook is divided into four parts: (1) Preprocessing; (2) *Totem*'s TI; (3) Visualization; and (4) Pseudotime. The Preprocessing section sets up the environment and imports and prepares data for *Totem*. The *Totem*'s TI section contains a two-step workflow which includes clustering and multiple spanning trees (MSTs), followed by smoothing the selected clustering/MSTs. The Visualization section of the notebook visualizes *Totem*'s cell connectivity, clustering, and trajectory's topology. Finally, the Pseudotime section performs re-rooting and compares *Totem*'s pseudotime estimates to those of *Palantir*.

The set of code blocks highlighted in the next subsections can be run by clicking over the start button icon that appears in every R chunk of the "QuickStart_Totem.Rmd" notebook, or by putting the mouse cursor over the line of code to be run and pressing CTRL+ENTER keys. Alternatively, the code can be copied and pasted into the R console.

3.3.1 Preprocessing

First, load the R packages required to run the analyses into the R environment:

```
library("dplyr")
library("Totem")
library("scater")
library("ggplot2")
library("SingleCellExperiment")
```

Set the seed to ensure reproducibility:

```
set.seed(1204)
```

Import the scRNA-seq data set of human CD34+ bone marrow cells as `SingleCellExperiment` R object:

```
sce <- readRDS(file = "../data/human_cd34_bm_rep1.rds")
```

The `sce` object contains the assays `counts`, `logcounts`, and `scaled`; the cell metadata `clusters`, `palantir_pseudotime`, `palantir_diff_potential`, `cell_types_long`, and `cell_types_short`; the dimensionality reduction `tsne`; and several metadata such as `cluster_colors`. In total, the data has 14,651 genes and 5,780 cells.

Prepare the data to Totem by filtering non-expressed genes:

```
sce <- PrepareTotem(object = sce)
```

3.3.2 Totem's TI

The mandatory input to Totem is a `SingleCellExperiment` object, which in this case is `sce`. This object needs to compulsorily contain the `logcounts` assay (which can be checked with `assayNames(sce)` function) and at least one dimensionality reduction (which can be checked with `reducedDimNames(sce)`). *Totem* uses the first dimensionality reduction result in `reducedDimNames(sce)` for clustering, which in this case is `tsne` (*see* **Note 1**). Thus, it is important to make sure that the first dimensionality reduction result is the right one to use for clustering.

The two-step `Totem`'s TI workflow consists of:

1. CLARA (*k*-medoids) clustering and multiple spanning trees (MSTs): `RunClustering()`
2. Smoothing selected clustering/MSTs with the principal curves algorithm: `SelectClusterings()` followed by `RunSmoothing()`

To run the two-step *Totem*'s TI workflow with default options (*see* **Note 2**), set the seed to maintain reproducibility and use the following code:

```
set.seed(123)
sce <- RunClustering(object = sce) %>%
    SelectClusterings(object = .) %>%
    RunSmoothing(object = .)
```

All the results are stored in the `sce` object as a list of elements (`cell.connectivity`, `all.clustering`, `all.clustering.scores`, `selected.clustering`, `dynwrap_trajectory`, `slingshot_trajectory`) in the metadata slot named totem (*see* **Note 3**).

3.3.3 Visualization

To explore the results and support decisions, *Totem* provides several visualization functions to explore cell connectivity (`VizCellConnectivity()`; *see* **Note 4**), clustering/MST (`VizMST()`), and trajectory and pseudotime (`VizSmoothedTraj()`).

To visualize the cell connectivities, retrieve the `tsne` dimensionality reduction using the `reducedDim()` function, and plot them using the `VizCellConnectivity()` function (Fig. 2):

```
dim_red <- reducedDim(sce, "tsne")
VizCellConnectivity(object = sce, viz.dim.red = dim_red)
```

To visualize the best selected clustering/MST result, retrieve its name and visualize it using the `VizMST()`function (Fig. 3a):

```
select.cluster
<- names(metadata(sce)$totem$slingshot_trajectory)
VizMST(object = sce, clustering.names = select.cluster,
 viz.dim.red = dim_red)
```

For comparison, the projection can also be visualized together with cell types from the original study (Fig. 3b):

```
plotReducedDim(sce, dimred = "tsne",
              color_by = "cell_types_short") +
    scale_color_manual(name= "Cell Types",
                   values=as.character(unlist(metadata(sce)
                   $cluster_colors)[!duplicated(levels(sce
                   $cell_types_short))])) +
    ggtitle("Cell types") +
 theme_void() +
 theme(legend.position = "bottom")
```

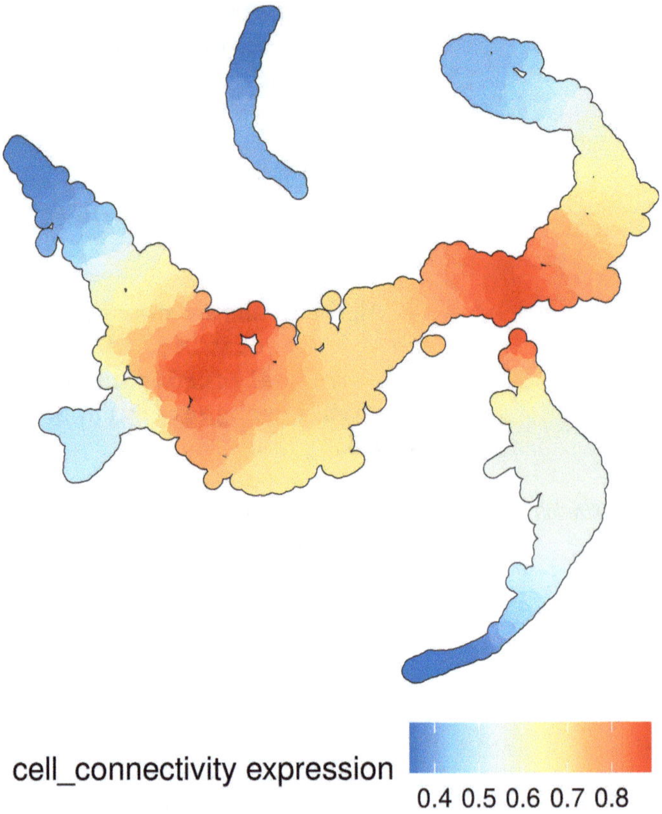

Fig. 2 Cell connectivity of the human CD34+ bone marrow cells projected onto t-distributed stochastic neighbor embedding (t-SNE). The cell connectivity highlights the averaged ratio of a cell cluster's connections to the number of clusters calculated across the minimum spanning trees obtained after clustering with `RunClustering()`. Each cell is represented by a dot. Higher connectivity values (red shade colors) indicate a farther distance from the leafs/end points of the trajectory and, thus, more likely represent trajectory branching points

3.3.4 Pseudotime

To visualize Totem's pseudotime, define the root of the cell trajectory to cluster 15, which is one of the clusters that correspond to the more immature HSCs (hematopoietic stem cells; *see* **Note 5**), and use the function `VizSmoothedTraj()` (Fig. 4a):

```
root.cluster <- 15
sce <- ChangeTrajRoot(object = sce,
 traj.name = select.cluster,
 root.cluster = root.cluster)
VizSmoothedTraj(object = sce, traj.names = select.cluster,
 viz.dim.red = dim_red, plot.pseudotime = TRUE)
```

To compare *Totem*'s pseudotime with that of Palantir from the original study, visualize Palantir's pseudotime using the function `plotReducedDim()` (Fig. 4b):

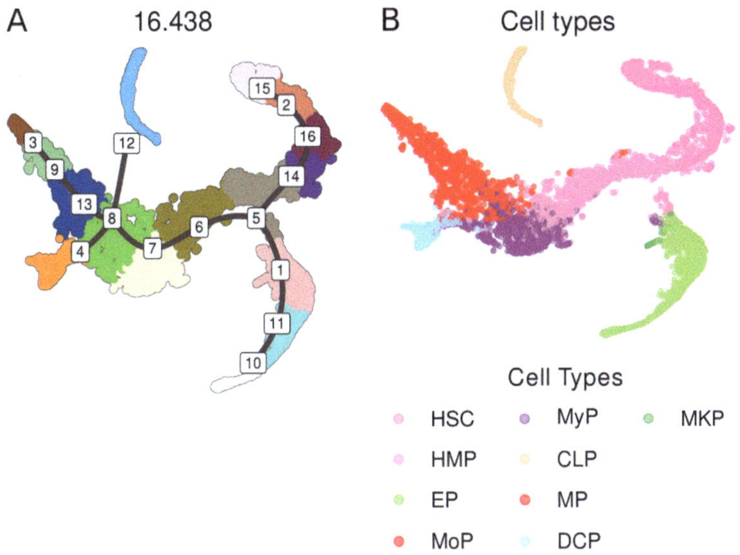

Fig. 3 Human CD34+ bone marrow cells projected onto t-distributed stochastic neighbor embedding (t-SNE), illustrating (**a**) the best minimum spanning tree (MST) result from *Totem* and (**b**) the cell types from the original study. HSCs (hematopoietic stem cells), HMP (hematopoietic multipotent progenitors), EP (erythroid progenitors), MoP (monocyte progenitors), MyP (myeloid progenitors), CLP (common lymphoid progenitors), DCP (dendritic cell progenitors), MKP (megakaryocyte progenitors)

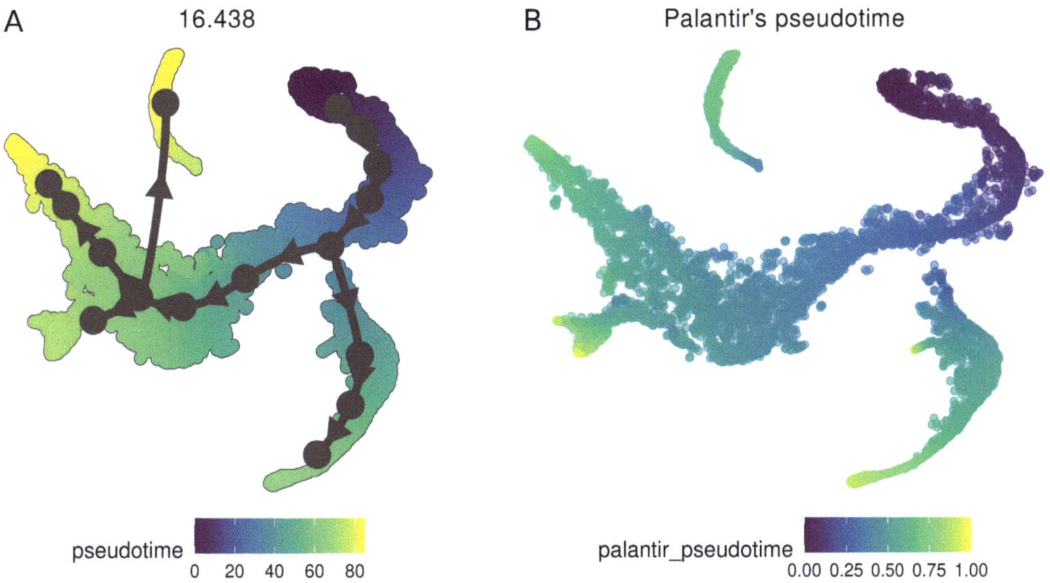

Fig. 4 Human CD34+ bone marrow cells projected onto t-SNE highlighted by the pseudotime obtained using (**a**) *Totem* and (**b**) Palantir used in the original study

```
plotReducedDim(sce, dimred = "tsne",
               colour_by = "palantir_pseudotime") +
               ggtitle("Palantir's pseudotime") +
               theme_void() +
               theme(legend.position = "bottom")
```

Despite the good correlation between the pseudotimes obtained using *Totem* and Palantir and the correct prediction of terminal states once the root was provided, the lymphoid lineage was wrongly predicted. CLP (common lymphoid progenitor) was predicted to diverge from MyP (myeloid progenitor) instead of HSCs (hematopoietic stem cells), as it should have been. One of the reasons that may explain this wrong prediction is that the low-dimensional representation used for clustering may not fairly represent the relationship between cell types. The CLP population appears farther apart from the HSC than the MyP cluster, making it more likely to branch from the latter than the former. In general, the user should be critical about lineages toward cell populations/clusters that do not show a continuous development in the dimensionality reduction projection used for clustering.

3.4 GuidedStart Protocol

To access the "Guided_Totem.Rmd" notebook in RStudio, navigate to the Files menu and open the notebooks folder. This notebook is divided into nine parts: (1) Prepare data; (2) Feature selection; (3) Dimensionality reduction; (4) Clustering (and MST); (5) Cell connectivity; (6) Select clusterings; (7) Smoothing MSTs; (8) Define a root; and (9) Pseudotime. Each one of these parts is summarized below:

1. Prepare data: Importing data to R and preparing it for analysis with *Totem*.
2. Feature selection: Selecting 2000 highly variable features for dimensionality reduction with *scran*.
3. Dimensionality reduction: Dimensionality reduction for clustering (PCA) and visualization (UMAP) using *Totem*.
4. Clustering (and MST): CLARA *k*-medoids clustering and MSTs using *Totem*.
5. Cell connectivity: Visualizing and interpreting cell connectivity estimated by *Totem*.
6. Select clusterings: Selecting and visualizing top six clustering and MST results using *Totem*.
7. Smoothing MSTs: Smoothing the selected MSTs with the principal curves algorithm using *Totem*.
8. Define a root: Defining the most probable root for the inferred trajectory.
9. Pseudotime: Visualizing *Totem*'s pseudotime and comparing it with Palantir's pseudotime.

3.4.1 Preprocessing

The first step of the GuidedStart protocol is exactly the same as the (1) Preprocessing step in the QuickStart protocol. However, the dimensionality reduction `tsne` is removed from the `sce` object to focus on exploring other dimensionality reduction methods for clustering and visualization:

```
set.seed(1204)
sce <- readRDS(file = "../data/human_cd34_bm_rep1.rds")
reducedDim(sce, "tsne") <- NULL
sce <- PrepareTotem(object = sce)
```

3.4.2 Feature Selection

The next step of the protocol is feature selection. It is not mandatory but highly recommended to avoid selection of uninformative genes, such as those with low abundance or invariant expression across different cells, and to increase the computational speed of downstream steps due to the reduced dimensionality. Selection of 2000 most highly variable genes (HVG) can be done using the `scran` R package:

```
var.genes <- scran::modelGeneVar(sce)
hvg <- scran::getTopHVGs(var.genes, n = 2000)
```

3.4.3 Dimensionality Reduction

After feature selection, perform a PCA for clustering with 50 principal components (PCs) using the `hvg` selected previously:

```
sce <- RunDimRed(object = sce,
 dim.red.method = "pca",
 dim.red.features = hvg,
 dim.reduction.par.list = list(ndim=50))
```

In order to see which PCs explain most of the biological variability in the data, an elbow plot can be created based on the standard deviation of the PCs along with visualization of the cells projected onto the first PCs (Fig. 5):

```
reducedDim(sce, "pca") %>%
 apply(X = ., MARGIN = 2, FUN = function(x) sd(x)) %>%
 as.data.frame(.) %>%
 `colnames<-`("Standard Deviation") %>%
mutate("PCs" = factor(1:nrow(.), levels = 1:nrow(.))) %>%
ggplot(data = ., mapping = aes(x = PCs,
y = `Standard Deviation`)) +
geom_point() +
theme_bw()
```

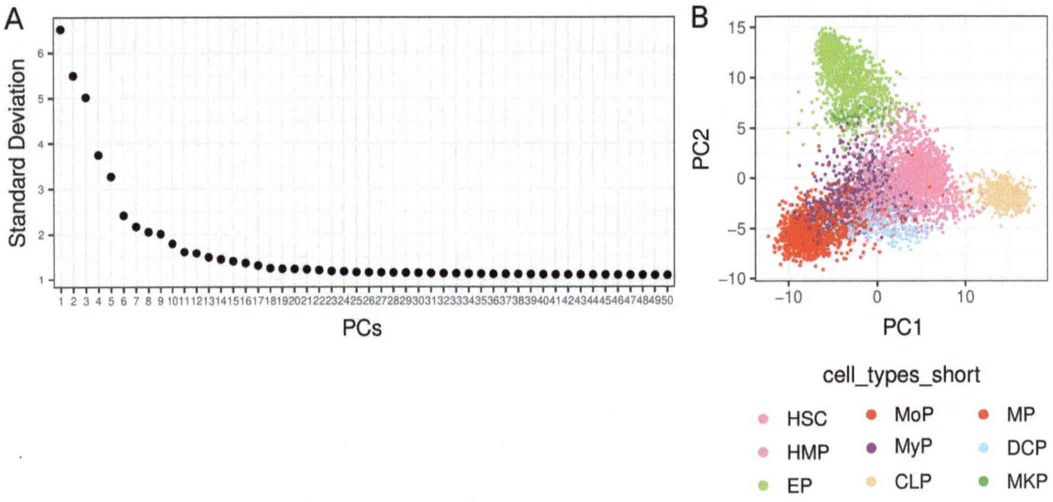

Fig. 5 Biological variation of human CD34+ bone marrow cells explored through principal component analysis (PCA). The elbow plot (**a**) shows the standard deviation of each principal component and can help selecting a suitable number of principal components (PCs) for downstream analysis. The PCA plot (**b**) visualizes the cells projected onto the first two PCs, with each cell colored according to the cell type from the original study. HSCs (hematopoietic stem cells), HMP (hematopoietic multipotent progenitors), EP (erythroid progenitors), MoP (monocyte progenitors), MyP (myeloid progenitors), CLP (common lymphoid progenitors), DCP (dendritic cell progenitors), MKP (megakaryocyte progenitors)

```
reducedDim(sce, "pca") %>%
 as.data.frame(.) %>%
 mutate("Cell_ID" = row.names(.)) %>%
 cbind(., colData(sce)) %>%
 ggplot(data = ., mapping = aes(x=comp_1,
 y=comp_2,
color=cell_types_short)
) +
 geom_point() +
 labs(x = "PC1", y = "PC2") +
scale_color_manual(values=as.character(unlist
(metadata(sce)$cluster_colors)[!duplicated(levels(sce$cell_-
types_short))])) +
theme_bw()
```

To perform clustering, the first six PCs were selected based on the previous visualizations (*see* **Note 1**):

```
pick.pcs <- 1:6
reducedDim(sce, "pca") <- reducedDim(sce, "pca")[, pick.pcs ]
```

To create a two-dimensional representation of the data for visualization, UMAP is selected as the dimensionality reduction method using the same options as with PCA and saved to the

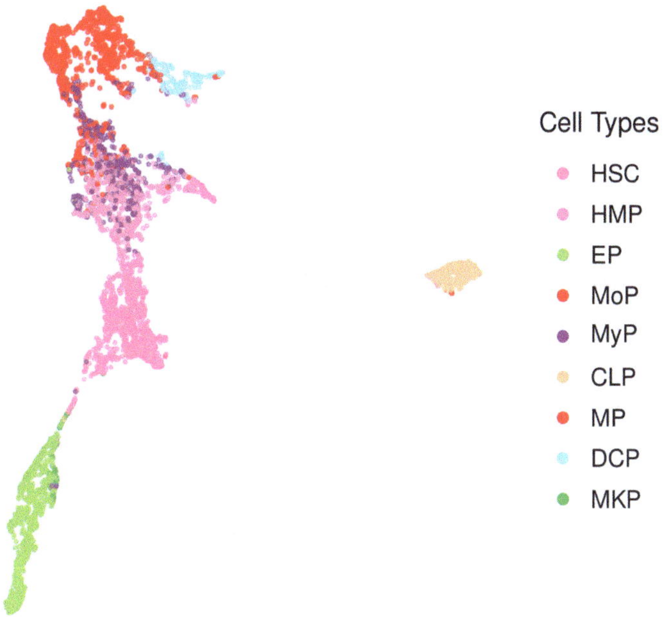

Fig. 6 Human CD34+ bone marrow cells projected onto uniform manifold approximation and projection (UMAP) highlighted by cell type. HSCs (hematopoietic stem cells), HMP (hematopoietic multipotent progenitors), EP (erythroid progenitors), MoP (monocyte progenitors), MyP (myeloid progenitors), CLP (common lymphoid progenitors), DCP (dendritic cell progenitors), MKP (megakaryocyte progenitors)

variable `dim_red`. The projection can then be visualized together with cell types from the original study (Fig. 6).

```
set.seed(123)
sce <- RunDimRed(object = sce,
 dim.red.method = "umap",
 dim.red.features = hvg,
 dim.reduction.par.list = list(ndim=2,
 pca_components = 6))
dim_red <- reducedDim(sce, "umap")
plotReducedDim(object = sce, dimred = "umap",
colour_by = "cell_types_short",
point_size=0.5) +
 scale_color_manual(name= "Cell Types",
values=as.character(unlist(metadata(sce)$cluster_colors)[!du-
plicated(levels(sce$cell_types_short))])) +
 theme_void()
```

Finally, the UMAP reduction is removed from the `sce` object to retain only `pca`:

```
reducedDim(sce, "umap") <- NULL
```

3.4.4 Clustering (and MST)

To cluster the single-cell data, use the CLARA algorithm (*k*-medoids) on the PCA-reduced data 10 K times (`N.clusterings = 10000`), ranging from 3 to 20 clusters (`k.range = 3:20`; *see* **Note 2**) and excluding clusters with less than 5 cells (`min.cluster.size = 5`):

```
set.seed(123)
sce <- RunClustering(object = sce, k.range = 3:20,
min.cluster.size = 5, N.clusterings
= 10000)
```

3.4.5 Cell Connectivity

To visualize the cell connectivity, the function `VizCellConnectivity()` is used, with the UMAP dimensionality reduction as input (stored in the variable `dim_red`):

```
VizCellConnectivity(object = sce, viz.dim.red = dim_red)
```

This information can be used to support the decision on the most likely trajectory (Fig. 7; *see* **Note 4**). Additionally, the cell connectivity can be used to select the clustering method (see the next section).

3.4.6 Select Clusterings

The next step is to select the top clustering results based on several metrics and inspected for further visualizations (*see* **Note 6**). Select the top six clustering results based on the variance ratio criterion and cell connectivity for MST calculation:

```
sce <- SelectClusterings(sce, selection.method = 3,
 selection.N.models = 6,
 selection.stratified = FALSE,
 prior.clustering = NULL)
```

After selecting the top six clustering results, they can be visualized using the `VizMST()` function (Fig. 8):

```
select.clusters <- ReturnTrajNames(sce)
VizMST(object = sce, clustering.names = select.clusters,
 viz.dim.red = dim_red)
```

3.4.7 Smoothing MSTs

Smooth the selected top six MSTs with the principal curves algorithm and visualize the smoothed trajectories:

```
sce <- RunSmoothing(sce)
smooth.msts.names <- ReturnTrajNames(sce)
VizSmoothedTraj(object = sce,
 traj.names = smooth.msts.names,
 viz.dim.red = dim_red,plot.pseudotime = FALSE)
```

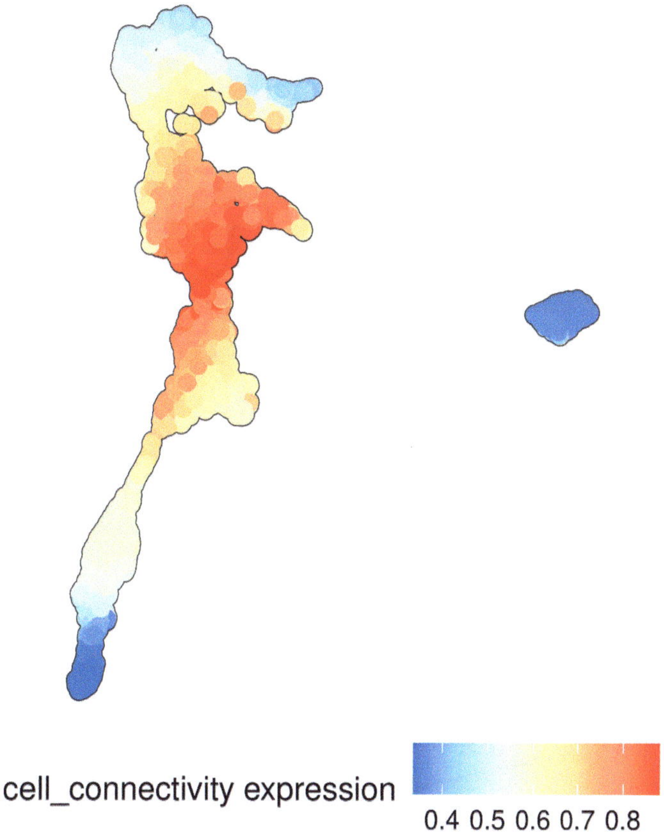

Fig. 7 Cell connectivity of the human CD34+ bone marrow cells projected onto uniform manifold approximation and projection (UMAP). The cell connectivity highlights the averaged ratio of a cell cluster's connections to the number of clusters calculated across the minimum spanning trees obtained after clustering with `RunClustering()`. Each cell is represented by a dot. Higher connectivity values (red shade colors) indicate a farther distance from the leafs/end points of the trajectory and, thus, more likely represent trajectory branching points

From the top six smoothed MST trajectories presented (Fig. 9), the only difference is in the MoP and DCP lineages and the branching from these lineages to the CLP (see Fig. 6). The CLP needs to diverge before the MyP, and MoP and DCP should diverge from MyP. The clustering/MST result `8.135` seems to meet these expectations and is selected below as the elected trajectory.

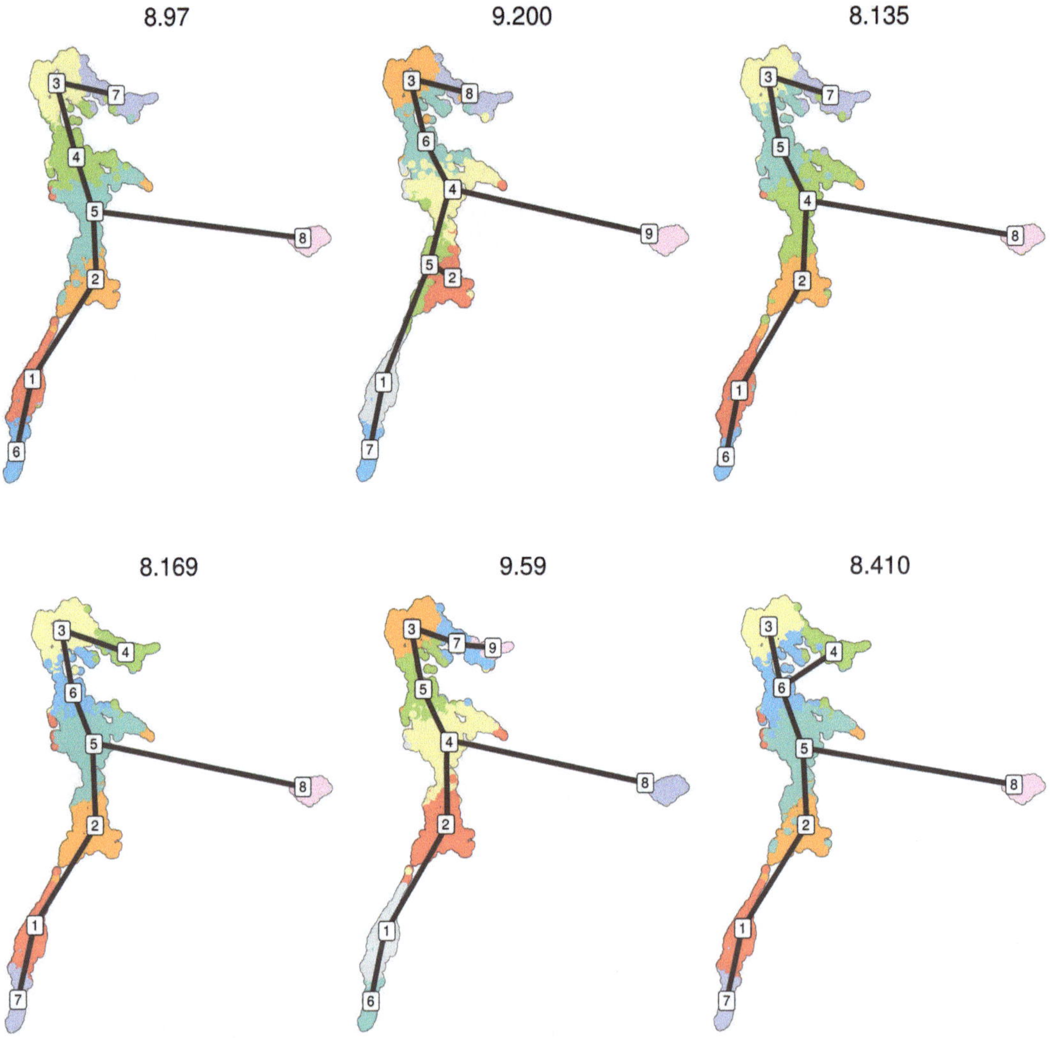

Fig. 8 Six top-ranked MST topologies based on the variance ratio criterion and cell connectivity, highlighted across the human CD34+ bone marrow cells projected onto uniform manifold approximation and projection (UMAP)

3.4.8 Define a Root

The following defines the root of the trajectory of the clustering/ MST result `8.135` as the cluster number 2, which corresponds to the most immature HSCs (*see* **Note 5**):

```
select.traj <- "8.135"
root.cluster <- 2
sce <- ChangeTrajRoot(object = sce, traj.name = select.traj,
 root.cluster = root.cluster)
```

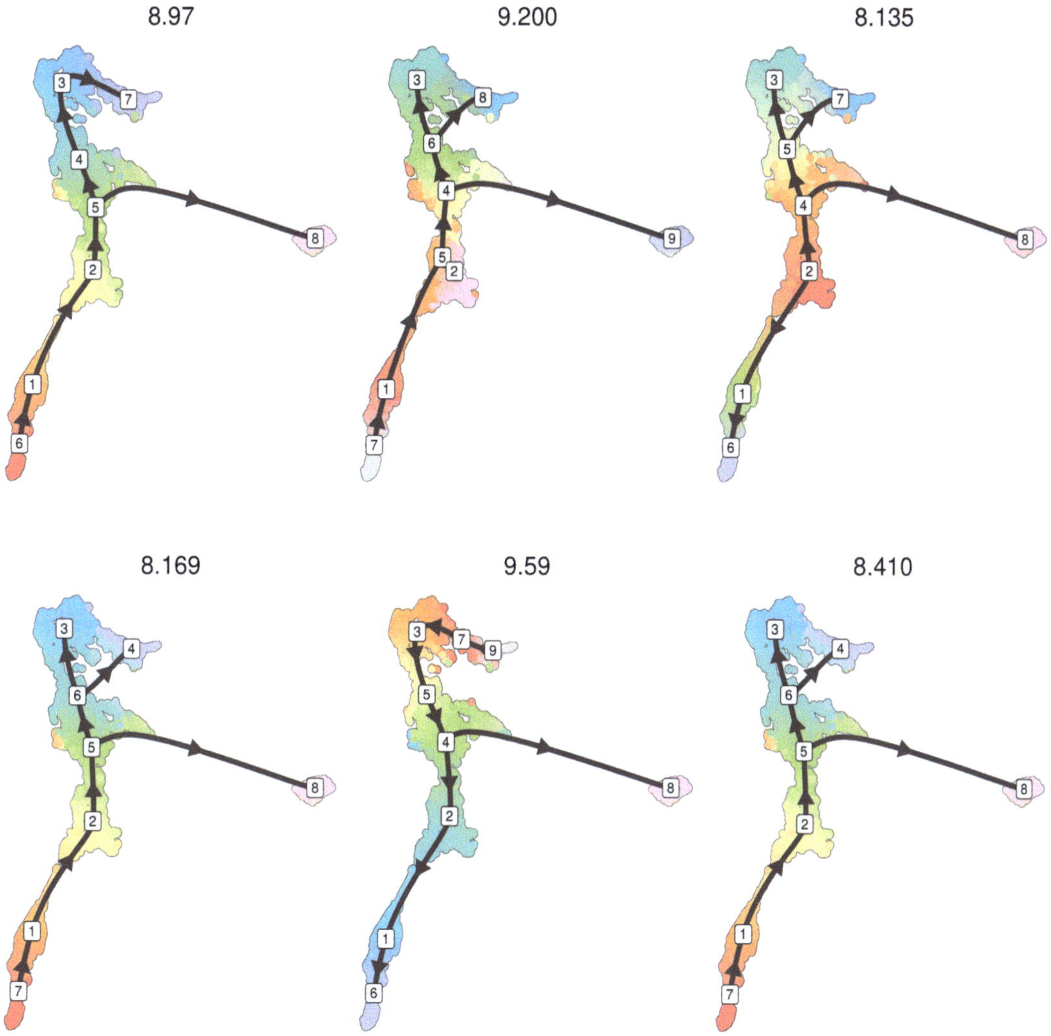

Fig. 9 Smoothed trajectories for the six top-ranked MST topologies across the human CD34+ bone marrow cells projected onto uniform manifold approximation and projection (UMAP)

3.4.9 Pseudotime

Plot the selected trajectory highlighting the pseudotime (Fig. 10a):

```
VizSmoothedTraj(object = sce,
 traj.names = select.traj,
 viz.dim.red = dim_red, plot.pseudotime = TRUE)
```

To compare *Totem*'s pseudotime (Fig. 10a) with *Palantir*'s pseudotime, visualize Palantir's pseudotime using the function `plotReducedDim()` (Fig. 10b):

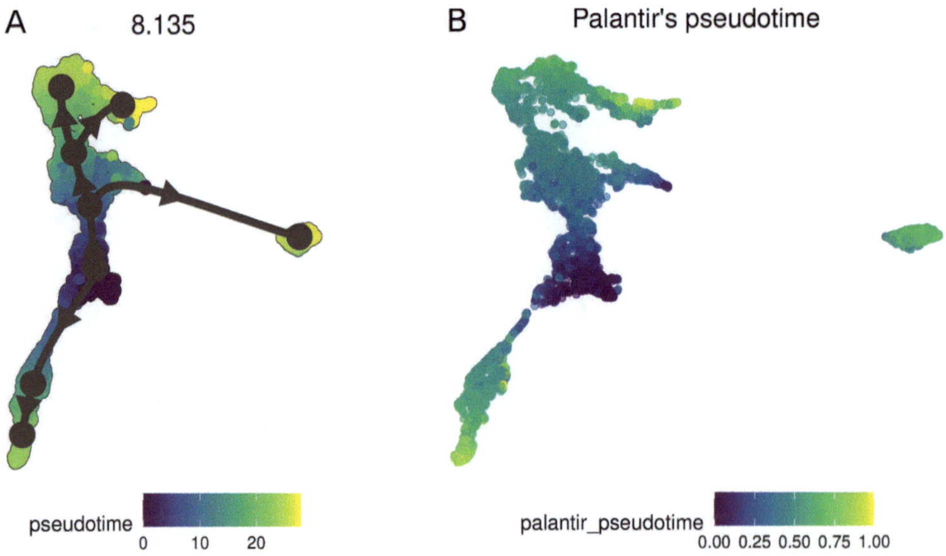

Fig. 10 Human CD34+ bone marrow cells projected onto uniform manifold approximation and projection (UMAP) highlighted by the pseudotime obtained using (**a**) *Totem* and (**b**) Palantir used in the original study

```
reducedDim(sce, "umap") <- dim_red
plotReducedDim(sce, dimred = "umap",
 colour_by = "palantir_pseudotime") +
 ggtitle("Palantir's pseudotime") +
 theme_void() +
 theme(legend.position = "bottom")
```

The pseudotime and trajectory obtained with *Totem* agree well with *Palantir*. Contrary to the inferred trajectory with the t-SNE projection in the QuickStart protocol, by using a more adequate dimensionality reduction projection, such as PCA, *Totem* was able to correctly predict the lineage related to the CLP (*see* **Note 1**).

The trajectory of interest obtained with Totem can be further used to perform downstream analyses such as the identification of genes supporting that same trajectory with third-party tools developed for this purpose. *Totem* allows converting the trajectory of interest to formats compatible with the *dyno* [10] and *tradeSeq* [21] for trajectory differential gene expression (*see* **Note 7**).

4 Notes

1. Dimensionality reduction to use for clustering and visualization

   The minimal requirement to run *Totem*'s TI method is a dimensionality reduction embedding. This can be obtained either from an upstream data analysis, such as integration or clustering as shown in QuickStart, or de novo from the normalized assay (`logcounts`), as shown in GuidedStart. In either case, the low-dimensional representation chosen for clustering should fairly represent the dynamic process being studied, such as cell cycle, activation, or differentiation, as it greatly impacts the inferred trajectories. Choosing the appropriate dimensionality reduction method and the number of dimensions is a difficult task. Typically, two or more dimensions need to be selected depending on the data set and the dimensionality reduction method chosen [6, 20]. The first dimension may represent the differentiation gradient and the second the cell cycle [6]. As demonstrated in the QuickStart protocol, it is important to note that using the same dimensionality reduction embedding for clustering and visualization might not be adequate. While t-SNE might be a proper method to visualize cell populations, other methods such as PCA might be more suitable for cell trajectory clustering [20]. *Totem* does not deal with disconnected trajectories, and, thus, the user should be very critical and doubtful about branchings toward milestones that appear segregated in the low-dimension projection, as it is the case of the common lymphoid progenitor population in both protocols.

2. Optimizing parameters: merge clusters

   By default, the Totem clustering function `RunClustering()` uses a target range of clusters from 3 to 20 (`k.range = 3:20`). However, this range may not be optimal for each data set, particularly if the data set involves unbalanced cell clusters. In such cases, it is recommended to provide a higher target range of clusters that can be merged later. More information can be found in the "Merging clusters" section at https://htmlpreview.github.io/?https://github.com/elolab/Totem-benchmarking/blob/main/Totem.html.

3. Exporting and accessing the *Totem* results

   Totem saves its results as a list under the metadata slot named `totem`. These can be accessed typing the following:

```
metadata(sce)$totem
```

The `sce` object with all the results stored on it can be exported using the function `saveRDS`:

```
saveRDS(object = sce, file = "../results/sce_quickstart.rds")
```

4. Supporting the decision of trajectory selection

   Cell connectivity can help support the decision about which topology to select among the top-ranked candidates. It is recommended that the user cross-checks the top inferred topologies with the cell connectivity values to ensure that they match the expectations, that is, if the highest and lowest connectivity values correspond to branching points and leafs/end points, respectively, in the inferred trajectory.

5. Optimizing parameters: root

   Totem randomly assigns a root to the trajectory during the inference, and the user should specify the root milestone or cell centroid afterward for a given topology to obtain a proper directed trajectory and pseudotime.

6. *Totem*'s limitations

   The aim of *Totem* is not necessarily to find the correct topology at first, but rather to identify a good set of candidate topologies that the user can evaluate based on their expectations and cell connectivity, as demonstrated in GuidedStart. It is important to note that *Totem* cannot infer topologies that include cycles and disconnected, converging, or diverging trajectories and, thus, any attempt to do so may result in incorrect trajectory predictions.

7. Downstream analyses: converting *Totem*'s trajectory to `dynwrap` and `PseudotimeOrdering` objects for differential gene expression

   The trajectory of interest obtained with *Totem* can be converted into a `dynwrap` (using the function `ReturnDynwrapObject()`) or `PseudotimeOrdering` (with `ReturnSlingshotObject()`) object in order to perform trajectory differential gene expression analysis with the third-party tools *dyno* [10] and *tradeSeq* [21], respectively:

```
dynwrap.obj
<- ReturnDynwrapObject(sce,traj.name=select.traj)
slingshot.obj <- ReturnSlingshotObject(sce,
 traj.name=select.traj)
```

The functions take as input the `SingleCellExperiment` object and the name of the trajectory (`traj.name` parameter; `select.traj` variable corresponds to the trajectory `"8.135"` in the GuidedStart protocol).

Acknowledgments

The authors thank the Elo lab for useful discussions and support, particularly Mats Perk and Tomi Suomi. AGGS has received funding from the European Union's Horizon 2020 research and innovation program under the Marie Skłodowska-Curie grant agreement No.: 955321. This research was also supported by the University of Turku, Åbo Akademi University, Turku Graduate School (UTUGS). LLE reports grants from the Academy of Finland (310561, 314443, 329278, 335434, 335611, and 341342) and Sigrid Jusélius Foundation during the conduct of the study. Our research is also supported by Biocenter Finland and ELIXIR Finland.

References

1. Pellin D, Loperfido M, Baricordi C et al (2019) A comprehensive single cell transcriptional landscape of human hematopoietic progenitors. Nat Commun 10(1):2395
2. Qiu C, Cao J, Martin BK et al (2022) Systematic reconstruction of cellular trajectories across mouse embryogenesis. Nat Genet 54(3):328–341
3. Dalton S (2015) Linking the cell cycle to cell fate decisions. Trends Cell Biol 25(10): 592–600
4. Kumar S, Fonseca VR, Ribeiro F et al (2021) Developmental bifurcation of human T follicular regulatory cells. Sci Immunol 6(59): eabd8411
5. Ding J, Sharon N, Bar-Joseph Z (2022) Temporal modelling using single-cell transcriptomics. Nat Rev Genet 23(6):355–368
6. Cannoodt R, Saelens W, Saeys Y (2016) Computational methods for trajectory inference from single-cell transcriptomics. Eur J Immunol 46(11):2496–2506
7. La Manno G, Soldatov R, Zeisel A et al (2018) RNA velocity of single cells. Nature 560(7719):494–498
8. Van der Maaten L, Hinton G (2008) Visualizing data using t-SNE. J Mach Learn Res 9(11)
9. McInnes L, Healy J, Melville J (2018) Umap: uniform manifold approximation and projection for dimension reduction. arXiv preprint arXiv:1802.03426
10. Saelens W, Cannoodt R, Todorov H et al (2019) A comparison of single-cell trajectory inference methods. Nat Biotechnol 37(5): 547–554
11. Smolander J, Junttila S, Elo LL (2023) Cell-connectivity-guided trajectory inference from single-cell data. Bioinformatics 39(9):btad515 https://doi.org/10.1093/bioinformatics/btad515
12. Maechler Martin (2019) Finding groups in data: cluster analysis extended Rousseeuw et al. R package version 2(0):242–248
13. Paradis E, Schliep K (2019) ape 5.0: an environment for modern phylogenetics and evolutionary analyses in R. Bioinformatics 35(3): 526–528
14. Street K, Risso D, Fletcher RB et al (2018) Slingshot: cell lineage and pseudotime inference for single-cell transcriptomics. BMC Genomics 19:1–16
15. Setty M, Kiseliovas V, Levine J et al (2019) Characterization of cell fate probabilities in single-cell data with Palantir. Nat Biotechnol 37(4):451–460
16. Zappia L, Lun A (2022) zellkonverter: conversion between scRNA-seq objects. R package version 1.8.0 https://github.com/theislab/zellkonverter. Accessed 6 Apr 2023
17. Stuart T, Butler A, Hoffman P et al (2019) Comprehensive integration of single-cell data. Cell 177(7):1888–1902
18. Hie B, Bryson B, Berger B (2019) Efficient integration of heterogeneous single-cell transcriptomes using Scanorama. Nat Biotechnol 37(6):685–691
19. Coifman RR, Lafon S, Lee AB et al (2005) Geometric diffusions as a tool for harmonic analysis and structure definition of data: diffusion maps. Proc Natl Acad Sci USA 102(21): 7426–7431
20. Sun S, Zhu J, Ma Y et al (2019) Accuracy, robustness and scalability of dimensionality reduction methods for single-cell RNA-seq analysis. Genome Biol 20(1):1–21
21. Van den Berge K, Roux de Bézieux H, Street K et al (2020) Trajectory-based differential expression analysis for single-cell sequencing data. Nat Commun 11(1):1201

Chapter 10

Zebrafish Thrombocyte Transcriptome Analysis and Functional Genomics

Weam Fallatah, Jabila Mary, and Pudur Jagadeeswaran

Abstract

Our laboratory is interested in investigating the maturation process of zebrafish thrombocytes, which are functional equivalents to human platelets. We have adopted the zebrafish model to gain insights into mammalian platelet production, or thrombopoiesis. Notably, zebrafish exhibit two distinct populations of thrombocytes in their circulating blood: young and mature thrombocytes. This observation is intriguing because maturation appears to occur in circulation, yet the precise mechanisms governing this maturation remain elusive. Our goal is to understand the mechanisms underlying thrombocyte maturation by conducting single-cell RNA sequencing (scRNA-Seq) on young and mature thrombocytes, analyzing these transcriptomes to identify genes specific to each thrombocyte population, and elucidating the role of these genes in the maturation process, by quantifying thrombocyte numbers after the piggyback knockdown of each of these genes. In this chapter, we present a comprehensive, step-by-step protocol detailing the multifaceted methodology involved in understanding thrombocyte maturation, which encompasses the collection of zebrafish blood, the separation of young and mature thrombocytes using flow cytometry, scRNA-Seq analysis of these distinct thrombocyte populations, identification of genes specific to young and mature thrombocytes, and subsequent validation through gene knockdown techniques.

Key words Thrombocyte, Single-cell RNA-Seq, Transcriptome, Bioinformatics, Piggyback knockdown, PANTHER, Primer 3, RFP+ thrombocytes, GFP+ thrombocytes

1 Introduction

Single-cell RNA sequencing (scRNA-Seq) has become a very effective tool in studying the gene expression of hematopoietic cells and their precursors in order to identify novel genes in their differentiation and development [1–4]. In our laboratory, we have harnessed the power of scRNA-Seq to gain insight into the gene expression patterns of hematopoietic cells and their precursors, particularly focusing on thrombocyte development and function in hemostasis in a zebrafish model [5–7]. Understanding the maturation and

Rajeev K. Azad (ed.), *Transcriptome Data Analysis*, Methods in Molecular Biology, vol. 2812,
https://doi.org/10.1007/978-1-0716-3886-6_10,

differentiation of thrombocytes requires unraveling novel genes involved in these processes [7].

Our journey began with a fortuitous discovery: the specific labeling of young thrombocytes with DiI-C18 (DiI) and the labeling of other blood cells with mepacrine [8]. This discovery allowed us to effectively separate young and mature thrombocytes by fluorescence-activated cell sorting (FACS) as well as by immunoselection [5, 9]. Subsequently, we developed a transgenic zebrafish line known as GloFli, in which young thrombocytes are labeled with red fluorescent protein (RFP+) and mature thrombocytes with green fluorescent protein (GFP+) [6]. This genetic tool further enhanced our ability to isolate these two thrombocyte populations through FACS [7].

With these meticulously separated thrombocyte populations in hand, we embarked on a comprehensive scRNA-Seq analysis using the 10× Genomics platform [7]. Our objective is to compare the transcriptomic data from young and mature thrombocytes, ultimately generating a list of genes that exhibit specific expression patterns in these two populations. This knowledge is pivotal in understanding the intricate process of thrombocyte maturation and function within the circulatory system of zebrafish.

Moving beyond transcriptomics, we are delving into functional genomics by employing a technique known as the "piggyback knockdown method." This method entails the delivery of three gene-specific antisense oligonucleotides (ASOs) into GloFli zebrafish cells via a VMO (Vivo-Morpholino oligomer) after intravenous (IV) injections to knock down three genes simultaneously. Forty-eight hours following this intervention, blood samples are harvested, and young and mature thrombocyte numbers are assessed, where RFP+ and GFP+ thrombocyte counts are used as indicators of thrombocyte quantities.

In summary, our research utilizes multiple methods, combining cutting-edge scRNA-Seq analysis to decipher the thrombocyte transcriptome with functional genomic investigations using the piggyback knockdown method. Through these methods, we aim to unravel the role of genes specifically expressed in young and mature thrombocytes, shedding light on their contributions to thrombocyte production and the broader field of hemostasis genetics in zebrafish.

2 Materials

Transgenic TG Glo fish (DsRed) and Fli fish (fli1:EGFP), wild-type (WT) zebrafish AB strain, E3 medium (0.17 mM KCl, 5 mM NaCl, 0.33 mM $CaCl_2$, 0.33 mM $MgSO_4$, pH 7.2, 0.1% methylene blue), MS-222 (0.1 to 0.15 mg/mL), fluorescence microscope, circulating water system supplemented with Instant Ocean, brine shrimp, fish flakes, paper towel, Kimtech wipes, Noyes Scissors 4.7, micropipette, polystyrene 5 mL conical tubes, 1x PBS (137 mM NaCl, 2.7 mM KCl, 10 mM Na_2HPO_4, 1.8 mM KH_2PO_4), thrombocyte buffer (1× PBS supplemented with 0.04% bovine serum albumin), BD FACSCanto flow cytometer, BD FACSDiva 8.0.2 software, next-generation sequencing (NGS), 10x Genomics Chromium Single Cell 3′ GEM library and Gel Bead Kits V3, Illumina Hiseq 2500 V3, Cell Ranger software (version 3.0.0), Ensembl and ZFIN databases, Protein Analysis Through Evolutionary Relationship (PANTHER) version 17.0, ASO, VMO, 10× ThermoPol® Reaction Buffer (500 mM NaCl, 10 mM Tris–HCl (pH 8.0), and 1 mM EDTA (pH 8.0)), PCR thermal cycler, 1 mL syringe with 27G × 1/2 gauge needles, 1.5 ml Eppendorf tube, 3.8% sodium citrate, BD Accuri™ C6 plus flow cytometer.

3 Methods

3.1 Zebrafish Husbandry

1. TG Glo fish and fli1:EGFP fish are kept in breeding tanks, and the next day, the embryos are obtained and kept in an E3 medium.
2. The 3dpf larvae are anesthetized by keeping them in MS-222 and observed under a fluorescence microscope to select the brightest larvae positive for both RFP and GFP (GloFli fish) and grow to adulthood.
3. GloFli fish and WT fish are maintained at 28 °C in the circulating system of deionized water supplemented with Instant Ocean. The fish are maintained at 10 h of dark and 14 h of the light cycle and fed with brine shrimp and flakes.

3.2 Blood Collection

1. The GloFli fish is positioned on a clean paper towel on its lateral side, and its skin is delicately cleaned using Kimtech wipes.
2. The fish is carefully dissected using scissors, cutting between the second and fourth black stripes in alignment with the ventral and dorsal fins.
3. Using a 10 μL pipette, 2 μL of blood welling out from the incision is collected and introduced into polystyrene 5 mL conical tubes containing 300 μL of freshly prepared ice-cold thrombocyte buffer (*see* **Note 1**) (Fig. 1).

3.3 Cell Sorting

1. To sort RFP+ thrombocytes and GFP+ thrombocytes from GloFli fish, the blood cells in the above polystyrene tube are loaded into BD FACSCanto flow cytometer. The thrombocytes are sorted into polystyrene 5 mL conical tubes filled with thrombocyte buffer.

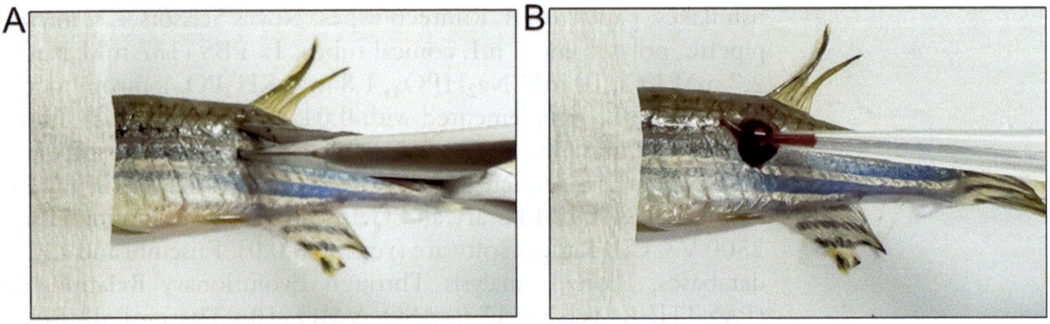

Fig. 1 Blood collection from adult zebrafish. Panels A and B show the position of the incision and the pipette tip with two microliters of blood collected, respectively

2. The FACSCanto flow cytometer is operated at 4 °C with a nozzle size of 100 for optimal performance.
3. To ensure high-quality analysis, cell viability is set within the range of 75% to 90%, and a minimum of 10,000 sorted cells is collected.
4. During the sorting process on the FACSCanto flow cytometer, samples are gated based on PE and GFP signals using BD FACSDiva software.
5. WT zebrafish samples are utilized as negative controls, while positive controls include Glo fish (DsRed) and TG (fli1:EGFP) fish blood samples.

3.4 Single-Cell RNA Sequencing

1. The sorted thrombocytes are maintained on ice and promptly utilized for RNA-Seq analysis by the NGS.
2. The sorted cells are immediately loaded onto the 10× Genomics Chromium Chip B. Bar-coded cDNA libraries are generated using the 10× Genomics Chromium Single Cell 3′ GEM library and Gel Bead Kits V3 as per the manufacturer's instructions (*see* **Note 2**).
3. After successful library preparation and quality control checks, sequencing of the libraries is performed on an Illumina Hiseq 2500 V3 platform, with a minimum depth of 20,000 read pairs per cell for each library. The Cell Ranger software is employed for barcode recovery.

3.5 Selecting Young- and Mature-Specific Protein-Coding Transcripts

1. The gene list from the scRNA-Seq is imported to an Excel spreadsheet to identify common genes expressed in RFP+ and GFP+ thrombocytes.
2. For this, RFP+ and GFP+ gene names are kept in red and green fonts, respectively, and then the list from both these thrombocytes is combined. In the combined list, the duplicated values are highlighted using the "conditional formatting" function,

followed by the elimination of duplicated genes using the "remove duplicates" function.

3. The remaining RFP+ and GFP+ unique genes are identified by their red and green color fonts using the "sort and filter by color" function.
4. To obtain protein-coding genes corresponding to human homologs, RFP+ and GFP+ thrombocyte unique genes are uploaded to PANTHER program by copying and pasting the list in the "Enter IDs" section. Then in the organism section, the "Homo sapiens" is selected. Subsequently, in the "Select Analysis" section, "Pie chart" is selected. After that, in the "Select Ontology," "Protein Class" is selected. Then, in the final output, pie chart lists human genes and categories of proteins where the gene names for each category are accessed by clicking on the appropriate icons. The PANTHER software enables the classification of proteins based on various functional criteria, including family and subfamily, molecular function, cellular component, biological process, and pathway, and filters out non-protein-coding genes, such as microRNA genes (Fig. 2).

1. The first step in this procedure involves obtaining zebrafish cDNA sequences for the above-identified group of protein-

Fig. 2 A representative PANTHER analysis of thrombocyte transcriptome data. When 397 unique genes expressed in young thrombocytes and 2153 unique genes expressed in mature thrombocytes are uploaded into the PANTHER program, and the human species is selected, human orthologous proteins are listed by functional category, including proteins unclassified by PANTHER. The program yielded a total number of protein hits of 189 and 1139 for these unique RFP+ and GFP+ genes, respectively. The percentage of each protein class for young and mature thrombocyte gene lists is shown by different colors in the pie chart, and their classification is provided with similarly coded colors

3.6 Evaluation of Protein-Coding Gene Transcripts from Young and Mature Thrombocytes by Screening with the Piggyback Knockdown Method for Thrombocyte Development

coding sequences in zebrafish corresponding to human orthologs from the Ensembl Genome Browser.

2. The cDNAs are divided into sets of three corresponding to three genes in the above list. For each set of three mRNAs predicted from these cDNAs, three 25 nucleotide length ASOs are designed.
3. To these ASOs, at their 3′-ends, three oligonucleotides are added: 5′-TATAAAT-3′, 5′-TAACTGA-3′, and 5′--TAAGAGG-3′, respectively (called ASO1, ASO2, and ASO3 in Fig. 3) to facilitate hybridization with VMO listed below.
4. For simultaneous three-gene knockdowns, a piggyback hybrid is prepared by mixing 2.5 μL of a 0.5 mM VMO 5′-CCTCT TACCTCAGTTACAATTTATA-3′ with 0.84 μL of each of the three ASOs (1.5 mM of each ASO). To this mixture, 0.5 μL of 10× ThermoPol® Reaction Buffer is added. Then, the piggyback hybrids are prepared by heating the mixture to 94 °C for 5 min and then slowly cooled to room temperature in a PCR thermal cycler. The piggyback hybrids enter the cells, and the ASOs form hybrids with the cognate mRNAs. The mRNAs are degraded by RNaseH (Fig. 3).

3.7 Zebrafish Injections for Knockdown

1. To prepare the zebrafish for the injections, an adult GloFli fish is positioned on its lateral side on a clean paper towel, and its head is covered with a damp Kimtech wipe.
2. The skin is carefully wiped using a dry Kimtech wipe. Subsequently, either 5 μL of the piggyback hybrid described earlier or 1× PBS is administered intravenously using a 27G × 1/2 needle (*see* **Note 3**) (Fig. 4).

3.8 Thrombocyte Quantification

1. In experiments involving the quantification of thrombocytes in both control and knockdown fish, the incision method is employed.
2. 2 μL of blood welling out from the incision is collected and introduced into 1.5 mL Eppendorf tube containing 0.5 μL of 3.8% sodium citrate. After mixing it well, 300 μL of 1× PBS is added followed by a thorough mixing.
3. The tube is then placed in BD Accuri™ C6 plus flow cytometer, and the cells are counted until it reaches 50,000 counts.
4. Then the young and mature thrombocyte population is analyzed by the red and green fluorescence in PE and FITC channels and quantified as the percentage of total cell counts.

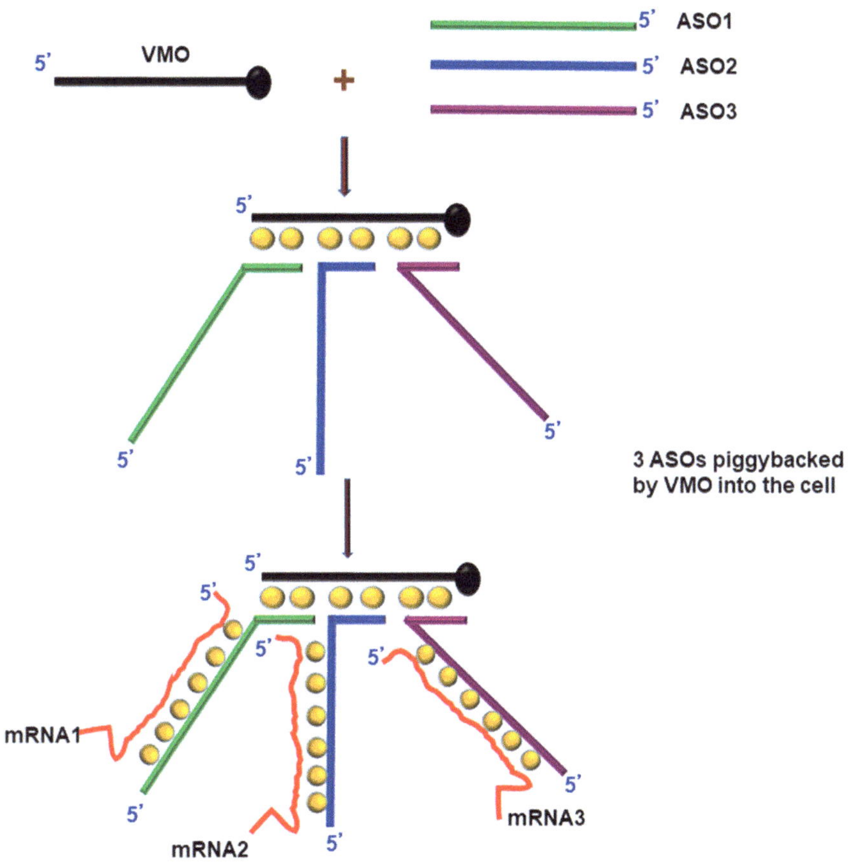

Fig. 3 Principle of piggyback knockdown method. Simultaneous knockdown of three genes. Note the hybrid formation between three antisense deoxyoligonucleotides, ASO1, ASO2, and ASO3, and one VMO at three locations. Also note the hybrid formation between three antisense deoxyoligonucleotides, ASO1, ASO2, and ASO3, and three mRNAs, mRNA1, mRNA2, and mRNA3. Yellow circles indicate base pairing, and the black circle represents octa-guanidine dendrimer. The top arrow leads to the formation of the hybrid between VMO and three ASOs in vitro. The bottom arrow points to the formation of the hybrid between three mRNAs and VMO/three ASOs (hybrid) in vivo after IV injections. RNase H degrades the mRNAs of the hybrids

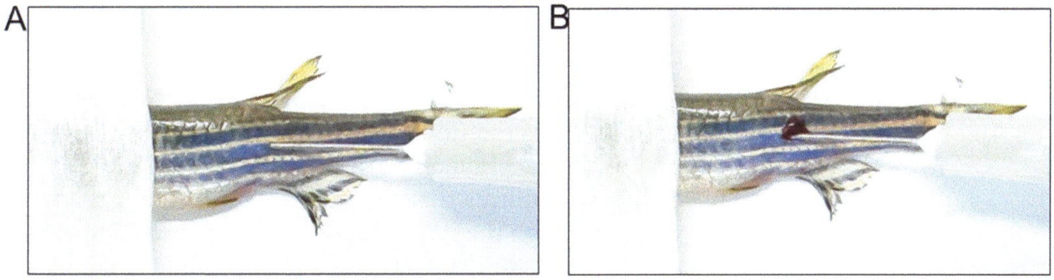

Fig. 4 IV injection of adult zebrafish. Panels A and B show the position of the needle before injection and a tiny amount of blood oozing out after injection, respectively. Note oozing blood is an indication of a successful injection

3.9 Secondary Screening

1. Once the entire genes are knocked down in sets of three, and the sets which yield thrombocyte numbers departing from the control fish are identified, individual genes from these positive sets are knocked down again to confirm which gene in these sets is truly functional in the thrombocyte development.
2. For individual gene knockdowns, three individual ASOs for each of the three genes from the set are designed. Here, the ASOs have the sequence 5′-TATAAATTGTAACTG-3′ added at its 3′-end, and three hybrids are prepared as described above and injected again into three sets of fish.
3. Thrombocytes are quantified after 48 h, as described above to select genes that participate in thrombocyte development (*see* **Note 4**).

In conclusion, the above procedure is useful not only for selecting thrombocyte-specific genes by scRNA-Seq and evaluating their function by large-scale knockdowns but could also be adapted to study expressed genes from other cell types followed by knockdowns if appropriate functional assays are available. For example, blood glucose levels could be measured if type II diabetes-related genes have to be evaluated by studying hepatocytes to understand their contribution.

4 Notes

1. Blood collection is the crucial part and requires a lot of practice. Since fish blood clots more rapidly than human blood, the collection should be fast. After collecting into the pipette tip, the blood should go to the citrate tube in less than a few seconds. Likewise, the pipette tip must be ready to collect the welling blood quickly.
2. Ideally, the FACS and the 10× Genomic machine should be in the same room. Unfortunately, this does not happen in the Core Facilities. Again, the speed is vital to take the sorted cells quickly to the 10× Genomic machine, and the sample must be kept on ice. Otherwise, cell death occurs, which is not useful for the sample loading.
3. IV injection is the tricky part of the method. Usually, the IV is successful 80% of the time. This is because the fish vessels are not visible, unlike human vessels. A tiny blood drop at the injection site confirms that the needle successfully penetrated the vein. Although the stripes on the fish body help, sometimes one may miss them because of the polymorphic variations in fish anatomy. That is why >6 fish must be injected for knockdown studies.

4. One cannot use this 48-h knockdown approach for long half-life proteins. This is an inherent defect in this method and all knockdown methods. However, it could be overcome by repeated injections every 48 h at least twice.

References

1. Macaulay IC, Svensson V, Labalette C et al (2016) Single-cell RNA-sequencing reveals a continuous spectrum of differentiation in hematopoietic cells. Cell Rep 14(4):966–977
2. Moore FE, Garcia EG, Lobbardi R et al (2016) Single-cell transcriptional analysis of normal, aberrant, and malignant hematopoiesis in zebrafish. J Exp Med 213(6):979–992
3. Carmona SJ, Teichmann SA, Ferreira L et al (2017) Single-cell transcriptome analysis of fish immune cells provides insight into the evolution of vertebrate immune cell types. Genome Res 27(3):451–461
4. Tang Q, Iyer S, Lobbardi R et al (2017) Dissecting hematopoietic and renal cell heterogeneity in adult zebrafish at single-cell resolution using RNA sequencing. J Exp Med 214(10): 2875–2887
5. Thattaliyath B, Cykowski M, Jagadeeswaran P (2005) Young thrombocytes initiate the formation of arterial thrombi in zebrafish. Blood 106(1):118–124
6. Fallatah W, De Silva IW, Verbeck GF et al (2019) Generation of transgenic zebrafish with 2 populations of RFP-and GFP-labeled thrombocytes: analysis of their lipids. Blood Adv 3(9): 1406–1415
7. Fallatah W, De R, Burks D et al (2022) Analysis of transcribed sequences from young and mature zebrafish thrombocytes. PLoS One 17(3): e0264776
8. Gregory M, Jagadeeswaran P (2002) Selective labeling of zebrafish thrombocytes: quantitation of thrombocyte function and detection during development. Blood Cell Mol Dis 28(3): 418–427
9. Kulkarni V, Kim S, Zafreen L et al (2012) Separation of young and mature thrombocytes by a novel immuno-selection method. Blood Cell Mol Dis 48(3):183–187

Chapter 11

Plant Transcriptome Analysis with HISAT–StringTie–Ballgown and TopHat–Cufflinks Pipelines

Xiaolan Rao

Abstract

The development of next-generation sequencing technology has led to a burst of data in a single assay. Management of a large dataset requires high demands on bioinformatic skills and computing resources. Here we present two popular pipelines for RNA-seq data analysis, using open-source software tools HISAT–StringTie–Ballgown and TopHat–Cufflinks. To meet the need of plant scientist, we describe in detail how to perform such comprehensive analysis beginning with raw RNA-seq reads and available reference genome. It allows biologists to align short reads to a reference genome, measure the transcript abundance, and analyze gene differential expression under two or more conditions. We also discuss other RNA-seq tools that are comparable or alternative to this protocol.

Key words Transcriptome, RNA-seq, Gene expression, Differential gene expression, HISAT, String-Tie, Ballgown, TopHat, Cufflinks

1 Introduction

A transcriptome refers to the total collection of message RNAs, noncoding RNAs, and small RNAs expressed by an organism. In the past decade, RNA sequencing (RNA-seq) has become a revolutionary tool to quantify the transcriptome [1–3]. RNA-seq offers accurate measurement of gene expression and detection of novel transcripts at high resolution. RNA-seq approach has been widely applied to determine the structure and dynamics of plant transcriptome [2, 4]. Accordingly, the increased RNA-seq applications have led to a fast-growing demand in bioinformatic skills on interpreting large amounts of data.

A typical protocol of RNA-seq analysis starts with trimming, alignment, quantification, and differential expression analysis [5, 6]. Millions of raw reads from 50 to 150 bp are generated using a high-throughput platform (e.g., Illumina HiSeq). In the initial step, the adapter sequences and poor-quality reads are

Rajeev K. Azad (ed.), *Transcriptome Data Analysis*, Methods in Molecular Biology, vol. 2812,
https://doi.org/10.1007/978-1-0716-3886-6_11,

removed. Then, the clean reads are mapped to a reference genome or transcriptome. The number of reads assigned to each gene or transcript is counted after normalization process. Finally, differentially expressed genes (DEGs) are accurately detected between two and more conditions. Although a plenty of approaches have been developed to analyze and interpret RNA-seq data [5–8], manipulation of RNA-seq tools remains a challenge to a bench scientist.

In this methods paper, we present a user-friendly guideline on plant RNA-seq analysis for a bioinformatic beginner. We perform two alternative methodological pipelines using a published *Arabidopsis* RNA-seq dataset. The two pipelines, HISAT–StringTie–Ballgown [7, 9, 10] and TopHat–Cufflinks [6, 11–13], consisted of three major steps in RNA-seq analysis including the reads mapping to the reference genome, read counts for each gene, and DEGs identification (Fig. 1). Additionally, we discuss other RNA-seq analysis software that are alternative or comparable to these tools for user design of customized pipelines. This paper will serve as a tutorial for plant scientists, who are in need of analyzing RNA-seq data on plant species with available reference genome.

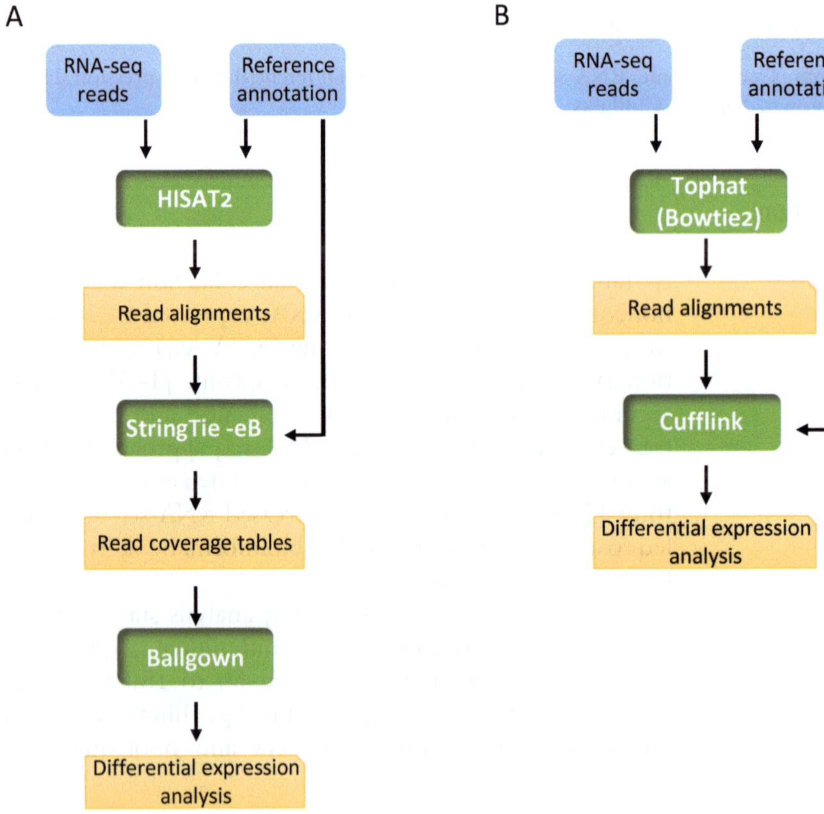

Fig. 1 An overview of the protocols. (**a**) HISAT–StringTie–Ballgown pipeline. (**b**) TopHat–Cufflink pipeline

2 Materials

2.1 RNA-seq Data

For this tutorial, we have selected the RNA-seq raw data from previous publication on *Arabidopsis thaliana* [14]. The study elucidated the coordinate defense between the receptor-like kinase FERONIA and multiple wall-associated kinases induced by lignin modification in plant cell walls. RNA-seq analysis was used to quantitate the changes in gene expression caused by loss of function of FERONIA. The raw data is deposited on the NCBI Sequence Read Archive (SRA) under the accession number GSE210734. We shall use wild-type replicates (GSM6436064 and GSM6436065) and *fer* mutant replicates (GSM6436070 and GSM6436071) as example data for the purpose of this demonstration. Subheading 3.1 will guide users to archive high-throughput sequencing reads from the SRA repository. Users can follow the similar steps with their own data.

2.2 Reference Genome

The *Arabidopsis thaliana* genome sequence and the reference-annotated genes (TAIR10) are used in this tutorial. The *A. thaliana* TAIR 10 package can be downloaded from Ensembl or Phytozome website.

2.3 Software and Tools

HISAT, StringTie, TopHat, and Cufflinks are run under Linux environment. Ballgown is run in an R interactive shell. The software used require a 64-bit version of the operating system, with at least 4 GB of RAM. In this tutorial, the example RNA-seq dataset is run on high-performance computing (HPC) clusters with 16 GB of RAM.

SRA Toolkit

SRA Toolkit can be downloaded from the website https://www.ncbi.nlm.nih.gov/home/tools/. This tutorial utilizes SRA Toolkit version 3.0.5.

SAMtools

SAMtools can be downloaded from the website http://www.htslib.org/. This tutorial utilizes SAMtools version 1.11.

FastQC

FastQC can be downloaded from the website https://www.bioinformatics.babraham.ac.uk/projects/fastqc/. This tutorial utilizes FastQC version 0.12.1.

HISAT

HISAT can be downloaded from the website http://ccb.jhu.edu/software.shtml. This tutorial utilizes HISAT2 version 2.2.1.

StringTie

StringTie can be downloaded from the website http://ccb.jhu.edu/software.shtml. This tutorial utilizes StringTie version 2.1.5.

Ballgown

Ballgown can be downloaded from the website http://ccb.jhu.edu/software.shtml. This tutorial utilizes Ballgown. In this demonstration, we use Bioconductor version 3.17 and R version 4.3.1 to install Ballgown version 2.32.0.

```
if (!require("BiocManager", quietly = TRUE))
    install.packages("BiocManager")
BiocManager::install()
BiocManager::install(c("ballgown", "genefilter", "dplyr",
"devtools"))
```

Bowtie

Bowtie can be downloaded from the website http://ccb.jhu.edu/software.shtml. This tutorial utilizes Bowtie2 version 2.3.4.3.

TopHat

TopHat can be downloaded from the website http://ccb.jhu.edu/software.shtml. This tutorial utilizes TopHat version 2.1.1.

Cufflinks

Cufflinks can be downloaded from the website http://ccb.jhu.edu/software.shtml. This tutorial utilizes Cufflinks version 2.2.1.

3 Methods

3.1 Datasets Acquisition

To simplify this demonstration, we have selected the first run for four samples (wild-type and *fer* mutant, two replicates). The datasets can be downloaded using SRA Toolkit. The command fastq-dump is applied to download the files with the corresponding NCBI accession number.

```
fasterq-dump SRR20966560
fasterq-dump SRR20966556
fasterq-dump SRR20966538
fasterq-dump SRR20966534
```

Each command will download a set of two files with suffixes _1.fq and _2.fq, which are forward and reverse reads in paired-end sequencing.

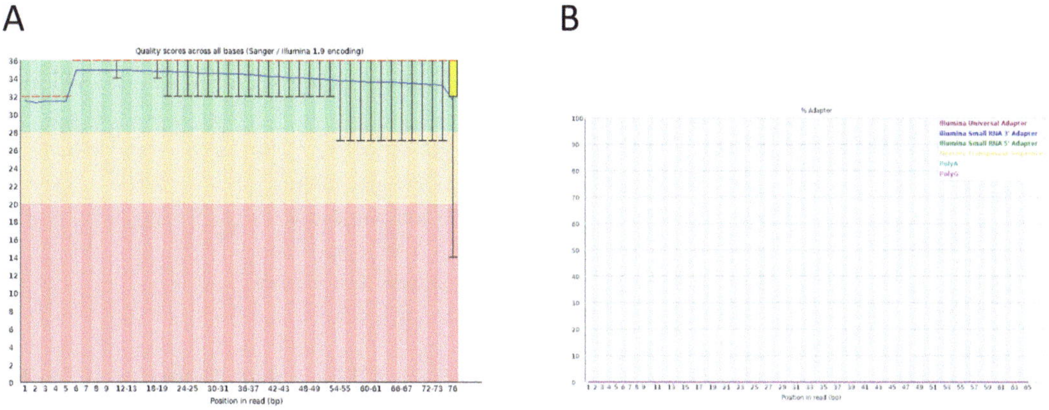

Fig. 2 Quality control results for RNA-seq data using SRR20966534_1.fq as an example. (**a**) per base quality content. (**b**) adapter content

3.2 Quality Control

After achieving the files, we use FastQC to check the quality of RNA-seq data. It provides an HTML report on a modular set of analyses. The example dataset shows a high quality (Fig. 2). If raw data is of poor quality, adapter trimming and quality filtering are needed. The tools for trimming reads are discussed in **Note 1**.

3.3 Pipeline One: HISAT–StringTie–Ballgown

3.3.1 Index the Reference Genome

Before mapping the raw RNA-seq reads to a reference genome, the genome needs to be indexed. The reference is built using hisat2-build function. Users shall follow the same command to build their own genomic indexes.

```
hisat2-build Arab_genome.fa Arab_genome
```

A set of new files is generated with the user-specified suffixes. Here we generate a set of eight files together as the index for *Arabidopsis* genome, including Arab_genome.1.ht2, Arab_genome.2.ht2, Arab_genome.3.ht2, Arab_genome.4.ht2, Arab_genome.5.ht2, Arab_genome.6.ht2, Arab_genome.7.ht2, and Arab_genome.8.ht2.

3.3.2 Alignment of the Reads to the Reference Genome Using HISAT

We use HISAT2 to align RNA-seq reads to the built index. We repeat this step on sample wt_1, wt_2, fer_1, and fer_2 with the corresponding output name.

```
hisat2 -x Arab_genome -1 SRR20966560_1.fastq -2 SRR20966560_2.
fastq -S wt_1.sam
hisat2 -x Arab_genome -1 SRR20966556_1.fastq -2 SRR20966556_2.
fastq -S wt_2.sam
hisat2 -x Arab_genome -1 SRR20966538_1.fastq -2 SRR20966538_2.
fastq -S fer_1.sam
hisat2 -x Arab_genome -1 SRR20966534_1.fastq -2 SRR20966534_2.
fastq -S fer_2.sam
```

These SAM files need to be converted into sorted BAM files using SAMtools. StringTie requires BAM file, which is encoded in binary file of SAM file. The differences between SAM and BAM files are shown in **Note 2**.

```
samtools sort -@ 8 -o wt_1.bam wt_1.sam
samtools sort -@ 8 -o wt_2.bam wt_2.sam
samtools sort -@ 8 -o fer_1.bam fer_1.sam
samtools sort -@ 8 -o fer_2.bam fer_2.sam
```

3.3.3 Transcript Quantification Using StringTie

The downloaded TAIR10 file is GFF format. It first needs to be converted into GTF format using gffread function. The definition of GFF and GTF format is shown in **Note 3**.

```
gffread Arab_genome.gene.gff -o Arab_genome.gene.gtf
```

We then use StringTie to compare each sample against the known transcripts. The BAM file and GTF file are taken as input. The transcript abundance is estimated in a count table, which fits requirement of Ballgown.

```
stringtie -p 8 -e -B -G Arab_genome.gene.gtf -o /ballgown/
Athaliana_wt_1/Athaliana_wt_1.gtf wt_1.bam
stringtie -p 8 -e -B -G Arab_genome.gene.gtf -o /ballgown/
Athaliana_wt_2/Athaliana_wt_2.gtf wt_2.bam
stringtie -p 8 -e -B -G Arab_genome.gene.gtf -o /ballgown/
Athaliana_fer_1/Athaliana_fer_1.gtf fer_1.bam
stringtie -p 8 -e -B -G Arab_genome.gene.gtf -o /ballgown/
Athaliana_fer_2/Athaliana_fer_2.gtf fer_2.bam
```

In this demonstration, we only consider the known transcripts (*see* **Note 4**). StringTie provides options to reconstruct the isoforms expressed from each gene. It allows users to assemble the novel transcripts from raw reads mapped to the reference genome.

3.3.4 Differential Expression Analysis Using Ballgown

Make sure that all essential packages have been successfully installed. Load the package in R console as below.

```
library(ballgown)
library(genefilter)
library(dplyr)
library(devtools)
```

Ballgown provides statistical tests for differential expression between groups. Before running Ballgown, we need to create a csv file containing phenotype information. In the tutorial, our samples are divided into two groups, wt and *fer*. Each group has two biological replicates. Users can create a similar csv file based on their experimental design.

ids	lines
Athaliana_fer_1	fer
Athaliana_fer_2	fer
Athaliana_wt_1	wt
Athaliana_wt_2	wt

Then, we use the following commands to load phenotype file and results generated from Ballgown.

```
pheno_data <- read.csv("phenodata.csv")
bg <- ballgown(dataDir = "ballgown", pData=pheno_data, sam-
plePattern = "Athaliana")
```

To filter genes with low variance across samples, we use subset function to extract genes with a variance above 1.

```
bg_filt = subset(bg, "rowVars(texpr(bg))>1" ,genomesubset=-
TRUE)
```

We use the stattest function to identify significantly differentially expressed genes between groups. Here the gene abundances are measured as FPKM values (*see* **Note 5**).

```
results_genes = stattest(bg_filt, feature="gene" , covariate
= "lines" , getFC = TRUE, meas = "FPKM")
```

Finally, genes with p-value can be written into a new csv file.

```
results_genes = arrange(results_genes,pval)
write.csv(results_genes, "gene_results.csv", row.names=FALSE)
```

3.4 Pipeline Two: TopHat–Cufflinks

3.4.1 Index the Reference Genome

First, we need to build reference genome with bowtie2-build function (*see* **Note 6**). For aligning reads to the known transcripts, we also need to create transcriptome sequences files using tophat -G/--GTF option (*see* **Note 4**).

```
bowtie2-build Arab_genome.fa Arab_genome
tophat -G Arab_genome.gene.gff --transcriptome-index=tran-
scriptome_known
```

3.4.2 Alignment of the Reads to the Reference Genome Using TopHat

Once the index is built, we use TopHat to align the raw reads to the reference genome and transcriptome.

```
tophat -p 8 -r 80 -o wt_1 --no-novel-juncs --transcriptome-
index= transcriptome_known Arab_genome SRR20966560_1.fastq
SRR20966560_2.fastq
tophat -p 8 -r 80 -o wt_2 --no-novel-juncs --transcriptome-
index= transcriptome_known Arab_genome SRR20966556_1.fastq
SRR20966556_2.fastq
tophat -p 8 -r 80 -o fer_1 --no-novel-juncs --transcriptome-
index= transcriptome_known Arab_genome SRR20966538_1.fastq
SRR20966538_2.fastq
tophat -p 8 -r 80 -o fer_2 --no-novel-juncs --transcriptome-
index= transcriptome_known Arab_genome SRR20966534_1.fastq
SRR20966534_2.fastq
```

3.4.3 Differential Expression Analysis Using Cufflinks

We directly run Cuffdiff to check differential expression between wild-type and *fer* mutant. The command will generate an output directory named as "diff_result," which contains a group of files and databases.

```
cp wt_1/accepted_hits.bam wt_1.bam
cp wt_2/accepted_hits.bam wt_2.bam
cp fer_1/accepted_hits.bam fer_1.bam
cp fer_2/accepted_hits.bam fer_2.bam
cuffdiff -p 8 -o cuffdiff -N Arab_genome.gene.gff wt_1.bam,
wt_2.bam fer_1.bam,fer_2.bam
```

Gene expression profiling can be found in genes.fpkm_tracking with FPKM counts across samples (*see* **Note 5**). The differential expression tests on genes between groups can be found in gene_exp.diff.

3.5 Results

After the process of HISAT–StringTie–Ballgown, 3075 genes show their p-value less than 0.05 between wild-type and *fer* mutant. Comparatively, 4142 genes are marked to be significantly differently expressed using TopHat–Cufflinks pipeline (*see* **Note 7**).

4 Notes

1. *Trimming tools*

   To eliminate adapter sequences and improve read quality, the trimming procedure can be carried out using FASTX-Toolkit (http://hannonlab.cshl.edu/fastx_toolkit/), Trimmomatic [15], Cutadapt [16], or sickle (https://github.com/

najoshi/sickle). The reads that pass the quality control will be used for downstream analysis.

2. *SAM and BAM formats*

   SAM file is a basic format containing the alignment information for reads. The general definition of SAM format can be found in the official SAM documentation (https://samtools.github.io/hts-specs/SAMv1.pdf). BAM file is the same format as SAM file except the binary encoding. The size of a BAM file is significantly smaller than that of the corresponding SAM file.

3. *GTF and GFF formats*

   GFF (General Feature Format) is a popular file format on various genomic features. Currently, genomic features for many organisms are stored in GFF3 format. The definition of GFF3 and the fields in each feature line can be found in the website (http://gmod.org/wiki/GFF3). GTF (Gene Transfer Format) is identical to GFF version 2 (http://gmod.org/wiki/GFF2).

4. *No novel isoform assembly*

   StringTie can assemble the input read alignments to produce novel isoforms that are absent in the given reference annotation. Using the -e option, StringTie will only estimate the expression levels of the given set of reference transcripts. Similarly, TopHat is designed to discover novel isoforms across with some junctions that are not specified in the given GFF/GTF file. Using the parameter "--no-novel-juncs," TopHat only detects reads across junctions indicated in the supplied GFF file.

5. *FPKM, RPKM, and TPM*

   Gene abundances in RNA-seq can be represented in TPM (transcripts per kilobase million), FPKM (fragments per kilobase million), or RPKM (reads per kilobase million). The differences among three terms are clearly described in previous publication [17–20]. Notably, Ballgown provides optional of TPM and FPKM, while Cufflinks provides FPKM.

6. *Bowtie 2 and Bowtie 1*

   Bowtie 2 is designed to support reads longer than 50 bp, while Bowtie 1 is more sensitive to shorter reads (less than 50 bp). The example RNA-seq in the tutorial is 75 bp paired-end reads. Thus, here we recommend Bowtie 2.

7. *Comparative analysis of different pipelines*

   In the tutorial, we detected 3075 and 4142 genes with FDR < 0.05 between wild-type and *fer* mutant using HISAT–StringTie–Ballgown and TopHat–Cufflinks, respectively. Varied results are due to different algorithms. It's consistent with previous studies that Cuffdiff generally detected the distinct number of gene expression changes [21, 22].

Here we have described two complete RNA-seq workflows, HISAT–StringTie–Ballgown and TopHat–Cufflinks. But there are a substantial number of alternative tools available at each step of RNA-seq analysis. For example, STAR is an ultrafast aligner in the genome-based alignment [23]; RSEM [24] and HTSeq [25] can be used to count the number of alignments mapped to each gene; DESeq2 [26] and edgeR [27] are efficient tools to obtain differentially expressed genes. Additionally, Salmon [28], Kallisto [29], or Sailfish [30] enables to encompass alignment, counting, and normalization in a single tool. Previous comparative studies have compared and evaluated the performance of different methods [18, 19, 21, 22, 31, 32]. Users can select an optimal workflow for RNA-seq data analysis based on their suitability.

Acknowledgments

This work was supported by the Youth Grant of Hubei University to X.R.

References

1. Wang Z, Gerstein M, Snyder M (2009) RNA-Seq: a revolutionary tool for transcriptomics. Nat Rev Genet 10(1):57–63. https://doi.org/10.1038/nrg2484
2. Lister R, O'Malley RC, Tonti-Filippini J, Gregory BD, Berry CC, Millar AH, Ecker JR (2008) Highly integrated single-base resolution maps of the epigenome in Arabidopsis. Cell 133(3):523–536. https://doi.org/10.1016/j.cell.2008.03.029
3. Stark R, Grzelak M, Hadfield J (2019) RNA sequencing: the teenage years. Nat Rev Genet 20(11):631–656. https://doi.org/10.1038/s41576-019-0150-2
4. Martin LB, Fei Z, Giovannoni JJ, Rose JK (2013) Catalyzing plant science research with RNA-seq. Front Plant Sci 4:66. https://doi.org/10.3389/fpls.2013.00066
5. Garber M, Grabherr MG, Guttman M, Trapnell C (2011) Computational methods for transcriptome annotation and quantification using RNA-seq. Nat Methods 8(6):469–477. https://doi.org/10.1038/nmeth.1613
6. Trapnell C, Roberts A, Goff L, Pertea G, Kim D, Kelley DR, Pimentel H, Salzberg SL, Rinn JL, Pachter L (2012) Differential gene and transcript expression analysis of RNA-seq experiments with TopHat and Cufflinks. Nat Protoc 7(3):562–578. https://doi.org/10.1038/nprot.2012.016
7. Pertea M, Kim D, Pertea GM, Leek JT, Salzberg SL (2016) Transcript-level expression analysis of RNA-seq experiments with HISAT, StringTie and Ballgown. Nat Protoc 11(9):1650–1667. https://doi.org/10.1038/nprot.2016.095
8. Williams J (2022) CyVerse for reproducible research: RNA-Seq analysis. Methods Mol Biol 2443:57–79. https://doi.org/10.1007/978-1-0716-2067-0_3
9. Kim D, Langmead B, Salzberg SL (2015) HISAT: a fast spliced aligner with low memory requirements. Nat Methods 12(4):357–360. https://doi.org/10.1038/nmeth.3317
10. Pertea M, Pertea GM, Antonescu CM, Chang TC, Mendell JT, Salzberg SL (2015) StringTie enables improved reconstruction of a transcriptome from RNA-seq reads. Nat Biotechnol 33(3):290–295. https://doi.org/10.1038/nbt.3122
11. Ghosh S, Chan CK (2016) Analysis of RNA-Seq data using TopHat and Cufflinks. Methods Mol Biol 1374:339–361. https://doi.org/10.1007/978-1-4939-3167-5_18
12. Kim D, Pertea G, Trapnell C, Pimentel H, Kelley R, Salzberg SL (2013) TopHat2: accurate alignment of transcriptomes in the presence of insertions, deletions and gene fusions. Genome Biol 14(4):R36. https://doi.org/10.1186/gb-2013-14-4-r36

13. Langmead B, Salzberg SL (2012) Fast gapped-read alignment with Bowtie 2. Nat Methods 9(4):357–359. https://doi.org/10.1038/nmeth.1923
14. Liu C, Yu H, Voxeur A, Rao X, Dixon RA (2023) FERONIA and wall-associated kinases coordinate defense induced by lignin modification in plant cell walls. Sci Adv 9(10):eadf7714. https://doi.org/10.1126/sciadv.adf7714
15. Bolger AM, Lohse M, Usadel B (2014) Trimmomatic: a flexible trimmer for Illumina sequence data. Bioinformatics 30(15): 2114–2120. https://doi.org/10.1093/bioinformatics/btu170
16. Martin M (2011) Cutadapt removes adapter sequences from high-throughput sequencing reads. EMBnetjournal 17:10–12
17. Mortazavi A, Williams BA, McCue K, Schaeffer L, Wold B (2008) Mapping and quantifying mammalian transcriptomes by RNA-Seq. Nat Methods 5(7):621–628. https://doi.org/10.1038/nmeth.1226
18. Dillies MA, Rau A, Aubert J, Hennequet-Antier C, Jeanmougin M, Servant N, Keime C, Marot G, Castel D, Estelle J, Guernec G, Jagla B, Jouneau L, Laloe D, Le Gall C, Schaeffer B, Le Crom S, Guedj M, Jaffrezic F, French StatOmique C (2013) A comprehensive evaluation of normalization methods for Illumina high-throughput RNA sequencing data analysis. Brief Bioinform 14(6):671–683. https://doi.org/10.1093/bib/bbs046
19. Li P, Piao Y, Shon HS, Ryu KH (2015) Comparing the normalization methods for the differential analysis of Illumina high-throughput RNA-Seq data. BMC Bioinform 16:347. https://doi.org/10.1186/s12859-015-0778-7
20. Evans C, Hardin J, Stoebel DM (2018) Selecting between-sample RNA-Seq normalization methods from the perspective of their assumptions. Brief Bioinform 19(5):776–792. https://doi.org/10.1093/bib/bbx008
21. Seyednasrollah F, Laiho A, Elo LL (2015) Comparison of software packages for detecting differential expression in RNA-seq studies. Brief Bioinform 16(1):59–70. https://doi.org/10.1093/bib/bbt086
22. Corchete LA, Rojas EA, Alonso-Lopez D, De Las Rivas J, Gutierrez NC, Burguillo FJ (2020) Systematic comparison and assessment of RNA-seq procedures for gene expression quantitative analysis. Sci Rep-Uk 10(1). https://doi.org/10.1038/s41598-020-76881-x
23. Dobin A, Davis CA, Schlesinger F, Drenkow J, Zaleski C, Jha S, Batut P, Chaisson M, Gingeras TR (2013) STAR: ultrafast universal RNA-seq aligner. Bioinformatics 29(1):15–21. https://doi.org/10.1093/bioinformatics/bts635
24. Li B, Dewey CN (2011) RSEM: accurate transcript quantification from RNA-Seq data with or without a reference genome. BMC Bioinform 12:323. https://doi.org/10.1186/1471-2105-12-323
25. Anders S, Pyl PT, Huber W (2015) HTSeq—a Python framework to work with high-throughput sequencing data. Bioinformatics 31(2):166–169. https://doi.org/10.1093/bioinformatics/btu638
26. Love MI, Huber W, Anders S (2014) Moderated estimation of fold change and dispersion for RNA-seq data with DESeq2. Genome Biol 15(12):550. https://doi.org/10.1186/s13059-014-0550-8
27. Robinson MD, McCarthy DJ, Smyth GK (2010) edgeR: a Bioconductor package for differential expression analysis of digital gene expression data. Bioinformatics 26(1): 139–140. https://doi.org/10.1093/bioinformatics/btp616
28. Patro R, Duggal G, Love MI, Irizarry RA, Kingsford C (2017) Salmon provides fast and bias-aware quantification of transcript expression. Nat Methods 14(4):417–419. https://doi.org/10.1038/nmeth.4197
29. Bray NL, Pimentel H, Melsted P, Pachter L (2016) Near-optimal probabilistic RNA-seq quantification. Nat Biotechnol 34(5): 525–527. https://doi.org/10.1038/nbt.3519
30. Patro R, Mount SM, Kingsford C (2014) Sailfish enables alignment-free isoform quantification from RNA-seq reads using lightweight algorithms. Nat Biotechnol 32(5):462–464. https://doi.org/10.1038/nbt.2862
31. Lin Y, Golovnina K, Chen ZX, Lee HN, Negron YL, Sultana H, Oliver B, Harbison ST (2016) Comparison of normalization and differential expression analyses using RNA-Seq data from 726 individual Drosophila melanogaster. BMC Genomics 17:28. https://doi.org/10.1186/s12864-015-2353-z
32. Gao D, Kim J, Kim H, Phang TL, Selby H, Tan AC, Tong T (2010) A survey of statistical software for analysing RNA-seq data. Hum Genomics 5(1):56–60. https://doi.org/10.1186/1479-7364-5-1-56

Chapter 12

Cotton Meristem Transcriptomes: Constructing an RNA-Seq Pipeline to Explore Crop Architecture Regulation

Róisín C. McGarry and Brian G. Ayre

Abstract

Plants stem cells, known as meristems, specify all patterns of growth and organ size. Differences in meristem activities contribute to diverse shoot architectures. As many architectural traits, such as branching patterns, flowering time, and fruit size, are yield determinants, meristem regulation is of fundamental importance to crop productivity. Cotton (*Gossypium* spp.) produces our most prevalent natural fiber that finds its way into products ranging from industrial cellulose, medical supplies, and paper currency, to a broad diversity of textiles, not least of which is our clothing. However, the cotton plant has growth habits that challenge management practices and limit harvest yield and quality. Unraveling and leveraging the genetic networks regulating meristem activities offers the potential to overcome these limitations. We use virus-based technologies in cotton to perturb signals regulating meristem fate and size. In this chapter, we describe our pipeline for altering cotton meristem dynamics and preparing, analyzing, and exploring the transcriptomes from isolated meristems.

Key words Cotton, Development, Meristem, RNA-Seq, Stem cells, Transcriptome, Virus, WGCNA

1 Introduction

Plant meristems are small pools of self-replenishing stem cells responsible for generating the whole plant body. To achieve this, meristems balance the proliferation of stem cells with the differentiation of emerging organs [1, 2]. Meristem homeostasis is exquisitely regulated—many mutations disrupting this balance increase meristem size and, consequently, the organs derived from these meristems are also larger [3–7]. This trait is called fasciation. In addition to this maintenance cycle, meristems perceive different signals, some of which initiate transient genetic programs that transform an indeterminate or vegetative meristem to a determinate, floral meristem to produce a flower [8]. Manipulating signals specifying meristem fate can alter flowering time and the placement of reproductive organs, potentially increasing yields. Thus, understanding these regulatory cascades and leveraging these genetic

Rajeev K. Azad (ed.), *Transcriptome Data Analysis*, Methods in Molecular Biology, vol. 2812,
https://doi.org/10.1007/978-1-0716-3886-6_12,

networks hold the promise to modify plant architecture and enhance crop productivity.

Cotton (*Gossypium* spp.) is our most important fiber crop and is also a valuable source of seed oil and protein for animal feed. Cultivated cotton has complex growth habits and yields bolls of variable sizes [9, 10]. Understanding and altering meristematic signals to increase and accelerate flowering, reduce vegetative branching, and generate larger cotton bolls could benefit crop yields and management practices. But, as cotton lacks extensive genetic tools, how can meristematic networks be efficiently probed and adjusted?

Viruses are powerful tools for plant biotechnology as they accelerate gene functional analysis and pose a modest, if any, risk of transmission to the next generation [11]. We used viruses, including the geminivirus *Cotton leaf crumple virus* (CLCrV) and tobravirus *Tobacco rattle virus* (TRV), to alter signals regulating meristem fate and size and dramatically impact cotton architecture [10, 12, 13]. To examine these changes quantitatively, we generated and analyzed the transcriptomes of meristems isolated from virus-infected cotton.

Probing meristem transcriptomes from virus-infected cotton merits further discussion. Virus-based approaches are transient and not every plant or cell is infected with the virus or virus-derived signal, meaning that the virus infection is not synchronous or homogenous in replicate plants. Thus, in contrast to stable transgenic lines wherein meristems from many genetically identical plants can be pooled, meristems from virus-infected plants need to be recognized as inherently variable, and, as such, we analyzed these as independent events [10]. Furthermore, isolating meristems from any plant is challenging. Within 0.5 cm of cotton apical tissues are multiple meristems, tightly compacted, enclosed by bracts, and surrounded by highly hirsute tissues (Fig. 1). Isolating these, and distinguishing among types of meristems, is difficult. Importantly, changes to meristematic networks precede phenotypes. For example, we analyzed target gene expression by RT-qPCR from meristems isolated from virus-infected plants at 7, 15, and 21 days post-inoculation (dpi). We detected differences in the levels of *SINGLE FLOWER TRUSS* (*SFT*) and *SELF-PRUNING* (*SP*) transcripts among virus-treated and control plants by 15 dpi, whereas the earliest visual phenotypes among the virus-treated plants presented at 27 dpi. With these considerations, we present our pipeline for investigating meristem transcriptomes following virus-based genetic manipulation in cotton (Fig. 1) [10]. We focus our discussion to dissecting meristems from virus-infected cotton seedlings, preparing RNA-Seq libraries from small quantities of isolated tissues, and analyzing the reads. We point out the cotton resources available to facilitate transcriptome analyses and visualize genetic networks.

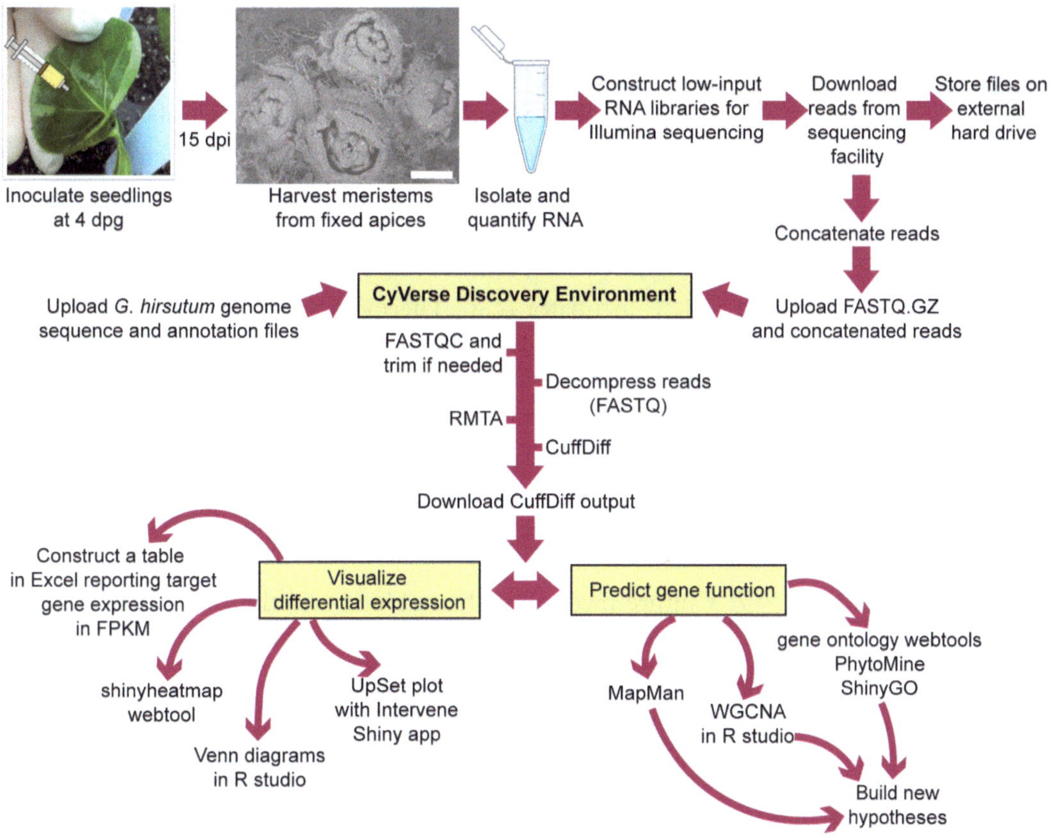

Fig. 1 Flowchart depicting the preparation and analyses of meristem transcriptomes from virus-infected cotton. Shown is the series of steps used, starting with *Agro*-inoculation (top left) and dissecting meristems from infected plants. Reads are processed largely in the CyVerse Discovery Environment. The output from the differential expression analysis is visualized using webtools, R Studio, and Excel worksheets. Lists of differentially expressed genes (DEGs) are analyzed with gene ontology and enrichment webtools and/or co-expression analysis (WGCNA). Scale bar is 500 μm

2 Materials

2.1 Plant Material and Growth Conditions

1. 3.5-inch square pots.
2. Flat plant trays (21.25 inches long × 11 inches wide).
3. Humidity domes, 7.5-inches tall, vented.
4. Coarse-cut growth mix, such as MetroMix 852 (Sungro Horticulture, Agawam, MA, USA) or similar.
5. Growth chamber at 25 °C and 16 h light/8 h dark photoperiod.
6. De-linted cotton seeds.

2.2 Agro-infiltration of Viral Vectors into Cotton Seedlings

1. Binary vectors harboring viruses: CLCrV-A and CLCrV-B genomes are cloned in pJRTAgroCLCrVA.008 and pJRTAgroCLCrVB13 [14], respectively; TRV RNA1 and RNA2 are cloned in pYL156 and pYL279 [15], respectively.
2. *Agrobacterium tumefaciens* GV3101 pMP90 electrocompetent cells.
3. Electroporator (e.g., Bio-Rad Gene Pulser Xcell Electroporation System).
4. 28 °C incubator (e.g., Thermo Scientific Heratherm™ Compact Microbiological Incubator).
5. 28 °C shaking (250 rpm) incubator (e.g., Barnstead/Lab-line MaxQ 4000 Orbital Incubator Shaker).
6. Filter-sterilized stock solutions: 1 M magnesium chloride (stored at 4 °C), 1 M MES [2-(N-morpholino)ethanesulfonic acid] (stored at 4 °C), 100 mM acetosyringone (stored at −20 °C in 1 mL aliquots).
7. Culture medium with antibiotic selection: LB broth and agar plates with kanamycin (30 μg/mL), rifampicin (50 μg/mL), and gentamycin (50 μg/mL).
8. Larger, overnight induction culture medium with antibiotic selection: LB broth with kanamycin (30 μg/mL), rifampicin (50 μg/mL), gentamycin (50 μg/mL), 10 mM MES, 20 μM acetosyringone.
9. Disposable 50 mL Falcon tubes.
10. Centrifuge (5000× *g*) with fixed angle rotor for 50 mL Falcon tubes (e.g., Sovall LYNX 4000).
11. Infiltration medium [15]: 10 mM MES, 10 mM magnesium chloride, 200 μM acetosyringone.
12. 1 mL disposable syringes.

2.3 Meristem Isolation

1. Glass plate.
2. Single-edged razor blades.
3. 2.0 mL screw-cap tubes.
4. Acetone.
5. Glass Pasteur pipettes.
6. Appropriate flammable 4 °C storage.
7. Stereomicroscope.
8. Fine-tipped (0.06 mm tip thickness) tweezers, such as Rubis #5 or Dumont #5 (Electron Microscopy Services, Hatfield, PA, USA).
9. Petri dishes, filter paper, wooden sticks.

2.4 RNA Isolation

1. BBs or other steel balls (2.0–4.5 mm).
2. Liquid nitrogen.
3. Retsch mill MM400.
4. Hot borate extraction buffer [16]: 0.2 M borax, 0.03 M EGTA, 1% (w/v) sodium dodecyl sulfate, 1% (w/v) sodium deoxycholic acid, 0.010 M DTT (added to solution before use), 1% (v/v) IGEPAL (added to solution before use), 2% (w/v) PVP-40 (added to solution before use).
5. Vortex mixer.
6. Microcentrifuge reaching 16,000× *g*.
7. Nuclease-free water.
8. Quick RNA Miniprep Plus columns (Zymo Research, Irvine, CA, USA) (*see* **Note 1**).
9. 1.5 mL Eppendorf tubes.

2.5 RNA Quantification

1. Nanodrop 2000c spectrophotometer (Thermo Fisher Scientific, Waltham, MA, USA), Synergy H4 hybrid reader (BioTek, Winooski, VT, USA), BioAnalyzer RNA 6000 Pico chip or High Sensitivity RNA ScreenTape Analysis (Agilent, Santa Clara, CA, USA).

2.6 Gene Expression Validation

1. Reverse transcription: reverse transcriptase such as Protoscript II (New England Biolabs, Ipswich, MA, USA), random hexamers and oligo $d(T)_{23}$ VN primers, dNTPs.
2. Gene-specific primers for qPCR (i.e., no more than 200 bp product) and primers for housekeeping genes, *GhUBQ10*, *GhGAPDH*, or *GhEF1α* (*see* **Note 2**).
3. Real-time thermocycler, such as ABI ViiA™7 (Thermo Fisher Scientific, USA).

2.7 RNA Library Construction, Quantification, and Sequencing

1. NEBNext® Single Cell/Low Input RNA Library Prep Kit for Illumina® (New England Biolabs, Ipswich, MA, USA).
2. NEBNext® Multiplex Oligos for Illumina® (Index Primers Set 1).
3. Agilent High Sensitivity D5000 ScreenTape (Agilent Technologies, Santa Clara, CA, USA).
4. Illumina sequencing platform (NextSeq 550) to fit your budget and generate ~30 million reads per sample.

2.8 Read Processing

1. Internet connection.
2. Free CyVerse account.
3. R Studio or R.
4. External hard drive or storage capabilities.

3 Methods

3.1 Plant Material and Growth Conditions

1. Gin and de-lint cotton seed [12]. Depending on the genetic background, some seed will require additional treatment to germinate [12]; Coker 312 seed can be sown directly.
2. Using coarse-cut growth mix, sow two cotton seeds per 3.5-inch pot. Fit pots into flat plant trays. Germinate seeds in a growth chamber set to 25 °C and 16 h day/8 h night photoperiod. Seeds will germinate by 4 days after sowing.
3. *Agro*-infiltration should be performed 3 or 4 days after the cotyledons first emerge from the potting mix. The cotyledons should be expanded but true leaves should be hardly visible. Infection efficiency drops quickly as the seedlings age.

3.2 Preparing *Agrobacterium* Cultures for Inoculation

1. Preparation for virus delivery via *Agro*-inoculation is well described [15], and included here for reference. Transform cloned viruses into *Agrobacterium tumefaciens* by electroporation and plate with appropriate selection (e.g., kanamycin at 30 μg/mL, rifampicin at 50 μg/mL, and gentamycin at 50 μg/mL). We use *A. tumefaciens* strain GV3101 pMP90. Incubate at 28 °C for ~48 h.
2. Use a single transformant to prepare a fresh streak plate, maintaining the same antibiotic selection, to obtain single colonies. Incubate at 28 °C for 48 h.
3. Set up 2 mL overnight starter cultures: Pipet 2 mL of sterile LB with appropriate antibiotics (e.g., kanamycin at 30 μg/mL, rifampicin at 50 μg/mL, and gentamycin at 50 μg/mL) into labeled 10 mL culture tubes. Inoculate each starter culture with a single colony from the freshly streaked plate of the *A. tumefaciens* strain harboring the viral vector. Culture at 28 °C overnight with shaking (250 rpm).
4. Inoculate larger, induction culture volumes. Use the 2 mL starter culture to seed, at a 1:100 dilution, a larger culture volume (10 mL, 50 mL, or larger volumes, *see* **Note 3**) of LB with antibiotics and supplemented with 10 mM MES and 20 μM acetosyringone. Incubate at 28 °C overnight with shaking (250 rpm).
5. Determine the OD_{600} of each culture, using a 1:5 dilution, with a spectrophotometer.
6. Transfer each culture to a sterile, labeled 50 mL Falcon tube. Pellet the cultures by centrifugation (Sorvall LYNX 4000) with a fixed angle rotor at 5000× *g* for 10 min at room temperature.

7. During centrifugation, calculate the volume of infiltration medium (10 mM $MgCl_2$, 10 mM MES, and 200 μm acetosyringone) required to resuspend each culture to an OD_{600} of 1.5. Prepare this volume of sterile infiltration medium.
8. When centrifugation is complete, decant the supernatant into suitable vessels and set aside for appropriate disposal of recombinant organisms. Resuspend the bacterial pellet with the sterile infiltration medium (i.e., the volume calculated in the preceding step). Leave cultures in the Falcon tubes, with caps loosened, in the dark at room temperature for 2–4 h for acetosyringone induction of the *Agrobacterium tumefaciens VIR* genes.

3.3 Virus Inoculation

1. CLCrV and TRV are bipartite viruses and both genomes are required to cause a systemic infection. After the cultures prepared in the last step incubate for 2–4 h, make the viral inoculum: for *Cotton leaf crumple virus*, pour equal volumes of the cultures harboring CLCrV-A with CLCrV-B in a 50 mL Falcon tube; for *Tobacco rattle virus*, pour equal volumes of the cultures harboring TRV RNA1 with TRV RNA2 in a 50 mL Falcon tube. Invert Falcon tubes to mix the inoculum. This inoculum is now ready to infiltrate into cotton cotyledons.
2. We infiltrate cotton cotyledons at 4 days post-germination, prior to the emergence of the first true leaf [13, 17]. Using a 1 mL disposable syringe, infiltrate both cotyledons from the abaxial (underside) surface, continuing until both cotyledons appear saturated. Repeat with replicate plants. It is critical in planning meristem transcriptome analyses to consider the number of plants to inoculate (*see* **Note 4**). We routinely inoculate six plants per treatment for standard gene functional analysis experiments [13]. However, we recommend inoculating a larger number of replicate plants per treatment to use for meristem isolation and phenotype validation. It is important to recognize that transcriptome changes precede phenotypic changes, and maintaining additional replicate plants for phenotype validation is a worthy investment before committing precious resources to expensive downstream efforts. For example, we routinely use six replicate plants for tissue isolation and maintain an additional six replicates in a growth chamber or greenhouse to analyze the resulting phenotype.
3. Following *Agro*-infiltration into cotton cotyledons, leave the infiltrated plants covered with a plastic humidity dome overnight at room temperature and out of direct light. After ~24 h, remove the dome and return the plants to the growth chamber.

3.4 Harvesting Apices and Dissecting Meristems

1. Harvest apices from seedlings and fix them in acetone. We target seedlings at 15 dpi based on a previous time-course analysis of target gene expression using RT-qPCR (*see* **Note 5**). Use a fresh, single-edged razor blade to remove the apex from each seedling. Then, working on a plate glass surface, remove leaves and as much stem tissue as possible. The final piece of tissue with the shoot apical meristem (and axillary meristems) is less than 1 cm in length. Place in a labeled 2 mL screw-cap tube containing 1 mL of 100% acetone. Repeat with the replicate samples, placing the apical tissue in separate, labeled screw-cap tubes. We use a new razor blade for each treatment, and repeat the process for all treatments.
2. After collecting all the samples, exchange the acetone in each screw-cap tube. Withdraw the acetone with a glass Pasteur pipette, and replace it with fresh acetone. Repeat for two changes of acetone before storing at 4 °C in an appropriate refrigerator (*see* **Note 6**). Ensure that the 2.0 mL screw-cap vial is now filled with acetone so that evaporative losses do not compromise tissue quality.
3. Working at a stereomicroscope, remove tissue from the acetone-filled 2.0 mL screw-cap tube. This is best achieved using wooden sticks, whittled on one end to a blunt, flat surface, so that the tissues can be removed from the screw-cap tube without being pierced or damaged. Blot the tissue on a piece of filter paper fitted within a petri dish. While visualizing with the aid of the stereomicroscope, use the tweezers to snap off unexpanded leaves and bracts to expose the meristems. Excise meristems by pinching each off with tweezers (Fig. 1; *see* **Note 7**) and return to the labeled screw-cap tube containing acetone. Return to 4 °C storage (*see* **Note 8**).

3.5 RNA Isolation from Meristems

1. Rinse BBs (4.5 mm steel balls) in acetone.
2. Withdraw acetone from each screw-cap tube using a Pasteur pipette and allow the samples in tubes to air-dry on the bench. Add two BBs to each 2 mL screw-cap tube and continue to air-dry (*see* **Note 9**).
3. Once dry, plunge the screw-cap tubes into liquid nitrogen for homogenization. We use a Retsch mill (MM400, Retsch USA, Newtown, PA, USA) with pre-chilled (−80 °C) adapters, and homogenize frozen tissues at 20 Hz for 15 s (*see* **Note 10**). Return the tubes to liquid nitrogen after homogenization.
4. Remove the sample tubes from liquid nitrogen and add pre-heated (80 °C) hot borate solution, as described [13, 16, 18]. Clean RNA using column purification with genomic DNA removal (Quick RNA Miniprep Plus, Zymo Research, Irvine, CA, USA) and elute with 30 μL water. Repeat the elution and pool both eluted volumes from the same sample (*see* **Note 11**).

5. Yields will be low from a single meristem, and we do not recommend attempting to visualize by gel electrophoresis. Quantify RNA with a BioAnalyzer RNA 6000 Pico chip or High Sensitivity RNA ScreenTape (Agilent, Santa Clara, CA, USA). If these sensitive methods of detection are unavailable, use a Nanodrop 2000c spectrophotometer (Thermo Fisher Scientific, Waltham, MA, USA), or Synergy H4 hybrid reader (BioTek, Winooski, VT, USA) for an estimate of RNA yield.

3.6 Library Construction from Low Input RNA Samples

1. Use equivalent amounts of RNA (~20 ng) from each meristem as starting material to prepare RNA-Seq libraries following the manufacturer's instructions. Because RNA yields from meristems are low, we use NEBNext® Single Cell/Low Input RNA Library Prep Kit for Illumina (New England Biolabs, Ipswich, MA, USA) without additional amplification steps.
2. Assess library quality and yield using Agilent High Sensitivity D5000 ScreenTape assay (Agilent Technologies, Santa Clara, CA, USA).
3. Libraries with concentrations ≥2 nM are suitable for sequencing. We use 75 cycles for single-end reads on an Illumina NextSeq 550 Sequencer (*see* **Note 4**).

3.7 Processing Reads

1. Download reads, as compressed fastq files (fastq.gz extension), from the sequencing facility, and save these original files to an external hard drive. To inspect and process the reads, we upload the fastq.gz files to the CyVerse Discovery Environment [19].
2. Determine read quality. We use FASTQC in the CyVerse Discovery Environment [19] to process the compressed fastq files (fastq.gz). There are multiple analysis modules reported, and we take particular note of the per base sequence quality. In these box-whisker plots, the read quality is visualized at each position of the read, with higher numbers on the y-axis indicating higher quality. We record the median quality scores, which are typically ≥32 for each nucleotide of 75 bp reads. While it is generally not necessary to trim reads for RNA-Seq experiments, gentle trimming may be used if sequence quality diminishes considerably over the length of the read.
3. Concatenate the reads. The NextSeq 550 has four lanes and the same library is loaded onto all lanes, meaning four sets of reads will be generated for each library (indicated with L001, L002, L003, L004 in the fastq.gz file names). The four sets of reads for each library can be concatenated. We use the Windows command line. We find it convenient to drag the four (or more if paired-end or re-sequenced) compressed sequence files (fastq.gz) for the same sample into a folder on the desktop. Using the command prompt, specify the directory. Then, use the copy function to concatenate all of the files in the folder:

Copy /b *.fastq.gz concatenated.fastq.gz.

This generates a single fastq.gz file for each sample, now named "concatenated.fastq.gz." Check the size of the new concatenated file. It should approximate the sum of all individual file sizes. Rename "concatenated.fastq.gz" with the sample name.

4. Upload the concatenated fastq.gz files to the CyVerse Discovery Environment [19] (*see* **Note 12**). Uncompress the zipped files using an app such as "Uncompress with gunzip 1.6-2" in the Discovery Environment. Uncompressed reads will have the fastq extension only.
5. Identify the *G. hirsutum* reference genome to use for mapping reads. There are now multiple cotton genomes sequenced and publicly available in Phytozome [20] and CottonGen [21], but most of these are the first versions of the assembly and annotation. We prefer to use the *G. hirsutum* v2.1 [22], and note that the third versions of the assembly and annotation (*G. hirsutum* v3.1) are now available. Download the compressed files from Phytozome or CottonGen, being sure to include the genome assembly in fasta format, annotation text file, and annotation/gene/transcript structural file in gtf format, and upload to CyVerse. Uncompress the uploaded files using an app such as "Uncompress with gunzip 1.6-2" in the Discovery Environment.
6. Map the reads to the *G. hirsutum* genome and assemble the transcripts. While there are several different approaches, we use RMTA_v2.6.3 (Reads Mapped, Transcripts Assembled) [23] in the CyVerse Discovery Environment. This workflow uses HISAT2 or Bowtie2 to map reads to the provided reference genome, assembles transcripts with Stringtie, and quantifies reads with FeatureCounts (*see* **Note 13**). This program requires a reference genome and annotation (as described in this section in point #5) and sequence-generated fastq reads (as described in this section in point #4) or sequence read archive (SRA) files from NCBI.
7. Determine the transcripts that are differentially expressed as a result of treatment. Different programs are available for differential expression analysis and we use Cuffdiff2 [24] in the CyVerse Discovery Environment. The outputs from this analysis include a "cuffdiffout" folder with FPKM tracking, and a "sorted data" folder, providing significantly differential expression for each transcript in pair-wise comparisons between treatments (*see* **Note 14**).
8. Report normalized expression data as fragments per kilobase of transcript per million mapped reads (FPKM). To quantify target gene expression following virus manipulation, we identify

the transcripts from the FPKM tracking file and visualize the expression in a table constructed in Excel. Then, we examine the pair-wise comparisons from the sorted data folder to determine whether expression is significantly different in the virus treatment compared with control treatments, and include these statistical differences in the table.

9. Prepare the sequencing data for sharing and uploading to NCBI GEO. Create an account with NCBI and complete the required metadata, which is an Excel file with information about the sample. The metadata, each original sequence read (as fastq.gz), and processed data with the Cuffdiff outputs will be uploaded together in submissions to NCBI GEO.

3.8 Interpreting and Presenting Differential Expression

1. Hypothesis testing: Is target gene expression altered following virus manipulation? The transcriptome data provides quantitative assessment of the effects of virus treatment on target gene expression. This is often presented in a table with statistically different levels of expression indicated (as described in *Processing Reads* point #7).
2. Tables, however, do not provide the strongest visual impact. Heatmaps can provide a more meaningful assessment of differential target gene expression among treatments, particularly if many target genes are being analyzed. These can be generated using accessible on-line tools, such as shinyheatmap [25] or Heatmapper [26] (source codes deposited in GitHub). Data, organized in Excel and saved as a CSV file, can be uploaded over the internet to http://shinyheatmap.com/ or http://heatmapper.ca/. Colors, clustering, scaling, and other options can be selected at these websites.
3. Venn diagrams are visual representations of the distributions of up- and downregulated genes among treatments. We construct these in R studio [27]. Prepare lists of the gene names (i.e., gene_ids) upregulated or downregulated by each treatment compared to the control, and save each list as a LS file. Construct Venn diagrams in R studio. With the following code (provided by Dr. Xiaolan Rao, Hubei University), we examined the sets of genes up- and downregulated when *SFT* was overexpressed or when *SP* was silenced compared with uninoculated plants in the DeltaPine 61 ("DP") and Texas 701 ("TX") genetic backgrounds [28].

```
workdir = "C:/My_Data"
setwd(workdir)
getwd()
library('grid')
```

```
library('VennDiagram')
TX_SFT_p = readLines("TX_SFT_up.ls")
TX_SP_p = readLines("TX_SP_up.ls")
DP_SFT_p = readLines("DP_SFT_up.ls")
DP_SP_p = readLines("DP_SP_up.ls")
VD = venn.diagram(width=3000,margin = .07,x = list(TX_vs_SFT
= TX_SFT_p, TX_vs_SP = TX_SP_p, DP_vs_SFT = DP_SFT_p, DP_vs_SP
= DP_SP_p), filename = 'overlap_up.tiff', fill = c('red','-
yellow','green','blue'), col = "transparent", alpha = 0.50)
TX_SFT_d = readLines("TX_SFT_down.ls")
TX_SP_d = readLines("TX_SP_down.ls")
DP_SFT_d = readLines("DP_SFT_down.ls")
DP_SP_d = readLines("DP_SP_down.ls")
VD = venn.diagram(width=3000,margin = .07,x = list(TX_vs_SFT
= TX_SFT_d, TX_vs_SP = TX_SP_d, DP_vs_SFT = DP_SFT_d, DP_vs_SP
= DP_SP_d), filename = 'overlap_down.tiff', fill = c('red','-
yellow','green','blue'), col = "transparent", alpha = 0.50)
```

4. Depending on the number of treatments, Venn diagrams can get very complicated very quickly. To consider the overlap of differentially expressed genes (DEGs) among a large number of treatments, we use UpSet plots where the intersections of DEGs shared among treatments can be visualized in a matrix and bar graph [10]. Several websites offer interactive shiny apps and we find Intervene [29] to be highly intuitive. Prepare a data file in Excel: include treatment names as headers, and list the DEGs for each treatment in the column; save the file in CSV format. Upload and adjust the settings and fonts and colors in the web tool. In the UpSet plot generated, the total number of DEGs for each target is indicated by "Set Size" along the *x*-axis. The *y*-axis shows the "Intersection Size," which is the number of DEGS in each treatment and shared among different treatments. The treatments with overlapping DEGs are illustrated in a matrix of connected dots. The plot can be downloaded in different file formats.
5. Understanding the functions DEGs is invaluable for testing the hypothesis and generating new hypotheses, but this can be overwhelming when a large number of DEGs are identified. Visualization tools that map DEGs to cellular or metabolic processes are helpful. One option toward understanding the function of the DEGs is to use MapMan ontology as an annotation framework [30]. This approach classifies DEGs into 35 bins, with each bin focusing mainly on metabolic pathways, such as lipid metabolism, secondary metabolism, major CHO metabolism, and more. Additional approaches to gene ontology are shared in point #7.

6. Comparative co-expression network analysis is a powerful strategy to uncover genetic signatures associated with a gene of interest. The premise to this approach is "guilt by association": highly connected genes sharing similar expression patterns are likely to be functioning in similar pathways. Large numbers of libraries, from multiple datasets, are needed to bolster this network analysis. We used the Weighted Gene Co-expression Network Analysis (WGCNA) package in R [31, 32], with default settings, to construct networks from 54 cotton libraries [28]. Among the outputs of this analysis are the sample dendrogram and trait heatmap and the module-trait relationships. The dendrogram shows the clustering of the different libraries, and the trait heatmap indicates target gene expression in each library. The module-trait relationship shows the different colored modules with the correlation coefficient and *p*-value of module gene expression compared to each target gene visualized in a heatmap. Importantly, WGCNA is a resource for future experiments: we leveraged this network analysis to identify a component of the *CLAVATA—WUSCHEL* meristem maintenance circuit significantly co-expressed with meristem maintenance gene *SELF-PRUNING* [10].
7. WGCNA constructs modules of relatedness. To determine the ontology of the genes comprising these modules, we use several different on-line web tools specifically for cotton. PhytoMine offers a convenient interface with Phytozome, and entering a list of gene identifiers can return descriptive information on each gene, and display the results from enrichment tests on Gene Ontology (GO) terms, protein domains, and pathways. However, this tool may not be updated as quickly as new cotton genomes are deposited in Phytozome. The Cotton Functional Genomics Database is another resource offering numerous tools and cotton genome sequences [33]. ShinyGo provides rapid gene ontology and enrichment analysis, and v0.80 includes cotton among the available species for analysis [34]. REViGO is a webtool that summarizes lists of GO terms graphically [35].
8. It may be helpful to validate expression of new genes of interest by RT-qPCR.
9. Ultimately, these analyses are used to design and test new hypotheses!

4 Notes

1. We use the genomic DNA removal columns and RNA purification columns from the Quick RNA Miniprep Plus kits (Zymo Research, Irvine, CA, USA) to complete the RNA extraction

with hot borate procedure [16]. We find these to be less expensive but of comparable quality to other kits.

2. We tested *GhUBQ10*, *GhGAPDH*, and *GhEF1α* as housekeeping genes in our gene expression analyses and find all to work well with minimal variation. We recommend testing two housekeeping genes in experiments. The primer sequences we use for cotton are as follows:
   (a) GhUBQ10 fwd: 5′-ggtttccgaatggtcttttg
   (b) GhUBQ10 rev: 5′-tttagccactgcatcacagc
   (c) GhGAPDH fwd: 5′-cattgttgccaatagctgga
   (d) GhGAPDH rev: 5′-gaaattgctgaagccgaaag
   (e) GhEF1α fwd (nt 802): 5′-ttcggaccttctggattgac
   (f) GhEF1α rev (nt 931): 5′-cacgcttgagatccttgaca
3. We prefer to culture an excess volume of each bacterial strain rather than limit the number of plants we can inoculate. To estimate the volume of overnight cultures to set up, we calculate as follows:

$$(2\text{ mL of culture per cotyledon}) \times (2\text{ cotyledons per plant}) \times (n\text{ replicate plants})$$

   We note that with these bipartite viruses, one genome (TRV RNA1 or CLCrV-B) will be used in every treatment, and therefore, larger volumes of overnight cultures are required for that strain. For example, to test four different TRV viruses with six replicate plants per treatment means 24 mL of each different TRV RNA2 culture and 96 mL of the TRV RNA1 culture.

4. Consider the number of plants to test. Our typical experiments use six replicate plants per virus treatment. For meristem isolations, we recommend increasing this to 12 or more plants per treatment. Harvest apices from six inoculated plants, but ensure that six replicate plants per treatment remain intact and are reserved for phenotype analysis. It is imperative to confirm the phenotype among the reserved plants: dissecting meristems and constructing libraries are laborious and expensive pursuits. At least three replicate plants per treatment will be used to construct independent libraries for RNA-Seq analysis. However, isolating meristems from cotton is not trivial, and it is too easy to lose samples in dissection or to obtain poor RNA or fail to generate libraries of sequencing quality. Having more experimental material from the outset provides much needed flexibility. We commonly isolate tissues and RNA from six replicate treatments and decide which three replicates will be used for library preparation after we have analyzed RNA quality by tape station.

5. Conduct preliminary time-course experiments to monitor target gene expression. These analyses will be invaluable to establish time points for harvesting experimental materials for library construction. Transcriptional changes precede phenotypic evidence of genetic modification. Positive controls are strongly recommended for any VIGS experiment, and to ensure that our viruses are working as expected, we routinely silence *magnesium chelatase I*, which results in chlorotic leaves by 7 dpi. However, this rapid visual assessment of altered accumulation of an otherwise abundant transcript in leaves does not correlate with targeted alterations of meristematic genes.
6. Apices stored in acetone at 4 °C will be stable for months. Alternatively, commercial RNA preservatives, such as RNA*later* (ThermoFisher Scientific, USA), can be used. We have not directly compared these reagents for differences in RNA preservation in side-by-side experiments.
7. It may be desirable to visualize the meristem using scanning electron microscopy (SEM) (Fig. 2). Samples dehydrated in acetone are ideally suited for directly mounting on a stub affixed with a conductive carbon adhesive tabs (Electron Microscopy Services, Hatfield, PA, USA) and viewing in a Hitachi TM1000 Tabletop SEM, which, conveniently, does not require additional preparation steps. However, after the meristem surface is bombarded by electron beams, it may be prudent to not use that specimen for RNA extraction and library construction.

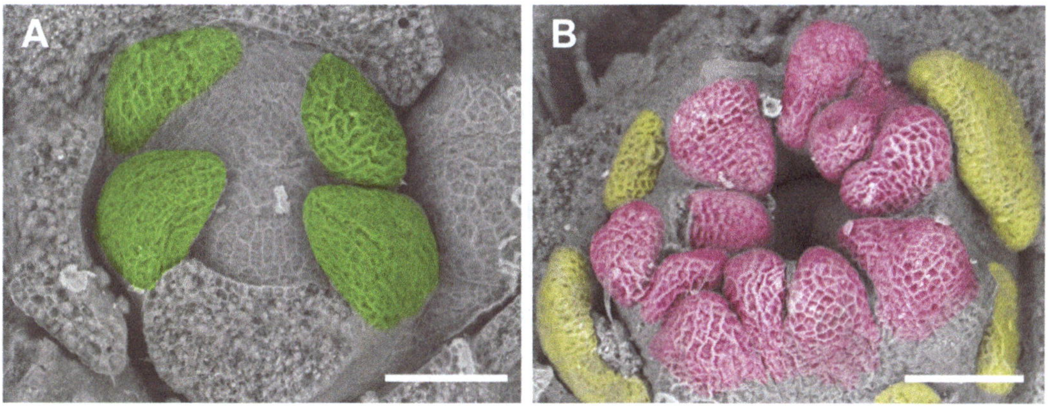

Fig. 2 Meristems isolated from cotton seedlings after treatments with different viruses. Seedlings were infiltrated with viruses at 4 days post-germination, and apices were harvested at 15 dpi. Apices were fixed in acetone, and meristems dissected. Isolated meristems were visualized using scanning electron microscopy. (**a**) A vegetative and indeterminate meristem (central dome) surrounded by emerging leaves (shown in green) is observed from a plant infected with a TRV virus. (**b**) A floral meristem, with emerging sepals shown in yellow and petals colored in pink, is shown from a plant infected with a different TRV virus. Scale bars are 100 μm

8. Isolating meristems from fixed apices is a delicate task not to be hurried. We usually isolate meristems from fixed tissues over a span of several days and return tissues in acetone to 4 °C until completed. We do not detect differences in RNA quality.
9. Allow the acetone to evaporate completely. Plunging samples into liquid nitrogen prematurely produces a block of frozen acetone which compromises homogenization.
10. We use BBs, which are inexpensive and disposable, to homogenize tissue in a Retsch mill. We recommend adding two BBs per 2 mL screw-cap tube and homogenize at 20 Hz for 15 s. Homogenizing at higher frequencies or for longer intervals can rupture the screw-cap tubes, leading to sample loss during RNA isolation. The tubes can also warm up with longer homogenization. If more homogenization is desired after 15 s, the tubes and sample holders should be cooled down again in liquid nitrogen before repeating the homogenization step.
11. Isolating RNA from cotton is challenging, and phenol- or guanidinium isothyocyante-based methods are not suitable. Protocols for isolating RNA by CTAB are very successful, particularly when the starting quantities of tissue are not limiting (i.e., leaves) [36]. However, the CTAB protocol involves multiple selective precipitation steps, which are daunting with minute amounts of difficult-to-obtain tissues. For these reasons, we prefer to use an established hot borate procedure followed by column purification [16]. We use the columns from Quick RNA Miniprep Plus Kits (Zymo Research, Irvine, CA, USA), and find these to be effective with respect to quality and cost.
12. For many wet-bench scientists, processing these data can be daunting because of the novelty or uncertainty in programming. The CyVerse Discovery Environment is an excellent platform to initiate RNA-Seq (and other) analyses. Importantly, the apps described include documentation for further clarification, and many insightful tutorials are available and are excellent resources. In addition, there are many web tools available to aid in DEG analyses. We also confer with bioinformaticians when analyses are beyond our capabilities.
13. The column headings of the reference gene.gff file and the "gene attribute" entered in the RMTA program must be the same. Discrepancies will cause the program to fail.
14. Many researchers will impose cut-off values for differential expression, using either a minimal FPKM or fold-change value. In our experiments, small changes in *SELF-PRUNING* expression, reported in FPKM, resulted in profound architectural changes [10, 13]. Thus, in the context of developmental

changes in meristems, small temporal or spatial differences can have appreciable impacts on cotton architecture, and we prefer not to implement cut-off values.

Acknowledgments

This research was supported by the United States–Israel Binational Agricultural Research and Development Fund BARD Project number US-5202-19 (BGA), Cotton Incorporated award 16-414 (RCM, BGA), and a BioDiscovery Institute award (RCM).

References

1. Fletcher JC (2020) Recent advances in Arabidopsis CLE peptide signaling. Trends Plant Sci 25(10):1005–1016. https://doi.org/10.1016/j.tplants.2020.04.014
2. Somssich M, Je BII, Simon R, Jackson D (2016) CLAVATA-WUSCHEL signaling in the shoot meristem. Development 143(18): 3238–3248. https://doi.org/10.1242/dev.133645
3. Chu Y-H, Jang J-C, Huang Z, van der Knaap E (2019) Tomato locule number and fruit size controlled by natural alleles of *lc* and *fas*. Plant Direct 3(7):e00142. https://doi.org/10.1002/pld3.142
4. Liu L, Gallagher J, Arevalo ED, Chen R, Skopelitis T, Wu Q, Bartlett M, Jackson D (2021) Enhancing grain-yield-related traits by CRISPR–Cas9 promoter editing of maize CLE genes. Nat Plant 7(3):287–294. https://doi.org/10.1038/s41477-021-00858-5
5. Rodríguez-Leal D, Lemmon ZH, Man J, Bartlett ME, Lippman ZB (2017) Engineering quantitative trait variation for crop improvement by genome editing. Cell 171(2): 470–480.e478. https://doi.org/10.1016/j.cell.2017.08.030
6. Wang X, Aguirre L, Rodríguez-Leal D, Hendelman A, Benoit M, Lippman ZB (2021) Dissecting cis-regulatory control of quantitative trait variation in a plant stem cell circuit. Nat Plant 7(4):419–427. https://doi.org/10.1038/s41477-021-00898-x
7. Xu C, Liberatore KL, MacAlister CA, Huang Z, Chu YH, Jiang K, Brooks C, Ogawa-Ohnishi M, Xiong G, Pauly M, Van Eck J, Matsubayashi Y, van der Knaap E, Lippman ZB (2015) A cascade of arabinosyltransferases controls shoot meristem size in tomato. Nat Genet 47(7):784–792. https://doi.org/10.1038/ng.3309
8. Meir Z, Aviezer I, Chongloi GL, Ben-Kiki O, Bronstein R, Mukamel Z, Keren-Shaul H, Jaitin D, Tal L, Shalev-Schlosser G, Harel TH, Tanay A, Eshed Y (2021) Dissection of floral transition by single-meristem transcriptomes at high temporal resolution. Nat Plant 7(6):800–813. https://doi.org/10.1038/s41477-021-00936-8
9. McGarry RC, Ayre BG (2021) Cotton architecture: examining the roles of *SINGLE FLOWER TRUSS* and *SELF-PRUNING* in regulating growth habits of a woody perennial crop. Curr Opin Plant Biol 59:101968. https://doi.org/10.1016/j.pbi.2020.10.001
10. McGarry RC, Kaur H, Lin Y-T, Puc GL, Eshed Williams L, van der Knaap E, Ayre BG (2023) Altered expression of *SELF-PRUNING* disrupts homeostasis and facilitates signal delivery to meristems. Plant Physiol. https://doi.org/10.1093/plphys/kiad126
11. Rössner C, Lotz D, Becker A (2022) VIGS goes viral: how VIGS transforms our understanding of plant science. Annu Rev Plant Biol 73(1):703–728. https://doi.org/10.1146/annurev-arplant-102820-020542
12. McGarry RC, Ayre BG (2012) Geminivirus-mediated delivery of florigen promotes determinate growth in aerial organs and uncouples flowering from photoperiod in cotton. PLoS One 7(5):e36746
13. McGarry RC, Prewitt SF, Culpepper S, Eshed Y, Lifschitz E, Ayre BG (2016) Monopodial and sympodial branching architecture in cotton is differentially regulated by the *Gossypium hirsutum SINGLE FLOWER TRUSS* and *SELF-PRUNING* orthologs. New Phytol 212: 244–258. https://doi.org/10.1111/nph.14037
14. Tuttle JR, Haigler CH, Robertson D (2012) Method: low-cost delivery of the cotton leaf crumple virus-induced gene silencing system.

Plant Methods 8:27. https://doi.org/10.1186/1746-4811-8-27

15. Burch-Smith TM, Schiff M, Liu Y, Dinesh-Kumar SP (2006) Efficient virus-induced gene silencing in Arabidopsis. Plant Physiol 142(1): 21–27. https://doi.org/10.1104/pp.106.084624
16. Wan C-Y, Wilkins TA (1994) A modified hot borate method significantly enhances the yield of high-quality RNA from cotton (*Gossypium hirsutum* L.). Anal Biochem 223(1):7–12
17. Tuttle JR, Idris AM, Brown JK, Haigler CH, Robertson D (2008) Geminivirus-mediated gene silencing from *Cotton leaf crumple virus* is enhanced by low temperature in cotton. Plant Physiol 148(1):41–50. https://doi.org/10.1104/pp.108.123869
18. Prewitt SF, Ayre BG, McGarry RC (2018) Cotton *CENTRORADIALIS/TERMINAL FLOWER 1/SELF-PRUNING* genes functionally diverged to differentially impact plant architecture. J Exp Bot 69(22):5403–5417. https://doi.org/10.1093/jxb/ery324
19. Goff S, Vaughn M, McKay S, Lyons E, Stapleton A, Gessler D, Matasci N, Wang L, Hanlon M, Lenards A, Muir A, Merchant N, Lowry S, Mock S, Helmke M, Kubach A, Narro M, Hopkins N, Micklos D, Hilgert U, Gonzales M, Jordan C, Skidmore E, Dooley R, Cazes J, McLay R, Lu Z, Pasternak S, Koesterke L, Piel W, Grene R, Noutsos C, Gendler K, Feng X, Tang C, Lent M, Kim S, Kvilekval K, Manjunath BS, Tannen V, Stamatakis A, Sanderson M, Welch SM, Cranston K, Soltis P, Soltis D, O'Meara B, Ane C, Brutnell T, Kleibenstein DJ, White JW, Leebens-Mack J, Donoghue MJ, Spalding EP, Vision TJ, Myers CR, Lowenthal D, Enquist BJ, Boyle B, Akoglu A, Andrews G, Ram S, Ware D, Stein L, Stanzione D (2011) The iPlant collaborative: cyberinfrastructure for plant biology. Front Plant Sci 2. https://doi.org/10.3389/fpls.2011.00034
20. Goodstein DM, Shu S, Howson R, Neupane R, Hayes RD, Fazo J, Mitros T, Dirks W, Hellsten U, Putnam N, Rokhsar DS (2011) Phytozome: a comparative platform for green plant genomics. Nucleic Acids Res 40(D1): D1178–D1186. https://doi.org/10.1093/nar/gkr944
21. Yu J, Jung S, Cheng C-H, Lee T, Zheng P, Buble K, Crabb J, Humann J, Hough H, Jones D, Campbell JT, Udall J, Main D (2021) CottonGen: the community database for cotton genomics, genetics, and breeding research. Plan Theory 10(12):2805
22. Chen ZJ, Sreedasyam A, Ando A, Song Q, De Santiago LM, Hulse-Kemp AM, Ding M, Ye W, Kirkbride RC, Jenkins J, Plott C, Lovell J, Lin Y-M, Vaughn R, Liu B, Simpson S, Scheffler BE, Wen L, Saski CA, Grover CE, Hu G, Conover JL, Carlson JW, Shu S, Boston LB, Williams M, Peterson DG, McGee K, Jones DC, Wendel JF, Stelly DM, Grimwood J, Schmutz J (2020) Genomic diversifications of five Gossypium allopolyploid species and their impact on cotton improvement. Nat Genet 52(5):525–533. https://doi.org/10.1038/s41588-020-0614-5
23. Peri S, Roberts S, Kreko IR, McHan LB, Naron A, Ram A, Murphy RL, Lyons E, Gregory BD, Devisetty UK, Nelson ADL (2019) Read mapping and transcript assembly: a scalable and high-throughput workflow for the processing and analysis of ribonucleic acid sequencing data. Front Genet 10:1361. https://doi.org/10.3389/fgene.2019.01361
24. Trapnell C, Williams BA, Pertea G, Mortazavi A, Kwan G, van Baren MJ, Salzberg SL, Wold BJ, Pachter L (2010) Transcript assembly and quantification by RNA-Seq reveals unannotated transcripts and isoform switching during cell differentiation. Nat Biotechnol 28(5):511–U174. https://doi.org/10.1038/nbt.1621
25. Khomtchouk BB, Hennessy JR, Wahlestedt C (2017) shinyheatmap: ultra fast low memory heatmap web interface for big data genomics. PLoS One 12(5):e0176334. https://doi.org/10.1371/journal.pone.0176334
26. Babicki S, Arndt D, Marcu A, Liang Y, Grant JR, Maciejewski A, Wishart DS (2016) Heatmapper: web-enabled heat mapping for all. Nucleic Acids Res 44(W1):W147–W153. https://doi.org/10.1093/nar/gkw419
27. R Development Core Team (2013) R: a language and environment for statistical computing. R Foundation for Statistical Computing, Vienna
28. McGarry R, Rao X, Li Q, van der Knaap E, Ayre B (2020) *SINGLE FLOWER TRUSS* and *SELF-PRUNING* signal developmental and metabolic networks to guide cotton architectures. J Exp Bot 71(19):5911–5923. https://doi.org/10.1093/jxb/eraa338
29. Khan A, Mathelier A (2017) Intervene: a tool for intersection and visualization of multiple gene or genomic region sets. BMC Bioinform 18(1):287. https://doi.org/10.1186/s12859-017-1708-7
30. Thimm O, Blasing O, Gibon Y, Nagel A, Meyer S, Kruger P, Selbig J, Muller LA, Rhee SY, Stitt M (2004) MAPMAN: a user-driven tool to display genomics data sets onto diagrams of metabolic pathways and other biological processes. Plant J 37(6):914–939.

https://doi.org/10.1111/j.1365-313X.2004.02016.x

31. Langfelder P, Horvath S (2008) WGCNA: an R package for weighted correlation network analysis. BMC Bioinform 9(1):559
32. Zhang BP, Horvath S (2005) A general framework for weighted gene co-expression network analysis. Stat Appl Genet Mol Biol 4:Article17
33. Zhu T, Liang C, Meng Z, Sun G, Meng Z, Guo S, Zhang R (2017) CottonFGD: an integrated functional genomics database for cotton. BMC Plant Biol 17(1):101. https://doi.org/10.1186/s12870-017-1039-x
34. Ge S, Jung D, Yao R (2019) ShinyGO: a graphical gene-set enrichment tool for animals and plants. Bioinformatics. https://doi.org/10.1093/bioinformatics/btz931
35. Supek F, Bošnjak M, Škunca N, Šmuc T (2011) REVIGO summarizes and visualizes long lists of gene ontology terms. PLoS One 6(7): e21800. https://doi.org/10.1371/journal.pone.0021800
36. Gambino G, Perrone I, Gribaudo I (2008) A rapid and effective method for RNA extraction from different tissues of grapevine and other woody plants. Phytochem Anal 19(6): 520–525. https://doi.org/10.1002/pca.1078

Chapter 13

Detecting Somatic Insertions/Deletions (Indels) Using Tumor RNA-Seq Data

Kohei Hagiwara and Jinghui Zhang

Abstract

Identification of somatic indels remains a major challenge in cancer genomic analysis and is rarely attempted for tumor-only RNA-Seq due to the lack of matching normal data and the complexity of read alignment, which involves mapping of both splice junctions and indels. In this chapter, we introduce RNAIndel, a software tool designed for identifying somatic coding indels using tumor-only RNA-Seq. RNAIndel performs indel realignment and employs a machine learning model to estimate the probability of a coding indel being somatic, germline, or artifact. Its high accuracy has been validated in RNA-Seq generated from multiple tumor types.

Key words Indel, Realignment, RNA-Seq, Cancer, Machine learning

1 Introduction

Identification of somatic variants in a cancer genome typically involves analyzing DNA sequence (DNA-Seq) data generated from the tumor and matched normal samples from the same patient. While not a routine practice, transcriptome sequencing (RNA-Seq) data can serve as an alternative or complementary data source for variation detection [1]. Specifically, RNA-Seq may be more effective than DNA-Seq for detecting subclonal variants in highly expressed genes [2], while expressed variants can be more relevant for cancer etiology and/or treatment as demonstrated in a study on glioblastoma multiforme [3]. However, utilizing RNA-Seq for somatic variant detection is challenging for several reasons. First, paired tumor-normal RNA-Seq data generated from the same tissue type is extremely rare, and this lack of a matching normal sample complicates the process of filtering out germline variants and common artifacts. Second, library preparation involves reverse transcription from RNA to cDNA, which introduces additional errors. Third, mapping RNA-Seq reads to genomic sequence

Rajeev K. Azad (ed.), *Transcriptome Data Analysis*, Methods in Molecular Biology, vol. 2812,
https://doi.org/10.1007/978-1-0716-3886-6_13,

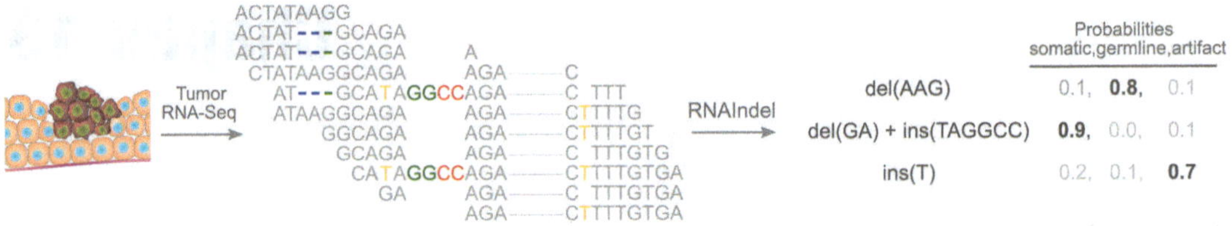

Fig. 1 RNA-Seq data prepared from a tumor sample contains both somatic and germline variants in addition to artifacts from library preparation, sequencing error, and inaccurate splice mapping. RNAIndel calls coding indels and estimates their probabilities of being somatic, germline, and artifactual. In the example read alignments, non-reference bases are colored, and spliced segments are rendered via solid lines

must consider splicing, which is more complex than DNA-Seq mapping. The difficulty of accurate mapping is a major source of false positive variants around splice junctions [1–3], particularly for indel analysis [4]. RNAIndel was developed to address these issues, implementing a machine-learning-based model trained on 765,475 indels from tumor samples profiled by both DNA-Seq and RNA-Seq [5]. Accepting a tumor RNA-Seq BAM file as input, this tool calls indels and classifies them into somatic, germline, or artifactual categories based on probability estimates (Fig. 1). Learning from the training data, indels with modest functional impact such as in-frame indels in non-conserved domains are more likely to be germline events, while sequence properties such as complex changes are predictive of somatic origin. Events in immediate proximity to splice junctions or error-prone regions such as homopolymers are likely artifacts. The program can also work with a PanelOfNormals (PON) [6] to further reduce false positive predictions.

2 Materials

2.1 Installation

RNAIndel can be run on Linux/Unix with Python 3.8 or higher, Java 8, and GNU C compiler preinstalled. The source code for RNAIndel can be downloaded freely (https://github.com/stjude/RNAIndel) and installed via the Python Package Index (PyPI). The most recent version of RNAIndel needs to be installed with indelPost (https://github.com/stjude/indelPost) [7], an indel realignment library, with the following command:

```
> pip install indelpost --no-binary indelpost --no-build-isolation
> pip install rnaindel
```

RNAIndel also requires a data package which contains pre-trained models and annotation files, which is available for both human reference assembly GRCh37 and GRCh38. Here, we show how to download and unpack the GRCh38 package:

```
> curl -L https://zenodo.org/records/10211317/files/data_dir_grch38.tar.gz
> tar -zxf data_dir_grch38.tar.gz
```

Pretrain the model, which takes 5–10 min. This step is required only once before the first run:

```
> rnaindel SetUp -d data_dir_grch38
```

Docker image of the latest version is also available (*see* **Note 1**). The pre-training step is not required to run the Docker image.

2.2 RNA-Seq Data

RNAIndel works best when the RNA-Seq read length is ≥75 bp (preferably 100 bp), which is suitable to effectively map indel-containing reads to the transcriptome. Aside from this requirement, RNAIndel is flexible to other variabilities in library construction protocols such as transcript enrichment method (e.g., ribosomal RNA depletion vs. poly-A capture). To maximize the sensitivity in indel detection, RNA-Seq reads should be mapped by STAR [8] in 2-pass mode. Another key parameter to consider is the mapping quality score. By default, STAR assigns a mapping quality of 255 to uniquely mapped reads, which is regarded as an invalid score by some variant analysis tools such as GATK [9]. To use the ensemble calling feature (*see* Subheading 3.2, Fig. 2) which integrates an external variant caller into the RNAIndel workflow, users should specify the score for uniquely mapped reads during RNA-Seq mapping that is compatible to their selected ensemble caller. As shown in the example below, uniquely mapped reads should be

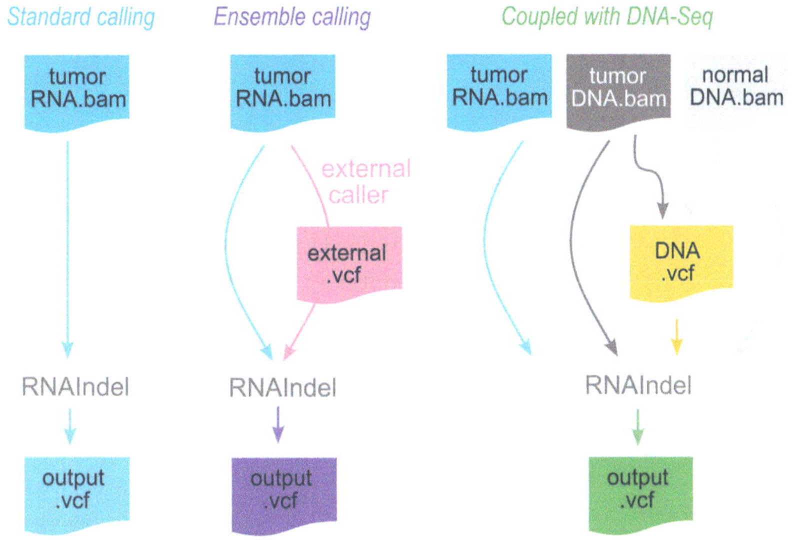

Fig. 2 Three analysis modes supported by RNAIndel. Standard calling (*left*) accepts a tumor RNA-Seq BAM file to analyze with the built-in caller. Ensemble calling (*middle*) takes as input a user-supplied VCF file generated by an external caller (external.vcf) along with the RNA-Seq BAM file, so that indel calls from both the external and built-in callers can be merged for improved sensitivity. The third mode enables discovery of indels using both DNA-Seq and RNA-Seq data (*right*). The input files include BAM files from both DNA-Seq and RNA-Seq well as a VCF file generated from tumor/normal-paired DNA analysis (DNA.vcf)

assigned with a score of 60 when running STAR mapping if GATK HaplotypeCaller [9] or MuTect2 [6] will be used as the ensemble caller:

```
> STAR --readFilesIn … --twopassMode Basic --outSAMmapqUnique 60
```

The output BAM file should be sorted by coordinate. Marking duplicate reads [10] is strongly recommended to improve analysis runtime performance:

```
> java -jar picard.jar MarkDuplicates I=in.bam O=out.bam M=metrics.txt
```

Further BAM file preprocessing is not required and in fact is discouraged as it may introduce unexpected behavior for downstream indel analysis.

3 Methods

3.1 Standard Calling

The built-in caller, a modified version of Bambino [11] with indel realignment [7], is used to locate indels in RNA-Seq data. A VCF file will be generated to report the detected indels with probabilities for each attribute (*left* in Fig. 2). Standard calling is invoked:

```
> rnaindel PredictIndels -i input.bam  \
                         -o output.vcf  \
                         -r reference_grch38.fa  \
                         -d data_dir_grch38  \
                         (-p 8 for faster analysis)
```

For faster analysis, increase the number of cores for multiprocessing (default: 1). On a server with 32-core AMD EPYC 7542 CPU @2.90 GHz, standard calling with 8 cores requires ~ 25 GB of memory and the analysis can be completed in <20 min.

3.2 Ensemble Calling

This option combines the indel calls from the built-in and an external caller which can improve sensitivity of indel detection (*middle* in Fig. 2):

```
> rnaindel PredictIndels -i input.bam  \
                         -o output.vcf  \
                         -r reference_grch38.fa  \
                         -d data_dir_grch38  \
                         -v external.vcf.gz  \
                         -q 60
                         (--include-all-external-calls)
```

The "*-v*" option allows users to supply a VCF file from a variant caller of the user's choice. In the output VCF, each indel will be annotated with its callers in the INFO field, in the format of

CALLER = *built_in*, *external*, or *both*. By default, RNAIndel integrates indels marked *PASS* in the externally called VCF file. Specify "*--include-all-external-calls*" to analyze all indels contained in the VCF file. A non-default mapping score (default: 255) might have been set for the compatibility with the external caller during RNA-Seq mapping (*2.2, RNA-Seq Data*) which will require the user to reset the score by specifying the "*-q*" option.

3.3 Coupled with DNA-Seq

When DNA-Seq data is available, users can integrate indels generated from these two data types (*right* in Fig. 2):

```
> rnaindel PredictIndels -i input.bam  \
                         -o output.vcf  \
                         -r reference_grch38.fa  \
                         -d data_dir_grch38  \
                         -t tumor_dna.bam  \
                         -n normal_dna.bam  \
                         (-v dna_paired_analysis.vcf.gz)
```

This integrated analysis discovers indels from both DNA-Seq and RNA-Seq. From the DNA perspective, this has the advantage of enhancing DNA-Seq analysis in regions with high expression and quantifying mutant allele expression level. From the RNA perspective, putative somatic indels detected in RNA-Seq can be matched to the DNA-Seq for validation via realignment, facilitating removal of false positive calls. A paired analysis from DNA-Seq may be supplied as an input file via the "*-v*" option. In this scenario, the DNA indels will be merged into the call set from RNA-Seq as in *Ensemble calling* and the output will include both the expression level of the mutant allele as well as additional indels detected exclusively in RNA-Seq.

3.4 Evaluating Putative Somatic Indels

To extract indels predicted as somatic (putative somatic indels), parse the VCF file for calls with a status of *PASS* in FILTER field and a value of *predicted_class* = *somatic* in INFO field. The somatic probability is a useful metric to remove artifact calls misclassified as somatic. Based on a pan-cancer TCGA application of RNAIndel (manuscript under preparation), we recommend a cutoff of 0.6 as a somatic probability threshold. Indels with probability <0.6 may be considered only if they occurred in driver genes relevant to the disease of interest. Regardless of the probability, indels with the *NEAR_EXON_BOUNDARY* flag, which is also in INFO field, are rarely true.

Ensemble calling (3.2) and *Coupled* with *DNA-Seq* (3.3) modes can further facilitate this evaluation. Indels should be prioritized for downstream analysis if they are called by both built-in and external callers (*CALLER* = *both*) and/or supported by DNA reads. Additionally, RNAIndel has an option to annotate possible recurrent artifacts when multiple samples are analyzed. This feature takes advantage of the fact that bona fide somatic indels rarely recur outside of a few known hotspots, which are already known in mutation databases such as COSMIC [12]. Hence, novel indels that recur in the sample cohort are potentially artifactual. The occurrence count utility can be run:

```
> rnaindel CountOccurrence --vcf-list vcf_paths.txt   \
                           -r reference_grch38.fa     \
                           (--out-dir /path/to/out_dir)
```

A text file listing the paths to the RNAIndel output VCF files (one path per line) is required. New VCF files with occurrence annotation for somatic prediction will be generated in the current or user-specified directory. Tips on manual review are given below (*see* **Note 2**).

4 Notes

1. If installation problems are encountered, use of a virtual environment can be helpful:

```
# create virtual environment
> python3 -m venv /path/to/dir_to_create_virtual_env
> source /path/to/dir_to_create_virtual_env/bin/activate

# installation
> pip install wheel
> pip install cython
> pip install python
> pip install indelpost
> pip install rnaindel

# run analysis

# To exit from the environment
> deactivate

# Re-entry
> source /path/to/dir_to_create_virtual_env/bin/activate
```

Alternatively, users can run the latest RNAIndel code via Docker:

```
> docker run --rm -v ${PWD}:/data ghcr.io/stjude/rnaindel:latest

# To run without sudo as in HPC systems
> singularity exec -B ${PWD}:/data \
  --containall docker://ghcr.io/stjude/rnaindel:latest rnaindel
```

2. When reviewing the results, use the domain knowledge of driver genes in the disease of interest. Also, note that the number of expressed coding somatic indels per sample is as few as <5 for many tumor types (excluding hypermutative cases such as microsatellite instability). When a number of normal RNA-Seq datasets are available, users can use the GATK toolkit to create a PanelOfNormals (PON) database in VCF format. RNAIndel can use a PON to remove false positives on the fly as:

```
> rnaindel PredictIndels -i… --pon user_defined_pon.vcf.gz
```

Acknowledgments

We would like to thank Mr. Michael Edmonson for his help with proofreading the manuscript.

References

1. Piskol R, Ramaswami G, Li JB (2013) Reliable identification of genomic variants from RNA-seq data. Am J Hum Genet 93:641–651. https://doi.org/10.1016/j.ajhg.2013.08.008
2. Mosen-Ansorena D (2019) Identification of mutated cancer driver genes in unpaired RNA-Seq samples. Methods Mol Biol 1878: 95–108. https://doi.org/10.1007/978-1-4939-8868-6_5
3. Coudray A, Battenhouse AM, Bucher P, Iyer VR (2018) Detection and benchmarking of somatic mutations in cancer genomes using RNA-seq data. PeerJ. https://doi.org/10.7717/peerj.5362
4. Sun Z, Bhagwate A, Prodduturi N, Yang P, Kocher JA (2017) Indel detection from RNA-seq data: tool evaluation and strategies for accurate detection of actionable mutations. Brief Bioinform 18:973–983. https://doi.org/10.1093/bib/bbw069
5. Hagiwara K, Ding L, Edmonson MN, Rice SV, Newman S, Easton J, Dai J, Meshinchi S, Ries RE, Rusch M, Zhang J (2020) RNAIndel: discovering somatic coding indels from tumor RNA-Seq data. Bioinformatics 36:1382–1390. https://doi.org/10.1093/bioinformatics/btz753
6. Cibulskis K, Lawrence MS, Carter SL, Sivachenko A, Jaffe D, Sougnez C, Gabriel S, Meyerson M, Lander ES, Getz G (2013) Sensitive detection of somatic point mutations in impure and heterogeneous cancer samples. Nat Biotechnol 31:213–219. https://doi.org/10.1038/nbt.2514
7. Hagiwara K, Edmonson MN, Wheeler DA, Zhang J (2022) indelPost: harmonizing ambiguities in simple and complex indel alignments. Bioinformatics 38:549–551. https://doi.org/10.1093/bioinformatics/btab601
8. Dobin A, Davis CA, Schlesinger F, Drenkow J, Zaleski C, Jha S, Batut P, Chaisson M, Gingeras TR (2013) STAR: ultrafast universal RNA-seq aligner. Bioinformatics 29:15–21. https://doi.org/10.1093/bioinformatics/bts635
9. DePristo MA, Banks E, Poplin R, Garimella KV, Maguire JR, Hartl C, Philippakis AA, del Angel G, Rivas MA, Hanna M, McKenna A, Fennell TJ, Kernytsky AM, Sivachenko AY, Cibulskis K, Gabriel SB, Altshuler D, Daly MJ (2011) A framework for variation discovery and genotyping using next-generation DNA sequencing data. Nat Genet 43:491–498. https://doi.org/10.1038/ng.806
10. Picard Tools. https://broadinstitute.github.io/picard/
11. Edmonson MN, Zhang J, Yan C, Finney RP, Meerzaman DM, Buetow KH (2011) Bambino: a variant detector and alignment viewer for next-generation sequencing data in the SAM/BAM format. Bioinformatics 27:865–

866. https://doi.org/10.1093/bioinformatics/btr032

12. Tate JG, Bamford S, Jubb HC, Sondka Z, Beare DM, Bindal N, Boutselakis H, Cole CG, Creatore C, Dawson E, Fish P, Harsha B, Hathaway C, Jupe SC, Kok CY, Noble K, Ponting L, Ramshaw CC, Rye CE, Speedy HE, Stefancsik R, Thompson SL, Wang S, Ward S, Campbell PJ, Forbes SA (2019) COSMIC: the catalogue of somatic mutations in cancer. Nucleic Acids Res 47(D1):D941–D947. https://doi.org/10.1093/nar/gky1015

Chapter 14

A Protocol for the Detection of Fusion Transcripts Using RNA-Sequencing Data

Fiza Hamid, Simran Arora, Pragya Chitkara, and Shailesh Kumar

Abstract

Fusion transcripts are formed when two genes or their mRNAs fuse to produce a novel gene or chimeric transcript. Fusion genes are well-known cancer biomarkers used for cancer diagnosis and as therapeutic targets. Gene fusions are also found in normal physiology and lead to the evolution of novel genes that contribute to better survival and adaptation for an organism. Various in vitro approaches, such as FISH, PCR, RT-PCR, and chromosome banding techniques, have been used to detect gene fusion. However, all these approaches have low resolution and throughput. Due to the development of high-throughput next-generation sequencing technologies, the detection of fusion transcript becomes feasible using whole genome sequencing, RNA-Seq data, and bioinformatics tools. This chapter will overview the general computational protocol for fusion transcript detection from RNA-sequencing datasets.

Key words Fusion transcript, Fusion gene, Tools, Chimeric RNAs, RNA-Seq

1 Introduction

Chimeric transcripts are the result of the fusion of two transcripts at the RNA level or formed from the fusion of two genes at the DNA level (Fig. 1) [1–3]. At the DNA level, gene fusion occurs through inter-chromosomal or intra-chromosomal rearrangement (Fig. 2a) that may arise as a result of translocation, inversion, insertion, tandem duplication, or deletion of genes (Fig. 2b) [2]. For example, the BCR-ABL1 fusion gene, a hallmark of CML, is generated from a translocation between chromosomes 9 and 22 [4]. The fusion gene BCR-ABL1 has a constitutive tyrosine kinase activity due to uncontrolled cell proliferation [5]. TMPRSS2-ERG, an oncogenic fusion gene found in prostate cancer, is formed due to the deletion of the intron between TMPRSS2 and ERG on chromosome 21 [6]. At the RNA level, gene fusion occurs via cis- or

Authors "Fiza Hamid" and "Simran Arora" have equally contributed to this chapter.

Rajeev K. Azad (ed.), *Transcriptome Data Analysis*, Methods in Molecular Biology, vol. 2812,
https://doi.org/10.1007/978-1-0716-3886-6_14,

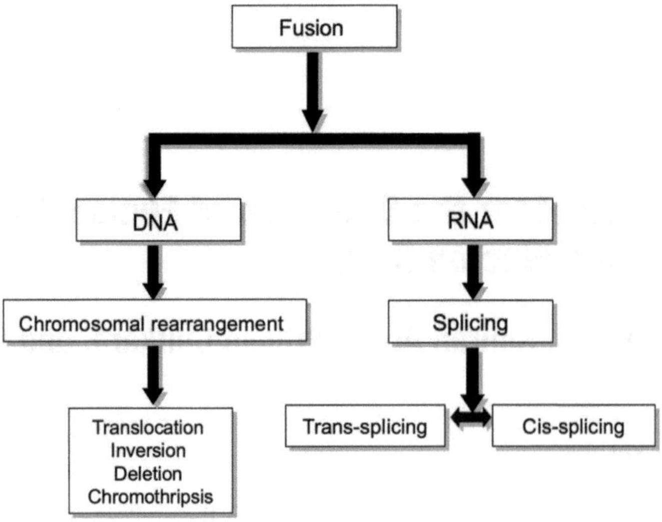

Fig. 1 Different ways of formation of fusion transcripts

trans-splicing (Fig. 2c). In trans-splicing, exons from mRNA of foreign genes, either on the same or different chromosome, fuse to produce a mature mRNA. For example, JAZF1-SUZ12 [7], a chimeric RNA found in normal endometrial stromal cells, is formed by trans-splicing the first three exons of the JAZF1 gene on chromosome 7 and the last 15 of the SUZ12 gene on chromosome 17. Fusion transcripts derived from cis-splicing may form from a single mRNA by transcriptional read-through, followed by splicing, or from two transcripts joined by splicing [2, 3]. For example, DUS4L–BCAP2 is a cis-spliced fusion that promotes gastric cancers [8, 9].

The functions of chimeric genes and RNAs depend on the location of the fusion junction. Chimeric RNAs in which the fusion junction is out-of-frame may function as long non-coding RNA and exhibit regulatory functions. In contrast, the in-frame chimeric transcript may code for a functional protein with a novel function or different cellular localization than the gene products [10]. For example, the Kua-UBE2Vl fusion protein is localized in the cytoplasm, while the parental gene, UBE2Vl, is found in the nucleus [11]. JAZF1-JJAZ1 is a chimeric protein that regulates cell proliferation and apoptosis [12]. Tyrosine kinase fusions, detected in various types of cancer, are found to cause the activation of signaling pathways that are involved in cell growth, survival, proliferation, and apoptosis. Fusion events have also been involved in important functionality in plants as well. For example, GN2, a chimeric gene formed from the fusion of LOC_Os02g45150 and LOC_Os02g56630 on chromosome 2, controls the grain number in Yuanjiang common wild rice (*O. rufipogon* Griff.) [13]. In rice (*O. japonica*), fusion genes Osjap07g28390 and Osjap09g15430

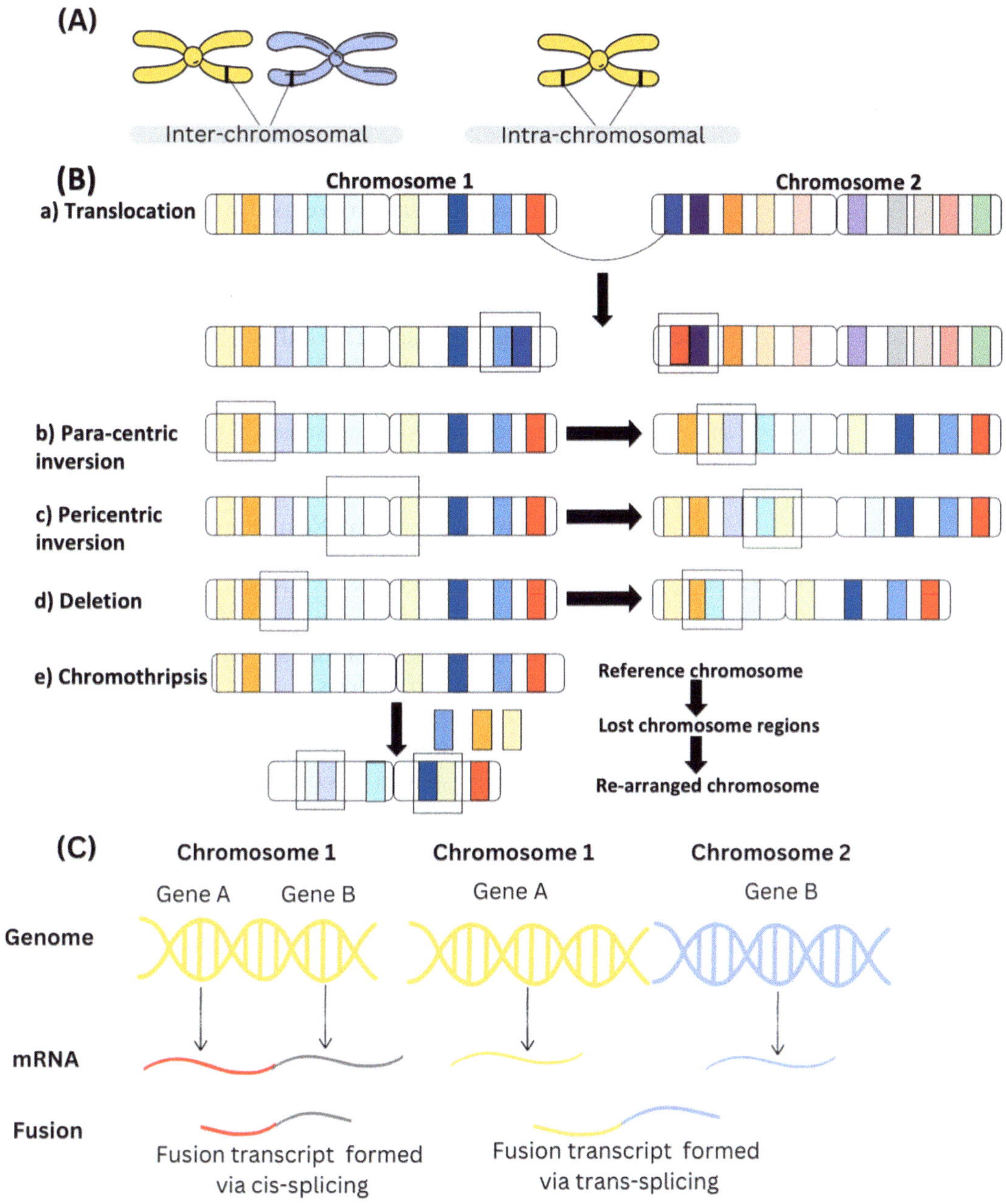

Fig. 2 Generation of different types of fusion transcripts. (**a**) Inter- and intra-chromosomal gene fusion, (**b**) fusion gene formation by chromosomal rearrangements, and (**c**) diagrammatic representation of fusion transcript formation at RNA level

regulate phenotype traits such as root length, shoot length, and seed germination, indicating the functional significance of these genes [14]. Much research has been going on to study the mechanism of formation and function of fusion genes, leading to better understanding, diagnostics, and targeted treatment of some cancers.

Studies of fusion events are essential for understanding the tumorigenesis, clinical biomarkers generation, designing drugs, and investigating other roles performed by fusion products. In plants, fusion identification is an untouched area that can be explored to understand the fusion events and their role in plants' physiology and functionality. Hence, fast and reliable identification of fusion genes is increasingly relevant for clinical and pharmaceutical applications. Various conventional methods have been used for detecting gene fusions, including, Southern blotting, FISH, and PCR-based approaches. The RT-PCR technique is widely used for detecting fusion transcripts and has applications in cancer diagnosis. Despite these techniques being sensitive to fusion detection, most conventional methods are labor-intensive and have low throughput.

Due to the advancement in Next-Generation Sequencing (NGS) technologies, whole genome sequencing (WGS), whole transcriptome sequencing (RNA-Seq), and bioinformatic tools are now used to detect fusion events. However, WGS, which detects fusion at the DNA level, is a computationally intensive analysis method that misses out on fusions formed at the RNA level, whereas RNA sequencing, which mainly focuses on the actively transcribed part of the genome, is a cost and time-effective method detecting intergenically spliced fusion and multiple alternative splice variants [15]. Hence, various fusion prediction tools have been developed utilizing RNA-sequencing datasets for identifying the full potential of chimeric RNAs. This chapter will provide a general protocol for fusion transcripts detection in RNA-Sequencing datasets using bioinformatics tools.

2 Requirements

- Operating System: 32-bit/64-bit Linux.
- FastQC [19] for quality assessment of RNA-Seq data.
- Fastp [21]. Not required if high-quality filtered reads are already available.
- EricScript [27] software and its dependencies (https://github.com/databio/ericscript).

3 Methods

3.1 Obtain Paired-End RNA-Seq Data of Human from NCBI SRA

The example paired-end RNA-seq data (SRR934947) can be downloaded from the NCBI SRA using following command:

- fastq-dump --split-files --origfmt SRR934947

3.2 Quality Assessment and Preprocessing of RNA-Seq Data

FastQC is a tool to check the quality of high-throughput sequencing datasets. It provides a comprehensive set of analyses and visualizations to assess the quality of sequence data. Moreover, FastQC creates an HTML report. The report includes different metrics and charts useful for pinpointing problems in the data.

To use FastQC, user generally need to download and install it on the system. Once installed, user can run it on the data files.

Download and Install FastQC

FastQC can be download from the official website: https://www.bioinformatics.babraham.ac.uk/projects/fastqc/

After downloading, follow the installation instructions provided on the website.

FastQC usage

- fastqc --outdir=/path/to/output_directory --threads=4 SRR934947_1.fastq SRR934947_2.fastq

The FastQC generates a list of output files, including an HTML report. User can open the HTML report in a web browser to visualize the quality metrics and assessments.

Consult the FastQC documentation for more details on how to interpret the results: https://www.bioinformatics.babraham.ac.uk/projects/fastqc/Help/

If the data is found to have low quality reads or adapter contamination, then we can use fastp. Fastp is a tool for quality control and preprocessing of high-throughput sequencing data. It can perform tasks such as adapter trimming, quality filtering, and other preprocessing steps. Use the following command for filtering your data using fastp:

- fastp --detect_adapter_for_pe -i SRR934947_1.fastq -I SRR934947_2.fastq -o SRR934947_1_trim.fastq -O SRR934947_2_trim.fastq --html=SRR934947.html --json=SRR934947.json

3.3 Fusion Transcripts Identification Using EricScript

According to the benchmarking study conducted by Kumar et al. [15], EricScript is one of the best tools for fusion detection, developed in with R, perl, and bash scripts. It uses the BWA aligner for mapping of reads on the transcriptome reference and samtools for dealing with SAM/BAM files.

Download EricScript with Git from GitHub

- git clone https://github.com/databio/ericscript.git

Once we downloaded EricScript, extract the package

- tar -xjf ericscript.tar.bz2

Before running for the first time EricScript, user need to make ericscript.pl executable:

- chmod +x /PATH/TO/ERIC/ericscript.pl

To get information about running EricScript, digit:

- /PATH/TO/ERIC/ericscript.pl --help

For fusion transcript detection, user need to download and build the Ensembl Database of the genome. To list the available genomes, type:

- /PATH/TO/ERIC/ericscript.pl --printdb

After a reference id is selected, user need to download and build the corresponding Ensembl Database. Here, it has shown how to prepare the database for homo sapiens.

- /PATH/TO/ERIC/ericscript.pl --downdb --refid homo_sapiens -db /PATH/TO/YOUR/DBFOLDER

We have to provide high-quality paired-end reads as input to EricScript for fusion transcript detection using the following command.

- /PATH/TO/ERIC/ericscript.pl -db /PATH/TO/YOUR/DBFOLDER --refid homo_sapiens -name SRR934947 -o /PATH/TO/OUTPUT/ SRR934947_1_trim.fastq SRR934947_2_trim.fastq

EricScript output:

- SRR934947.results.total.csv file in the output folder contains all the predicted gene fusions. SRR934947.results.filtered.csv contains the predicted gene fusions with EricScore > 0.50.

4 Methodology for Fusion Transcript Identification

A wide variety of tools (Table 1) utilizing RNA-Seq datasets are available to identify the potential chimeric transcripts from single-end and paired-end read. However, the paired-end reads have an advantage due to the availability of connectivity data between the sequenced ends. A large number of freely available online databases, such as TCGA [16], NCBI SRA, and GTEx [17], are available to download raw sequencing data [18]. This downloaded raw data is then processed for analyzing the fusion transcripts.

4.1 Quality Control Processing of RNA-Seq Datasets

The raw RNA-Seq data must be processed by applying various quality control software packages to ensure the quality of the reads. Most updated tools for this purpose are:

Table 1
Different bioinformatics tools to identify fusion transcripts along with their methodologies

Method	Brief overview of methodology
ChimeraScan	Align the paired-end reads to a reference, select discordant reads and re-aligned. The breakpoint was resolved by aligning to putative fusions.
deFuse	Align the paired-end reads to spliced and unspliced reference, resolves ambiguous discordant alignments, resolves junctions using a novel dynamic programming algorithm.
EricScript	BWA is used to align reads to the genome. Discordant reads are used to identify candidate gene fusions. BLAT is to identify precise fusion breakpoints by aligning to fused exons.
InFusion	Align the reads to the reference transcriptome using Bowtie to detect potential split reads and discordantly mapped read pairs which examined further to define candidate fusions.
JAFFA-Direct	Align the reads to remove introns and map resultant reads on transcriptome using BLAT and putative fusion candidates analyzed for junction detection.
MapSplice	A second-generation algorithm for splice junction detection. It can be applied on long as well as short reads. It involves segmentation of reads, exonic and spliced alignment, segments assembly, and junction detection for candidate alignments.
nFuse	Detects complex genomic rearrangements (CGRs) from WGS-seq and RNA-seq that leads to fusion transcripts predicted from high-throughput cDNA sequencing.
STAR-Fusion	Using STAR, it align RNA-seq reads to the reference for chimeric detection, followed by extensive filtering to remove false-positive results.
TopHat-Fusion	Uses Bowtie to find reads that align within exons and creates a set of partial exons from these alignments with hypothetical boundaries, re-align the initially unmapped reads, and find fusion events.

- *FastQC:* This tool is highly accessible for quality check on raw data, e.g., FASTQ reads, from high-throughput sequencing platforms. The main aim of this tool is to determine whether data has any problem that should be resolved before doing further analysis. The BAM, SAM, or FASTQ formats are used as input files, which, as a result, provide summary graphics and tables to evaluate the datasets. It allows exporting of data as an HTML-based permanent report. This tool is available at https://www.bioinformatics.babraham.ac.uk/projects/fastqc [19].
- *ClinQC:* It is a user-friendly and robust pipeline used for conversion of format, quality control, trimming, and filtering raw sequencing from Sanger and NGS sequencing methods. First, this tool converts the input file to a standard format, i.e., FASTQ, which is the most accepted format for predicting chimeric transcripts. Then, this file further undergoes adapter removal, PCR priming, duplicate filtering, contamination removal, and quality check and generates a QC report, which

is directly used in downstream analysis. This tool is available at https://sourceforge.net/projects/clinqc [20].

- *fastp:* An ultra-fast FASTQ preprocessor used to generate clean data for downstream analysis. It performs quality control, filtering, adapter trimming, and many other functions. This tool is implemented in Python and Java, and is freely available at https://github.com/OpenGene/fastp [21].

4.2 Fusion Transcript Prediction

Following the quality control step, the high-quality reads are further used as input for fusion transcript prediction tools that directly give the predicted chimeras as output. Many tools are available for predicting fusion transcripts that employ different kinds of algorithms. Despite this, the fundamental procedure for detecting fusion remains the same and can be divided into three steps:

(a) *Reads mapping and filtering*

Most of the tools are based on the principle of mapping as their initial step. Each read is first mapped on the reference genome or transcriptome in this step to find the concordantly mapped, discordantly mapped, and unmapped reads (Fig. 3). The discordantly mapped reads can be classified into two types:

- Split reads—Reads those have fusion junctions are known as split reads. The alignment of these reads helps in analyzing the fusion events.
- Spanning reads—These are those reads in which the sequence matches two different sequences with neither bearing a fusion junction [22].

In single-end reads, only split reads can be used for fusion detection, while for paired-end reads, spanning reads can also be used as the junction lies in the insert sequence. After mapping, each read is evaluated, and the reads irrelevant to fusions (concordant reads) are removed. It is known as Split mapping. Some tools like TopHat-Fusion and FusionMap utilizes "Split Mapping" to filter out all the mapped reads, while some tools, such as SnowShoes-FTD, follow the principle of "spanning reads" in which all the odd reads are preserved without filtration [23]. Those reads that are less favorable for fusion detection are removed by most of the tools, as in FusionSeq, more than ten filters are present to filter out the illicit fusions [24]. In addition, tools like FusionMap, FusionHunter, ShortFuse, SnowShoes-FTD, and TopHat-Fusion use a threshold value of the distance between the fusion partners to eliminate the read-through transcript.

(b) *Fusion junction detection*

As shown in Fig. 3, the second step of the procedure is to detect the fusion junction through the split and spanning

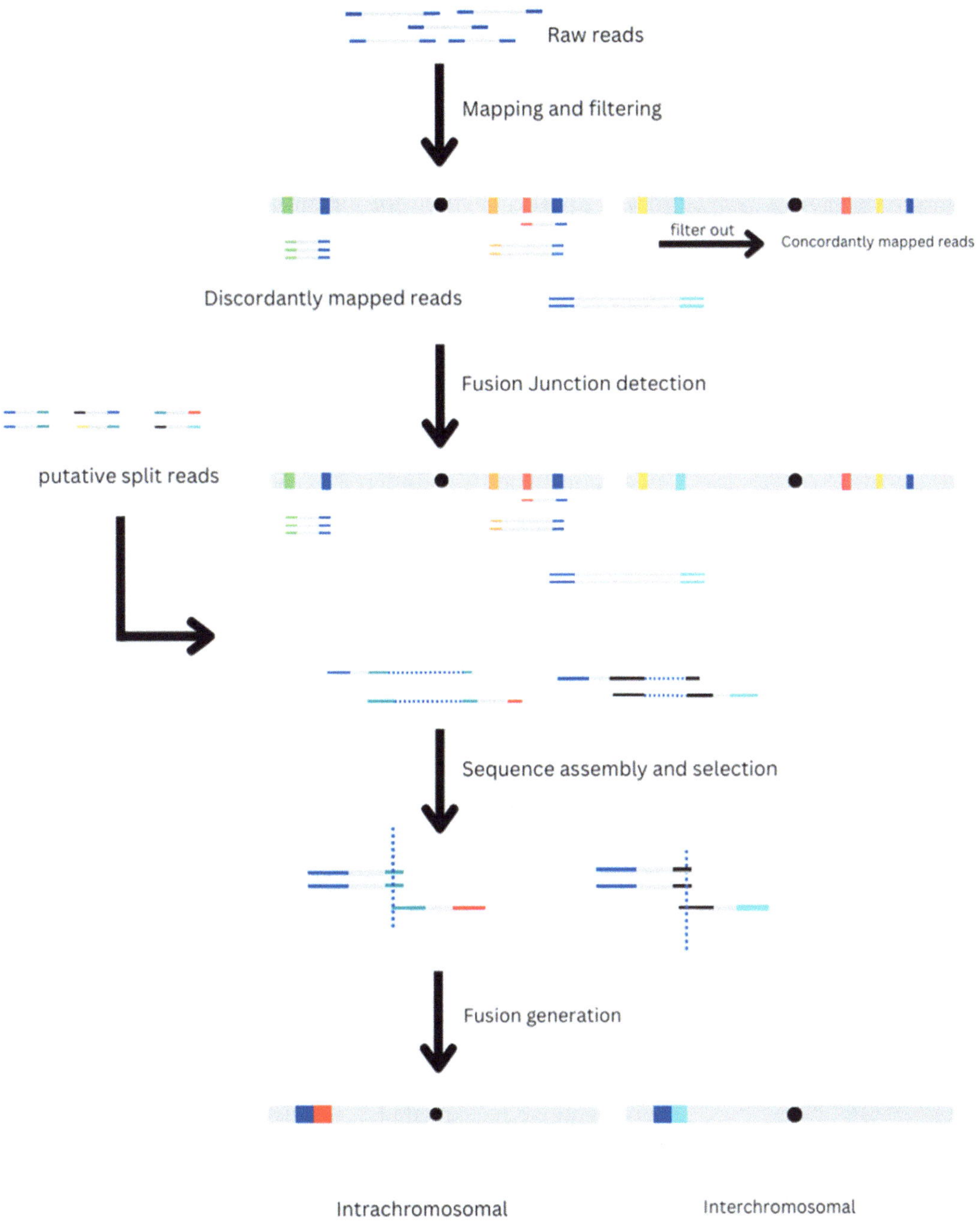

Fig. 3 Steps involved in the detection of fusion through "split reads" and "spanning reads." (i) Mapping of raw reads to the reference genome followed by filtering of concordantly mapped reads. (ii) Discordantly mapped reads are further used for the identification of fusion junction detection. The dotted line represents the expected junction of the fusion gene. (iii) Assembly of the expected fusion sequence and one with the highest possibility is selected. The vertical dotted line represents the fusion breakpoint

reads mapping. In the case of "split read" mapping, unmapped reads generated from the previous step are broken into smaller fragments, and the first and last segments of each "split read" are again aligned independently. After recognizing the alignment pattern, the boundaries of the original fragments are adjusted, and this realignment helps to detect fusion junctions. Another method for junction identification is through "spanning reads" followed by extracting the candidates confirmed by "split reads." First, discordant alignments are collected with a maximal set of reads and the same pair of breakpoints and grouped into clusters. Next, the identification of each candidate fusion junction's boundary region and fusion junction loci are predicted through its cluster. In the last step, unmapped reads are aligned on predicted fusion transcripts. The one with the maximum number of aligned unmapped reads is considered a candidate fusion [24].

(c) *Fusion assembly and selection*

In the final step, those two parental genes involved in fusion formation are joined together to form a fusion sequence using the predicted junction (Fig. 3). The previously unmapped reads are then aligned on these candidate fusions. Such mapped reads in this step are known as supporting reads. These reads help to eliminate false candidates; however, the risk of removing those fusions expressed at low-level increases. To deal with this problem, some tools have a scoring function based on features, including mapping quality, read depth, and several supporting reads. In addition, tools like FusionSeq generate a score via empirical analysis, while defuse generates a score via machine learning modalities [25, 26].

5 Some Basic Tools for Fusion Identification

1. *EricScript (ChimERIC TranSCRIPT detection algorithm):* It is a computational framework that detects chimeric transcripts in paired-end RNA-Seq data [27]. This tool is based on the four alignment processes for detecting candidate fusion: reads mapping against transcriptome using BWA, discordant alignment identification, exon junction reference identification (using BLAT by aligning them to fused exons) and analysis, and last is Scoring and filtering the potential fusion. According to Kumar et al., as compared to sensitivity, EricScript is good for its balance between time and memory requirements [15]. In addition, this tool reduces the significant number of false-positive calls.
2. *ChimeraScan:* This tool utilizes discordant Bowtie genome alignment to detect candidate fusion. First, alignment of

paired-end reads is done on the genome as well as transcriptome reference. Then discordantly aligned reads are trimmed and re-aligned. This step improves sensitivity for the identification of chimeras. After that, ChimeraScan detects the reads spanning the chimera junction, and based on an abundance of fusion-supporting reads; it filters out false-positive chimeras. As a result, it generates a tabular text file explaining each chimera [28]. It is a user-friendly, highly preferential tool for predicting fusions from longer reads that provide better coverage of fusion junctions.

3. *JAFFA:* This algorithm is a transcript-centric approach that utilizes either raw RNA-Seq reads or preassembled transcripts for detecting gene fusions. It performs well for read lengths 50 bp or greater and works for single- and paired-end reads. The name, location, and cDNA sequence of the candidate gene fusion breakpoint are provided as output. JAFFA aligns the reads to a reference transcriptome, and those reads that are aligned to different genes are considered candidate fusion reads. The alignment step is evaluated by re-aligning the reads to a reference and clustering the breakpoint positions on exon-exon boundaries [29]. The major disadvantage of this tool is that it is insensitive to the fusion incorporated into the intronic and intragenic sequence as this tool considers only exonic sequences as reference. It is open-source and available at https://github.com/Oshlack/JAFFA/wiki.
4. *TopHat:* The latest version of TopHat is TopHat-Fusion which allows RNA-Seq reads to align across fusion junctions. It does not rely on gene annotation to discover fusion deriving from the known gene, unknown gene, and unannotated spliced variants of a known gene. This tool finds candidate fusion, aligns reads across them, and then filters out false fusion by post-processing steps. It uses Bowtie to find the reads that align within exons. Following this, it creates a set of partial exons from these alignments with hypothetical boundaries, re-aligns the initially unmapped reads, and finds fusion events. As this tool does not rely on gene annotation and discordantly mapping paired reads, so increases the sensitivity and helps in finding novel genes and splice variants fusions of known genes. This tool is available at http://tophat-fusion.sourceforge.net/ [30].
5. *SOAPfuse:* This tool utilizes paired-end RNA-Seq data to identify fusion transcript. This software is implemented in Perl and is only executable on Linux. This tool has higher detection efficiency and uses fewer computational resources than other tools. This tool utilizes discordant, and junction reads to confirm the sites. SOAPfuse constructs a library of fusion junction sequences, which undergoes a series of filters to filter out false-

positive fusions. The output of this software provides information on fusion location, junction sequence, and diagram representing the location of reads concerning the junction sequence and expression level of the exon, which is very helpful for further analysis. It is available at http://soap.genomics.org.cn/soapfuse.html [31].

6. *FusionSeq:* This tool also utilizes paired-end RNA-Seq to detect fusion transcript. FusionSeq considers all possible exons involved in the junction sequence and generates a junction library from all possibilities. The output provides a list of fusion candidates with high confidence and is scored so they can easily undergo validation studies. The disadvantage of using this is that it requires high CPU time and memory, mainly when multiple samples are analyzed in parallel. In addition, it removes false fusion generated by misalignment and random pairing [32].
7. *STARChip (STAR Chimeric Post):* As per studies, it was found that across the genome, circular isoforms of RNA are expressed and may be responsible for the disease. This software applies the STAR aligner to chimeric alignments to generate annotated circRNA and fusions. A significant advantage of this tool is that it is very effective for high-dimensional datasets and provides better performance and time efficiency [33].
8. *STAR-Fusion (Component of CTAT):* This method is fast and efficient for detecting fusion transcript from RNA-Seq data. This tool is implemented in Perl, and takes RNA-Seq data as input and provides a list of candidate fusion transcripts as output. It utilizes chimeric reads found by STAR into chimeric.out.junction file to find candidate fusion, followed by filtering to remove false positives. It is an open-source software freely available at http://star-fusion.github.io (GitHub-STAR-Fusion/STAR-Fusion: STAR-Fusion Codebase n.d.) [34].
9. *MapSplice:* Similar to TopHat, this tool is also based on Bowtie indexing [18]. The primary purpose of this tool is to demonstrate the exon splice junction in the sampled mRNA by using long and short reads and also detect the alignment of mRNA sequences tag to reference genome. This pipeline follows two phases: tag alignment phase and splice inference, involving segmentation of reads, exonic and spliced alignment, segments assembly, and junction detection for candidate alignments. As a result, it has a high specificity and sensitivity with CPU and memory efficiency. Furthermore, this software does not depend on features like intron length, and splice site so it can detect both canonical and non-canonical splices. In addition,

MapSplice has the advantage of increased accuracy because of the quality and diversity of read alignment [35].

10. *FusionHunter:* This tool is implemented in C and C++ and based on discordant read pairs to identify the candidate fusion and follows Bowtie for the alignment of reads on the genome. First, FusionHunter maps the reads to the reference genome to detect the putative exons and transcript, followed by resolving false positives caused by homology to detect the candidate fusion. Then by using the Pseudo-reference of each candidate, it detects the reads that precisely span the junction. As output, they generate a SAM format file which is further used to report fusion transcript [36].
11. *TrinityFusion* [22]*:* This tool utilizes "Trinity" for de novo transcriptome assembly using RNA-Seq Data. GMAP is used to identify candidate fusion, and Bowtie2 is utilized for capturing those reads that support the fusion. According to the characteristics of the fusion gene partners and evidence, candidate fusions are filtered. Based on the type of reads utilized for de novo assembly and fusion identification, TrinityFusion has three types:
    - *TrinityFusion-C* utilizes only the chimeric read.
    - TrinityFusion-*UC* utilizes both chimeric as well as those reads which are not mapped to the genome.
    - TrinityFusion-*D* utilizes all input reads.
12. *Arriba:* This tool is based on the STAR aligner and identifies gene fusion from the RNA-Seq data. High sensitivity and short runtimes are important features of this tool. This algorithm produces reusable alignments that are further utilized for gene expression quantification. Unlike other tools which are designed on STAR, this does not need to reduce the STAR parameter for the detection of fusion that emerged from local deletions. This feature affects the mapping of reads with long introns and may interfere quantification of expression. Besides fusion detection, this tool can detect another structural rearrangement that is clinically relevant [22].
13. *ChimPipe:* This pipeline identifies fusion genes and chimeric transcript using Illumina paired-end RNA-Seq data. This tool utilizes split reads and discordant paired-end reads to detect fusions, and the identified Chimeras can be verified in vitro with high accuracy. First, RNA-Seq reads are mapped using GEMtools, and those reads which are not mapped are passed to GEM RNA-mapper to extract reads that split map to a different chromosome. Then by ChimSplice gathered, split reads are given to generate consensus junctions. Further, ChimPE associates each chimeric junction with their discordant PE reads by splitting them into consistent and inconsistent

junctions. Finally, ChimFilter filters out the false positive and generates reliable chimeric junctions [37].

6 Factors Affecting Fusion Detection

Various factors affect fusion detection by RNA-Seq data, some of which are listed below.

- *Type of reads:* The sequencing format of the reads, including single-end sequencing reads, sequenced from one side, and paired-end reads that are generated when sequencing is done from both ends, influences fusion detection. It has been shown that paired-end reads provide better fusion detection [24].
- *Effect of reference sequence:* It was found that the majority of the detection tool need both genome and transcriptome sequences. Some tools require transcriptome as reference genomes, such as EricScript and JAFFA, whereas Bellerophontes, Dissect, MapSplice, TRUP, and CRAC require genome sequences as reference [23].
- *Assembly and mapping of reads:* RNA-Seq based approaches used two methods for the fusion junction prediction: "Mapping first" and "Assembly first." In the "Mapping first" approach, the split and spanning reads are first aligned to the reference genome, while in the "Assembly first," the reads are first assembled into reads and contigs, and then these are aligned back to the reference to detect the fusion breakpoint [38]. According to Kumar et al. the "mapping first" approach is less time-consuming and better for detecting fusion candidates using RNA-Seq [24]. Tools like JAFFA and BreakFusion utilize a combination of both approaches.
- *Effect of read length:* Read length also affects fusion detection, with the majority of tools having greater accuracy with longer reads compared to shorter reads, while some tools such as SOAPfuse and FusionHunter also show better accuracy with shorter reads. Some tools like PRADA show similar result despite variable reading length [22].

7 Conclusion

With the advancement of NGS, a large amount of data has been generated. This data can be utilized for various kinds of study using bioinformatic tools. In this chapter, we discussed the detection of fusion transcript using RNA-Seq data with the help of various bioinformatics tools. The tools discussed in this chapter are useful for accurately predicting candidate fusion transcripts, but much more work needs to be done in this area. For future tools, some fundamental characteristics like user-friendly, high sensitivity, high specificity, time effectiveness, and greater computational memory should be taken into consideration.

References

1. Singh A, Zahra S, Das D, Kumar S (2019) AtFusionDB: a database of fusion transcripts in Arabidopsis thaliana. Database (Oxford) 2019. https://doi.org/10.1093/DATABASE/BAY135
2. Dai X, Theobard R, Cheng H et al (2018) Fusion genes: a promising tool combating against cancer. Biochim Biophys Acta Rev Cancer 1869:149–160. https://doi.org/10.1016/J.BBCAN.2017.12.003
3. Li Z, Qin F, Li H (2018) Chimeric RNAs and their implications in cancer. Curr Opin Genet Dev 48:36. https://doi.org/10.1016/J.GDE.2017.10.002
4. Rowley JD (1973) Letter: a new consistent chromosomal abnormality in chronic myelogenous leukaemia identified by quinacrine fluorescence and Giemsa staining. Nature 243: 290–293. https://doi.org/10.1038/243290A0
5. Shtivelman E, Lifshitz B, Gale RP, Canaani E (1985) Fused transcript of abl and bcr genes in chronic myelogenous leukaemia. Nature 315: 550–554. https://doi.org/10.1038/315550A0
6. Hollenhorst PC, McIntosh LP, Graves BJ (2011) Genomic and biochemical insights into the specificity of ETS transcription factors. Annu Rev Biochem 80:437–471. https://doi.org/10.1146/ANNUREV.BIOCHEM.79.081507.103945
7. Hrzenjak A (2016) JAZF1/SUZ12 gene fusion in endometrial stromal sarcomas. Orphanet J Rare Dis 11. https://doi.org/10.1186/S13023-016-0400-8
8. Kim HP, Cho GA, Han SW et al (2014) Novel fusion transcripts in human gastric cancer revealed by transcriptome analysis. Oncogene 33:5434–5441. https://doi.org/10.1038/ONC.2013.490
9. Tang Y, Qin F, Liu A, Li H (2017) Recurrent fusion RNA DUS4L-BCAP29 in non-cancer human tissues and cells. Oncotarget 8:31415–31423. https://doi.org/10.18632/ONCOTARGET.16329
10. Babiceanu M, Qin F, Xie Z et al (2016) Recurrent chimeric fusion RNAs in non-cancer tissues and cells. Nucleic Acids Res 44:2859–2872. https://doi.org/10.1093/NAR/GKW032
11. Thomson TM, Lozano JJ, Loukili N et al (2000) Fusion of the human gene for the polyubiquitination coeffector UEV1 with Kua, a newly identified gene. Genome Res 10:1743–1756. https://doi.org/10.1101/GR.GR-1405R
12. Li H, Wang J, Mor G, Sklar J (2008) A neoplastic gene fusion mimics trans-splicing of RNAs in normal human cells. Science 321: 1357–1361. https://doi.org/10.1126/SCIENCE.1156725
13. Chen H, Tang Y, Liu J et al (2017) Emergence of a novel chimeric gene underlying grain number in rice. Genetics 205:993–1002. https://doi.org/10.1534/GENETICS.116.188201/-/DC1
14. Zhou Y, Zhang C, Zhang L et al (2022) Gene fusion as an important mechanism to generate new genes in the genus Oryza. Genome Biol 23:1–23. https://doi.org/10.1186/S13059-022-02696-W/FIGURES/8
15. Kumar S, Vo AD, Qin F, Li H (2016) Comparative assessment of methods for the fusion transcripts detection from RNA-Seq data. Sci Rep 6:21597. https://doi.org/10.1038/srep21597
16. Weinstein JN, Collisson EA, Mills GB et al (2013) The cancer genome atlas pan-cancer analysis project. Nat Genet 45(10): 1113–1120. https://doi.org/10.1038/ng.2764
17. Lonsdale J, Thomas J, Salvatore M et al (2013) The genotype-tissue expression (GTEx) project. Nat Genet 45(6):580–585. https://doi.org/10.1038/ng.2653
18. Narayan A, Pahwa B, Kumar S (2022) Computational tools and databases for fusion transcripts: therapeutic targets in cancer. In: Systems biomedicine approaches in cancer research, pp 115–127. https://doi.org/10.1007/978-981-19-1953-4_6
19. Babraham bioinformatics—FastQC a quality control tool for high throughput sequence data. https://www.bioinformatics.babraham.ac.uk/projects/fastqc/. Accessed 7 Dec 2022
20. Pandey RV, Pabinger S, Kriegner A, Weinhäusel A (2016) ClinQC: a tool for

quality control and cleaning of Sanger and NGS data in clinical research. BMC Bioinform 17. https://doi.org/10.1186/S12859-016-0915-Y
21. Chen S, Zhou Y, Chen Y, Gu J (2018) fastp: an ultra-fast all-in-one FASTQ preprocessor. Bioinformatics 34:i884–i890. https://doi.org/10.1093/BIOINFORMATICS/BTY560
22. Haas BJ, Dobin A, Li B et al (2019) Accuracy assessment of fusion transcript detection via read-mapping and de novo fusion transcript assembly-based methods. Genome Biol 20. https://doi.org/10.1186/S13059-019-1842-9
23. Wang Q, Xia J, Jia P et al (2013) Application of next generation sequencing to human gene fusion detection: computational tools, features and perspectives. Brief Bioinform 14:506–519. https://doi.org/10.1093/BIB/BBS044
24. Kumar S, Razzaq SK, Vo AD et al (2016) Identifying fusion transcripts using next generation sequencing. Wiley Interdiscip Rev RNA 7:811–823. https://doi.org/10.1002/WRNA.1382
25. Lee CH, Wbin O, Mariño-Enriquez A et al (2012) 14-3-3 fusion oncogenes in high-grade endometrial stromal sarcoma. Proc Natl Acad Sci USA 109:929–934. https://doi.org/10.1073/PNAS.1115528109
26. Steidl C, Shah SP, Woolcock BW et al (2011) MHC class II transactivator CIITA is a recurrent gene fusion partner in lymphoid cancers. Nature 471(7338):377–381. https://doi.org/10.1038/nature09754
27. Benelli M, Pescucci C, Marseglia G et al (2012) Discovering chimeric transcripts in paired-end RNA-Seq data by using EricScript. Bioinformatics 28:3232–3239. https://doi.org/10.1093/BIOINFORMATICS/BTS617
28. Iyer MK, Chinnaiyan AM, Maher CA (2011) ChimeraScan: a tool for identifying chimeric transcription in sequencing data. Bioinformatics 27:2903–2904. https://doi.org/10.1093/BIOINFORMATICS/BTR467
29. Davidson NM, Majewski IJ, Oshlack A (2015) JAFFA: high sensitivity transcriptome-focused fusion gene detection. Genome Med 7:1–12. https://doi.org/10.1186/S13073-015-0167-X/COMMENTS
30. Kim D, Salzberg SL (2011) TopHat-Fusion: an algorithm for discovery of novel fusion transcripts. Genome Biol 12:1–15. https://doi.org/10.1186/GB-2011-12-8-R72/FIGURES/6
31. Jia W, Qiu K, He M et al (2013) SOAPfuse: an algorithm for identifying fusion transcripts from paired-end RNA-Seq data. Genome Biol 14:1–15. https://doi.org/10.1186/GB-2013-14-2-R12/FIGURES/10
32. Sboner A, Habegger L, Pflueger D et al (2010) FusionSeq: a modular framework for finding gene fusions by analyzing paired-end RNA-sequencing data. Genome Biol 11:1–19. https://doi.org/10.1186/GB-2010-11-10-R104/FIGURES/5
33. Akers NK, Schadt EE, Losic B (2018) STAR Chimeric Post for rapid detection of circular RNA and fusion transcripts. Bioinformatics 34:2364–2370. https://doi.org/10.1093/BIOINFORMATICS/BTY091
34. GitHub—STAR-Fusion/STAR-Fusion: STAR-Fusion codebase. https://github.com/STAR-Fusion/STAR-Fusion. Accessed 6 Dec 2022
35. Wang K, Singh D, Zeng Z et al (2010) MapSplice: accurate mapping of RNA-Seq reads for splice junction discovery. Nucleic Acids Res 38:e178. https://doi.org/10.1093/NAR/GKQ622
36. Li Y, Chien J, Smith DI, Ma J (2011) FusionHunter: identifying fusion transcripts in cancer using paired-end RNA-Seq. Bioinformatics 27:1708–1710. https://doi.org/10.1093/BIOINFORMATICS/BTR265
37. Rodríguez-Martín B, Palumbo E, Marco-Sola S et al (2017) ChimPipe: accurate detection of fusion genes and transcription-induced chimeras from RNA-Seq data. BMC Genomics 18:1–17. https://doi.org/10.1186/S12864-016-3404-9/TABLES/3
38. Deng W, Murugan S, Lindberg J et al (2022) Fusion gene detection using whole-exome sequencing data in cancer patients. Front Genet 13:287. https://doi.org/10.3389/FGENE.2022.820493/BIBTEX

Chapter 15

GAN Learning Methods for Bulk RNA-Seq Data and Their Interpretive Application in the Context of Disease Progression

Yongjin Kim and Mookyung Cheon

Abstract

A generative adversarial network (GAN) is a generative model that consists of two adversarial networks, a discriminator and a generator, usually in the form of neural networks. One of the useful things about applying GANs is that they can synthesize two states to produce an intermediate output that implies a semantic feature. When applied to omics data that determine phenotypes of a disease, GANs can be used to associate these intermediate outputs with the progression of the disease. In this chapter, to realize the above idea, we will introduce the application of GAN methods to bulk RNA-seq data, which cover data preprocessing, training, and latent interpolation between different phenotypes describing disease progression.

Key words Generative adversarial networks (GANs), Bulk RNA-seq, Disease progression, Alzheimer's disease

1 Introduction

Alzheimer's disease (AD) is a complex neurological disorder that leads to dementia, and in addition to amyloid-beta and tau pathology, several molecular mechanisms, such as neuroinflammation, oxidative stress, mitochondrial dysfunction, synaptic plasticity impairment, impaired protein clearance, and vascular dysfunction, contribute to its development [1–3]. These modifiers, or predisposing factors, can interact with the primary amyloid and tau proteinopathies to affect the incidence of the disease, severity of symptoms, and rate of progression. These factors can come from a variety of sources, including genetic mutations, environmental factors, and lifestyle habits [1, 4]. Given the complex interactions between major proteinopathies and their modalities, it is expected that more innovative approaches will need to be developed in multiple research and clinical areas to accurately treat the disease

Rajeev K. Azad (ed.), *Transcriptome Data Analysis*, Methods in Molecular Biology, vol. 2812,
https://doi.org/10.1007/978-1-0716-3886-6_15,

[5]. In particular, personalized and precise dementia diagnostics will require breakthrough technologies to better understand why individual patients develop dementia or why some people are resistant to it.

Multi-omics data, including genomics, epigenetics, transcriptomics, proteomics, and metabolomics, are critical to comprehensively understand the molecular changes in individual AD patients [6]. However, much of these omics data need to be taken from brain tissue, not human blood, to be analyzed, which is impossible to do on a living person for diagnostic and therapeutic purposes [7]. The practical difficulty of obtaining tissue samples from patients motivated the development of an AI-powered prediction system. Recent notable achievements include predicting tissue-specific transcriptome [8] and epigenetics data from genomics data [9]. In a different direction from these studies, we developed a disease progression prediction system, which generates intermediate states by a generative model, based on phenotypic omics data such as transcriptomes. Generative models including Variational AutoEncoder (VAE) [10], Generative Adversarial Networks (GAN) [11], Transformer [12], and diffusion probabilistic models [13] might be suitable for such disease progression prediction systems. However, when dealing with transcriptomic data, the sample size is often much smaller compared to the training data used in typical image generation AI methods. Therefore, it is crucial to consider network architectures that can effectively learn transcriptome features within the limitations of these generative models.

The initial application of GANs to transcriptomic data focused on single-cell RNA-seq data [14]. Despite their brilliant idea, the similarity between the generated data and the real data was low due to the noisy nature of the single-cell data itself. Because our goal is to track and predict changes at the gene level during disease progression, we designed a pipeline that utilizes bulk RNA-seq data rather than single-cell data with large variation in gene expression.

This chapter describes the GAN application which we used on bulk RNA-seq data from two mouse models of amyloid-beta pathology (5xFAD) [15] and tau pathology (TPR50-P301S) [16], respectively. The two datasets, which encompass both wild-type (WT) and disease phenotypes (AD), exhibit variations in the number of age groups, and as a result, employ distinct methods to analyze disease progression. Given that the 5xFAD dataset has three age groups (2 months (2M), 4 months (4M), and 7 months (7M)) and the TPR50-P301S dataset has two age groups (3 months (3M) and 6 months (7M)), we will discuss both methods of applying GAN in this chapter. The GAN application includes data preprocessing, data augmentation, normalization, learning process, optimization, latent interpolation, gene clustering, and gene set enrichment.

2 Materials: Data Preprocessing and GAN Training

2.1 Download and RNA-Seq Analysis

1. Sequence read archive files of public AD-associated data were downloaded from the NCBI Gene Expression Omnibus (GEO) database (accession numbers GSE104775 for 5xFAD [17], GSE90693 for TPR50-P301S [18]), and raw fastq files were generated using the fastq-dump command in the NCBI SRA toolkit [19].
2. We followed a commonly used RNA-seq analysis pipeline. Data cleaning and adapter trimming were performed using Trimmomatic [20]. Alignment to the mouse reference genome GRCm38 was performed using HISAT2 [21]. Gene counts were quantified by using HTSeq-counts [22] and filtering to remove low read counts (average count < 100). Differential gene expression analysis was performed using the DESeq2 package [23]. There are differences in the program versions used in the two datasets, as the versions employed were the latest available at the time of analysis (*see* **Note 1**).

2.2 Gene Selection and Data Augmentation

The selection of the gene list on which to train a GAN to monitor disease progression should be carefully considered in advance, depending on the purpose of the study. While it is desirable to select as many genes as possible, in our experienced opinion, selecting a large number of genes from training data with a small number of samples can be problematic to ensure the robustness of the results. In other words, the number of genes must not be so large that small changes in the hyperparameters of the network will not significantly alter the pattern of gene expression changes.

1. We only selected 1208 significantly differentially expressed genes (DEGs) between AD7M and WT7M of 5xFAD mice and a union set of 3767 DEGs from four comparative pairs for TPR50 mice: WT maturation (WT6M vs. WT3M), AD maturation (AD6M vs. AD3M), early phenotype changes (AD3M vs. WT3M) and late phenotype changes (AD6M vs. WT6M).
2. We processed the gene expression counts by transforming into the expression profile as regularized logarithmic data (RLD) utilizing the DESeq2 method. Leveraging 36 (5xFAD) or 20 (TPR50) authentic sample data, segregated into six or four phenotype-age groups, we developed a data augmentation procedure for RLD, thereby generating a total of 846 or 380 samples, respectively (*see* **Note 2**).

2.3 Normalization

Based on our empirical observations, the normalization process of real input gene expression profiles should take into account the characteristics of the activation function employed in the final layer

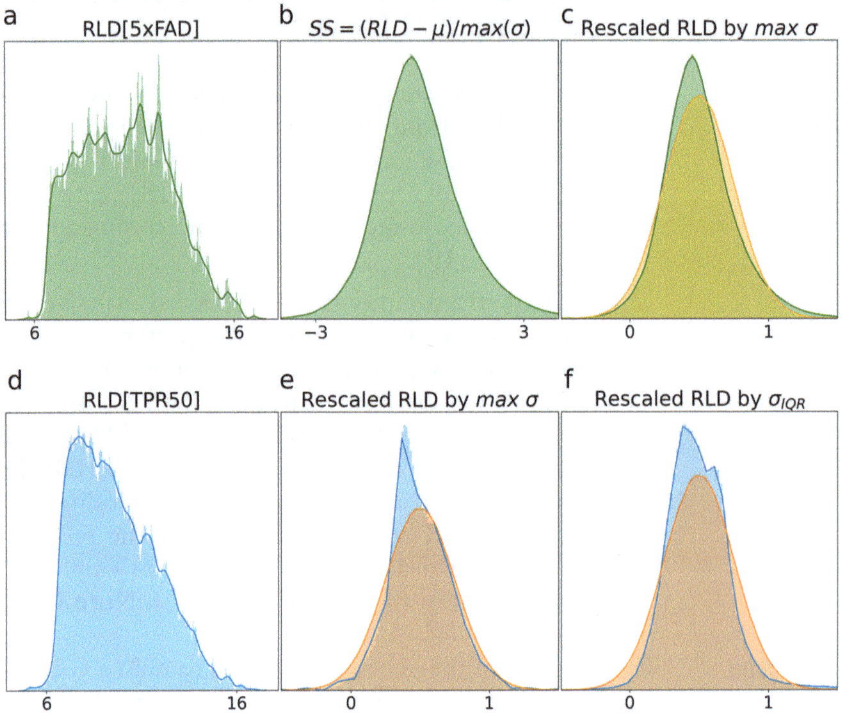

Fig. 1 (**a**–**c**) Plots showing the distributions of the 846 augmented and real samples from the 5xFAD dataset: (**a**) the distribution of original RLD values, (**b**) the distribution (SS) after performing standard scaling with the denominator max($\sigma[k, j]$), (**c**) the distribution after rescaling SS as SS/(3.918σ_{SS}) + 0.5. (**d**–**f**) Plots illustrating the distributions of the 380 augmented and real samples from the TPR50 dataset: (**d**) the distribution of original RLD values, (**e**) the distribution of rescaled RLD values obtained using the denominator max($\sigma[k, j]$) in standard scaling, (**f**) the distribution of rescaled RLD values obtained using the denominator σ_{IQR} in standard scaling

of the GAN's generator. After experimenting with various activation functions, we found that the leaky ReLU activation function, as demonstrated in the study by Ghahramani et al., is well suited for GAN training. Nonetheless, employing this activation function results in generated values that are either less than zero or exceed one. Therefore, to maintain consistency between the real data and the generated data, it is necessary to adjust the normalization of the real data based on the distribution pattern exhibited by the generated data.

1. The DESeq2 normalized data (RLD) of the selected genes were subjected to rescaling such that 95% of the values fell within the [0, 1] interval. Furthermore, it is desirable for the rescaled distribution of all gene expression values to resemble a Gaussian normal distribution (Fig. 1 and *see* **Notes 3** and **4**). This step is beneficial in ensuring that the generated values for each gene fall within the range of the actual values. Thus, the

normalization processes applied to the 5xFAD and the TPR50 datasets exhibit slight differences.

2.4 Training Data Set

1. For the 5xFAD dataset, the 846 augmented samples were divided into a training set and a test set. Specifically, 762 samples were randomly selected for the training set, while the remaining 84 samples were allocated for the test set. The objective of the test set data was to evaluate the generator's ability to produce realistic synthetic samples on data that were not included in the training process. We did not employ a validation set for hyperparameter optimization due to the tricky nature of determining an appropriate scoring measure in the context of the unsupervised generative model utilized to depict disease progression.
2. However, for the TPR50 dataset, we employed the entire set of 380 augmented samples as the training set. This decision was driven by our primary focus on generating highly accurate synthetic data to effectively capture disease progression through latent interpolation, rather than validating the network architecture and methodology using a separate test set.

2.5 Network Architecture, Hyperparameters, and GAN Training

1. We utilized the Wasserstein GAN with a gradient penalty, comprising two networks: the generator and the discriminator. The networks were implemented using fully connected layers (FC) and a leaky rectified linear unit (ReLU) activation function. TensorFlow with Keras in Python was utilized for the implementation (*see* **Note 5**).
2. After conducting multiple trials with different pairs of hidden layer units, we determined the number of hidden layer units for the generator and discriminator to be 250 and 150, respectively, for the 5xFAD dataset, and 450 and 270, respectively, for the TPR50 dataset (*see* Fig. 2).
3. We initialized the random weight parameters in the networks within the range of [−0.3, 0.3], and generated random variables in the latent space following the distribution of the rescaled RLD profile data. By selecting random variables in the latent space from the distribution of real data, the generated samples seem to be adhere to the same manifold structure, thereby eliminating the need to fine-tune parameters based on a combination of Gaussian and Poisson distributions. This approach not only ensures a more stable learning process in practice but also proves notably beneficial in enhancing the training performance of the network.
4. The weights were updated through learning based on the loss of the Wasserstein GAN with gradient penalty (gradient penalty = 10), using the Adam optimizer (learning rate = 1e−5)

a

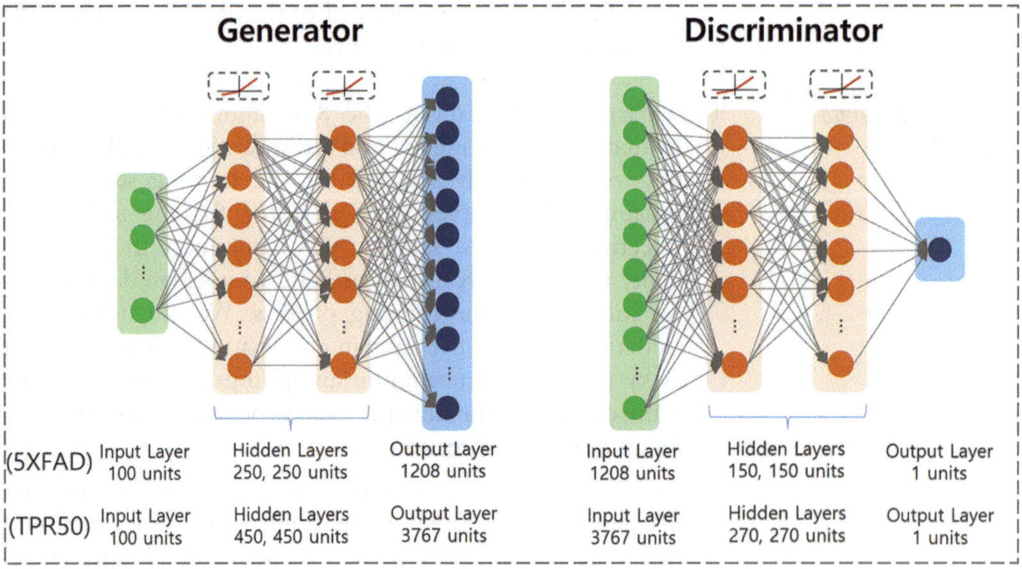

b

Hyperparameter	Setting	
	5XFAD	TPR50
Initial learning rate	1e-5	1e-5
Number of layer units (Generator)	100, 250, 250, 1208	100, 450, 450, 3767
Number of layer units (Discriminator)	1208, 150, 150, 1	3767, 270, 270, 1
Activation	Leaky ReLU	
Optimizer	ADAM	
Batch size	32	
Epochs	200,000	300,000
Loss	$E_{\hat{x}\sim P_g}[D(\hat{x})] - E_{x\sim P_r}[D(x)] + \lambda E_{\hat{x}\sim P_x}[(\|\nabla_{\hat{x}} D(\hat{x})\|_2 - 1)^2]$, $\lambda = 10$	

Fig. 2 (**a**) The network architecture and (**b**) hyperparameters for training the 5xFAD and the TPR50 datasets

for 200,000 (5xFAD) or 300,000 (TPR50) epochs, and employing a minibatch size of 32.

5. We assessed the performance of the generator output at various checkpoints using t-distributed stochastic neighbor embedding (t-SNE) and by examining the correlation between real and generated samples (Fig. 3a, b for t-SNE).

3 Methods: Implementation of the Concept of Disease Progression

3.1 Realistic Fakes and Latent Interpolation

1. At each epoch, we randomly generated 10,000 latent vectors (z) and obtained corresponding gene expression profile vectors (x) by applying the generator G. The distributions of correlations between these generated fake and real samples were plotted in Fig. 3c, d. We obtained the top ten profile vectors with the highest correlation ($G(z)_{\text{high}}$) for each augmented sample. From these vectors, we calculated the average latent space

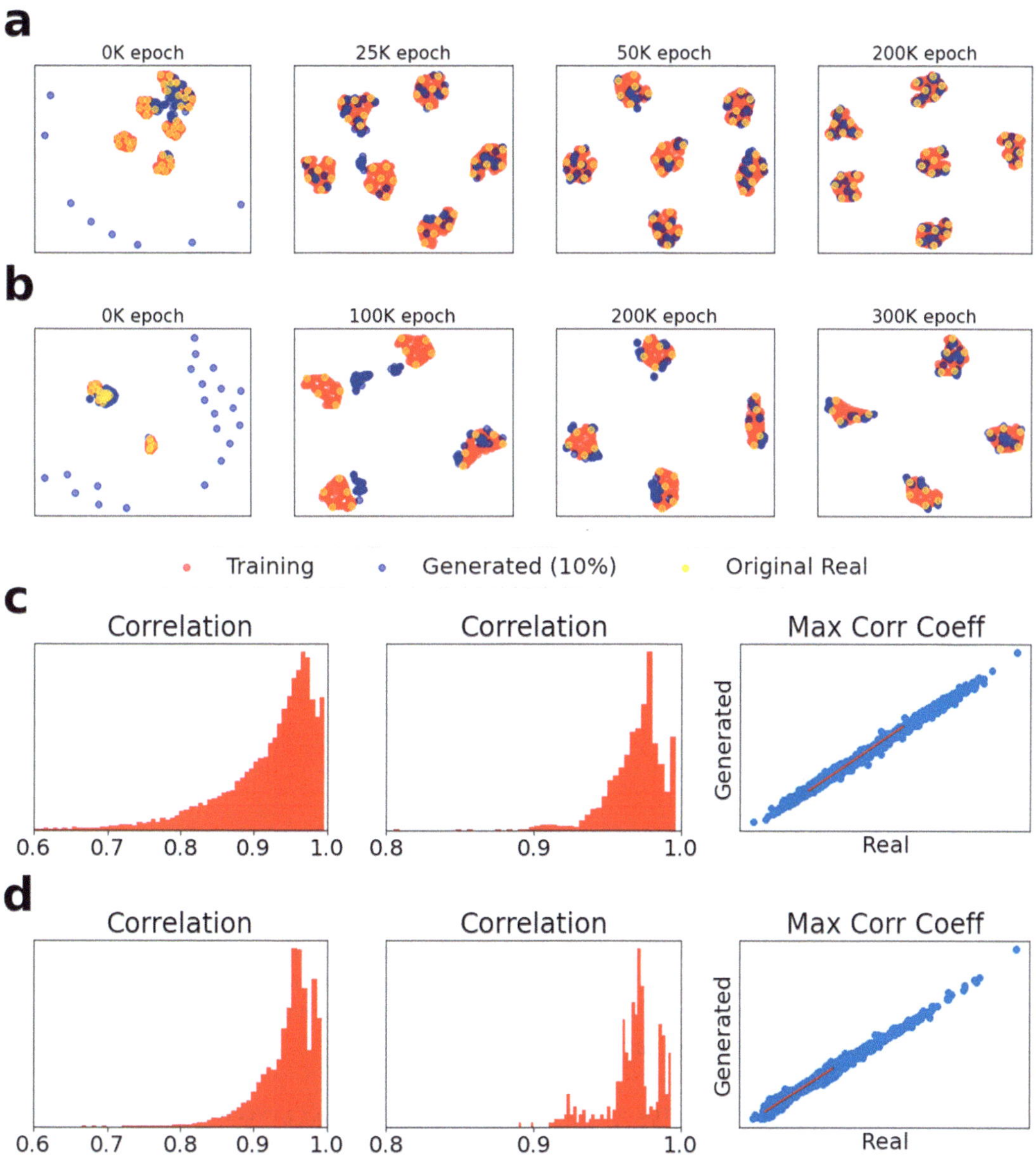

Fig. 3 (**a**, **b**) Temporal plots demonstrate the good match between the generated samples and (**a**) the six groups (5xFAD) and (**b**) the four groups (TPR50). t-SNE plots at four epochs are presented, where different colored dots represent the training samples (red), the original real samples (yellow), and samples generated at a rate of 10% of the training data (blue). (**c**, **d**) Plots depicting the quality of the generated data for (**c**) the 5xFAD and (**d**) the TPR50 datasets. (left) Distribution of the highest correlation among the real-and-augmented samples for each of the 10,000 generated data. (middle) Distribution of the highest correlation among the 10,000 generated data for each of the real-and-augmented samples. (right) Scatter plot showing the pair with the highest correlation coefficient between the averaged top ten correlated samples and the original real samples

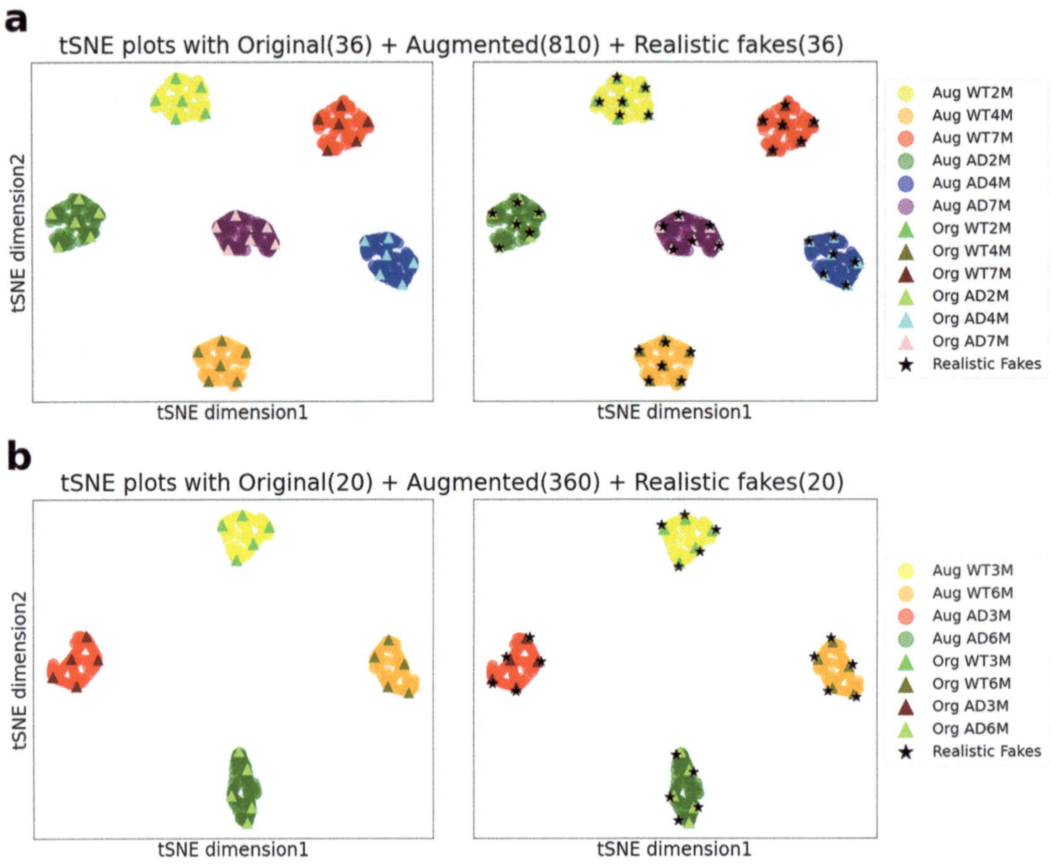

Fig. 4 The t-SNE plots of the combined data are shown for (**a**) the original (36), augmented (810), and realistic fake (36) samples for the 5xFAD dataset, and (**b**) the original (20), augmented (320), and realistic fake (20) samples for the TPR50 dataset. The t-SNE plots demonstrate excellent clustering of the original and augmented samples within their respective groups. Additionally, the t-SNE plot of the combined data, which includes the realistic fake samples (represented by black stars), exhibits noticeable overlaps between the original samples and the realistic fake samples

vector ($<z>$) over the ten $G(z)_{\text{high}}$ vectors, and used it to generate an averaged profile ($G(<z>)$).

2. Using this approach, we generated 846 realistic fake gene profiles ($G(<z>)$) for the 5xFAD dataset, corresponding to the 846 augmented data points in each epoch, resulting in 114 realistic fakes for each group and 36 realistic fakes for the original real (Fig. 4a).
3. Similarly, for the TPR50 dataset, we generated 380 realistic fake gene profiles, corresponding to the 380 augmented data points in each epoch, yielding 95 realistic fakes for each group 20 realistic fakes for the original real (Fig. 4b).
4. A difference vector ($\Delta(i, j)$ was calculated by subtracting the averaged latent vectors between two groups (i and j) as

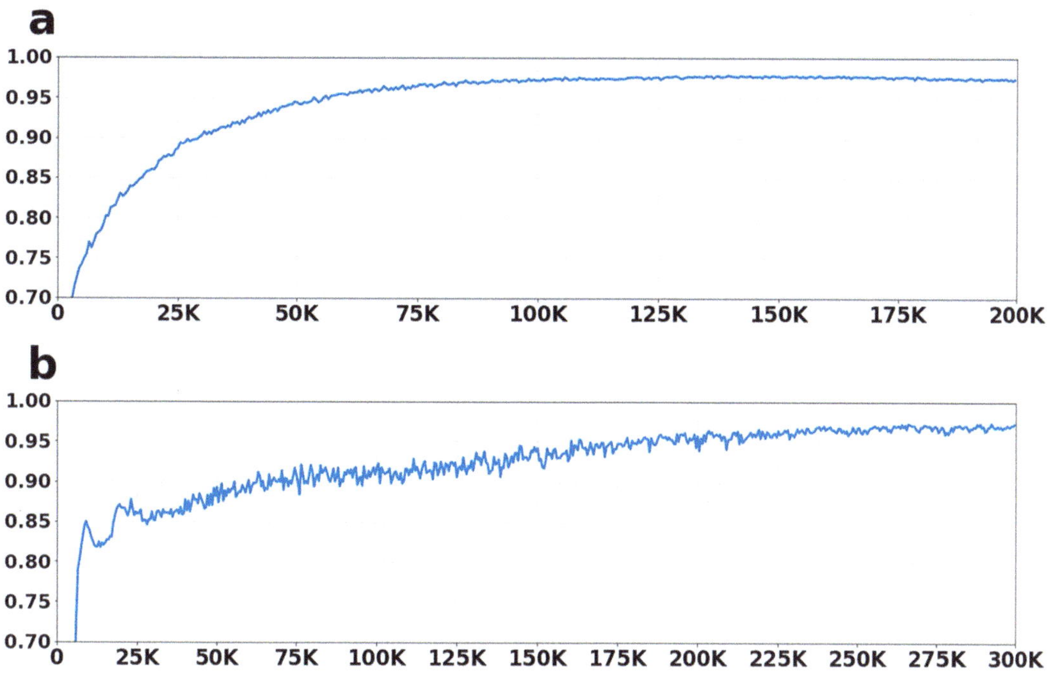

Fig. 5 The average curves of correlation values between the realistic fake and the corresponding original real samples for (**a**) the 5xFAD dataset and (**b**) the TPR50 dataset

$\Delta(i,j) = \sum_k^n PG(i,k)/n - \sum_k^n PG(j,k)/n$, where n represents the number of all realistic fakes for each phenotype group (PG) such as $n = 141$ for the 5xFAD dataset and $n = 95$ for the TPR50 dataset. For instance, to identify the transitional state between 6-month WT and AD in the TPR50 dataset, we estimated the simulated transitional states using the arithmetic equation $z(t, i) = z_{WT6M(i)} + t\Delta$ and the generator $G(z(t, i))$, where t ranges from 0 to 1 and $z_{WT6M(i)}$ represents n latent vectors for WT6M.

5. We calculated the transition curves (TCs), denoted as $T(t)$, by averaging the values for 1208 or 3767 genes, where $T(t) = \sum_k^n G(z(t,k))/n$.
6. Finally, we obtained the average transition curves ($<T(t)>$) by averaging the TCs over 100 epoch points within 50k epochs (75k ~ 125k for the 5xFAD dataset or 225k ~ 275k for the TPR50 dataset). The interval for performing the averaging was determined based on the average curve of correlation values between the realistic fake and the corresponding original real samples (Fig. 5).

3.2 Transition Curves Describing Disease Progression

We postulate that the gene transition curves (TCs) extracted from the trained generative model capture phenotypic changes that could be observed during disease progression. Indeed, we observed a strong correlation in the curve patterns of genes that are known to be highly related or share similar names. This finding suggests that co-expressed genes involved in causal biological processes exhibit similar patterns as they traverse the simulated disease progression trajectory. In the 5xFAD mouse study, which encompassed three age groups (2, 4, and 7 months), we focused on a single comparative set of differentially expressed genes (AD7M vs. WT7M) to estimate TCs from WT7M to AD7M for analyzing simulated disease progression. Given the incorporation of data from 2-month and 4-month time points during GAN training, it is anticipated that the latent space would encompass the semantic features pertaining to early expression changes that subsequently manifest in the 7-month phenotype alteration. However, based on our preliminary pilot study with only two age groups (3 and 6 months) for the TPR50 dataset, it was observed that the phenotypic changes at 6 months inadequately capture early changes and primarily depict later-stage alterations. Therefore, in order to analyze disease progression by examining patterns of TCs, we adopted one-way TCs for the 5xFAD dataset, which included three age groups. In contrast, for the TPR50 dataset consisting of two age groups, we employed four-way TCs.

3.3 One-Way TCs for the 5xFAD Dataset, Gene Association to Patterns, and Pathway Analysis

1. Using the one-way TCs (from WT7M to AD7M) of the 1208 genes (7M DEGs), we categorized them into six distinct patterns (Fig. 6). Each transition curve can be associated with multiple predefined patterns by assessing the correlations with the six predefined patterns (represented by the red lines in Fig. 6). The patterns are classified based on whether the curve changes primarily occur during early stages (patterns 1 and 4), gradual progression (patterns 2 and 5), or late stages (patterns 3 and 6). The first three patterns (patterns 1–3) represent transition curves for upregulated genes, while the last three patterns (patterns 4–6) correspond to downregulated genes.
2. Instead of using traditional clustering methods to group genes into six patterns, we intentionally assigned genes to predefined patterns based on their high correlation, which allows some genes to be associated with multiple patterns. This deliberate approach ensured that a sufficient number of genes displaying early-stage changes were included, which proved beneficial for gene set enrichment analysis encompassing gene ontology and pathways, thereby facilitating a comprehensive understanding of disease-related molecular processes.
3. Using the gene sets associated with the six patterns, we conducted gene ontology and pathway analysis. Through assessment of adjusted p-values and enrichment ratios, we identified molecular events occurring early or late in the disease

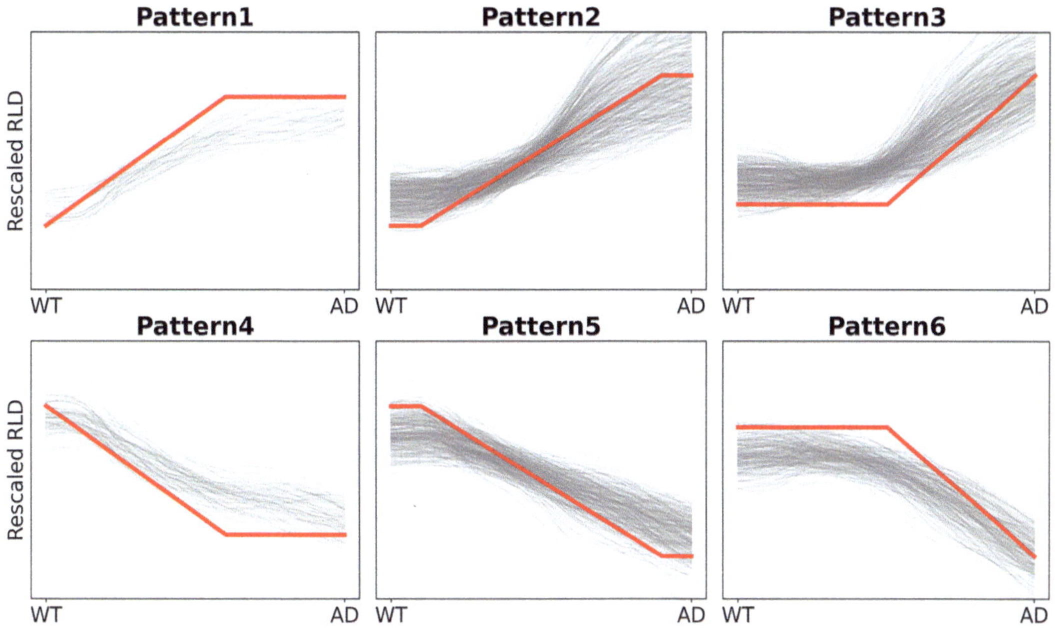

Fig. 6 The one-way TCs of genes from WT7M to AD7M were presented alongside the six predefined patterns represented by red lines. A curve is assigned to a pattern when its correlation coefficient (*r*) is higher than 0.95, or when the maximum correlation coefficient (*r*) is above 0.90

progression. For example, we observed that pattern 1 had the highest −log10(adj.p) and log2(ER) values for cholesterol biosynthesis compared to other patterns, while pattern 3 exhibited the highest values for cholesterol metabolism, response to lipoprotein particle, and immunity-associated pathways.

4. Based on these findings in the amyloidopathy mouse model, we can infer that cholesterol biosynthesis occurs earlier than cholesterol metabolism, as well as other lipid metabolism processes related to amyloid pathology and immune-associated pathways. Given the primary objective of this paper, which is to highlight the application of GANs in disease progression, we have provided only brief descriptions of gene ontologies and pathways to elucidate the uncovering of insights into disease progression.

3.4 Four-Way TCs for the TPR50 Dataset, Gene Clustering into Patterns, and Pathway Analysis

1. As mentioned previously, because one-way simulated disease progression between late stages exhibited limited sensitivity to initial changes, we estimated four-way TCs for 3767 genes (Union DEGs). Here, the four-way TCs were defined as follows: TC1 (late phenotype changes, WT6M → AD6M), TC2 (WT maturation or aging, WT3M → WT6M), TC3 (early phenotype changes, WT3M → AD3M), and TC4 (AD maturation or aging, AD3M → AD6M) (Fig. 7).

a

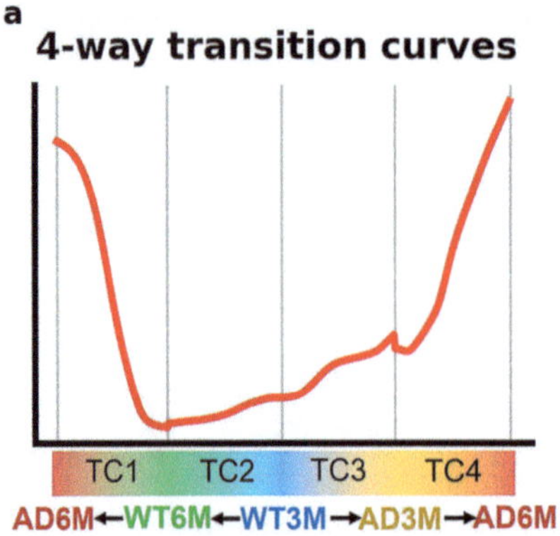

b

	TC1	TC2	TC3	TC4	Number of genes
P1	▲	▲	△	▲	643
P2	▲	–	–	▲	567
P3	▲	▽	▲	▲	240
P4	△	▲	▽	▲	477
P5	▼	▲	▼	▲	48
P6	▼	▼	▲	▼	871
P7	▼	▽	▽	▼	335
P8	▼	▲	▼	▼	200

Fig. 7 (**a**) Representation of the four-way TCs starting from WT3M to AD6M, either via WT6M or AD3M. (**b**) The eight pattern groups are described by trends of four TCs using the symbols ▲(up), ▼(down), △ (even/up), ▽(even/down), and –(even)

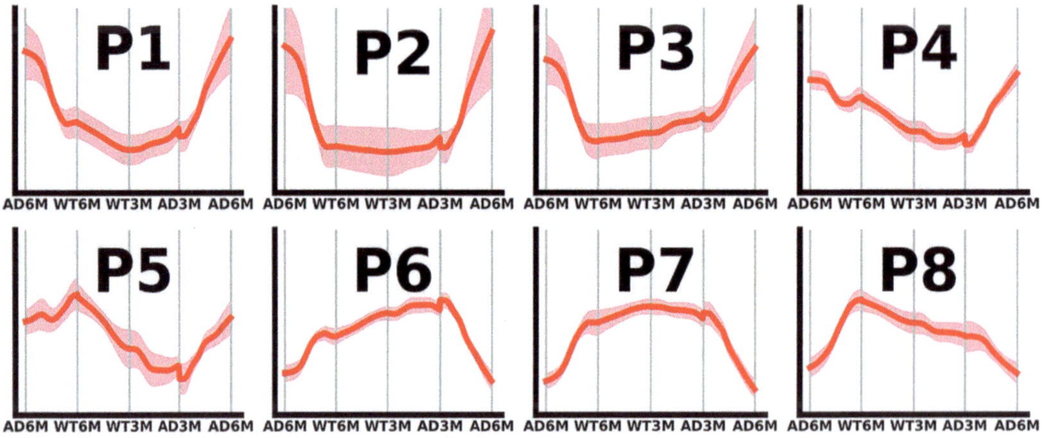

Fig. 8 The four-way TCs of the eight pattern groups are shown, with the red lines representing the average TCs and the variations indicated by one standard deviation

2. The 3767 TCs were merged into 56 clusters using affinity propagation clustering (APC). While certain clusters exhibited high similarity, they were shown to be separate clusters due to the distinct scales of expression levels (*y*-axis). Among the 56 clusters, we categorized the 46 clusters containing 3381 TCs into eight pattern groups (P1 to P8) based on their direction (upward or downward) for each TC (Figs. 7b and 8). The remaining ten clusters, consisting of 388 TCs, could not be merged as they exhibited undefined patterns.
3. Using the eight pattern groups, we also conducted gene ontology and pathway analysis. After analyzing the molecular mechanisms associated with the genes in each pattern, we specifically focused on two groups of patterns that revealed early molecular mechanisms. The P3 group was found to be

involved in both early and late pathological processes but not associated with WT maturation (Figs. 7b and 8), indicating pure pathological features in early tauopathy. The primary Gene Ontology Biological Process (GOBP) annotations included vasculature development and blood vessel morphogenesis, and the associated genes were enriched in brain endothelial cells. The smallest group, P5, consisting of only 48 genes, exhibited a unique pattern characterized by upregulation during maturation processes and downregulation during phenotype changes (Figs. 7b and 8). The dominant Gene Ontology Biological Process (GOBP) annotations for this group were related to mitochondrial metabolic processes, including oxidative phosphorylation, ATP synthesis-coupled electron transport, and ATP metabolic processes. Consequently, mitochondrial metabolic processes were found to be downregulated in both early and late tauopathy, while being upregulated during WT maturation. Vascular or mitochondrial dysfunction has been identified in several studies as an early molecular mechanism associated with Alzheimer's disease [24, 25].

4. Similar to our approach using the simulated four-way TCs by GANs, molecular pathway analysis based on gene expression patterns can be performed using the Weighted Gene Co-expression Network Analysis (WGCNA) method [26]. However, in our study, the WGCNA analysis using only the original samples did not provide the resolution needed to separate distinct molecular mechanisms, such as P1 (gliosis), P2 (immune response), and P3 (vascular).

3.5 Further Analysis and Variations

Both of our studies encompassed gene network analysis for each pattern, predictions of cell type-specific expression, and validation using human postmortem brain tissue data. These methods provide robust validation for the viability of our proposed analysis pipeline, further supporting its applicability in the context of our research. As mentioned earlier, in line with the primary objective of showcasing the application of GANs, we intentionally omitted specific biological validation procedures.

Indeed, one of the challenges encountered in the GAN-based analysis of bulk RNA-seq data is the establishment of a standardized analysis pipeline. Factors such as the sample size, number of groups, and data distribution of the RNA-seq data must be taken into consideration when determining the appropriate network architecture, hyperparameters, and other settings. Additionally, the optimization of hyperparameters using a validation set necessitates the definition of well-defined measures aligned with the research objectives. Moreover, the selection between one-way or multi-way transition curves (TCs) as means to visualize disease progression is a decision that researchers must make based on their specific requirements. Consequently, the methods outlined in this chapter may necessitate various methodological adaptations to suit individual needs. Furthermore, it is plausible to explore disease progression using vector arithmetic in the latent space beyond the scope of GAN-based generative models. Emerging generative models, such as transformer or diffusion probabilistic models, are predicted to hold promise in leveraging this concept to comprehensively investigate disease progression.

4 Notes

1. The preprocessing software versions used for the TPR50-P301S dataset (GSE90693) are as follows: Trimmomatic-v.0.38, HISAT2-v.2.1.0, HTSeq-counts-v.0.12.4, and DESeq2-v.1.28.1.
2. The original dataset for 5xFAD (TPR50) comprises a total of 36 (20) samples, divided into six (four) phenotype-age groups, each containing six (five) samples. The data augmentation procedure was initiated by considering pairwise combinations within each group, each of which comprised six (five) samples. Within a set of six (five) samples, fifteen (ten) pairwise combinations can be generated. Within the context of a combination between two original samples (S_1 and S_2), we generated nine augmented samples (S_{aug}) employing linear interpolation ($S_{\mathrm{aug}} = xS_1 + (1 - x)S_2$), where x ranges from 0.1 to 0.9 in increments of 0.1. Thus, through the product of 6 (4) groups, 15 (10) pairs, and 9 linear interpolations, we generated 810 (360) augmented samples. Combined with the original 36 (20) samples, this provided a comprehensive total of 846 (380) samples for GANs training.
3. For the 5xFAD dataset, we employed the standard scaling operations for a given sample (i), gene (j), and group (k) as follows: $\mathrm{SS}[i, j] = (\mathrm{RLD}[i, j] - \mu[j]) / \max(\sigma[k, j])$, where $\mu[j]$ represents the average expression of gene j across all samples, and $\sigma[k, j]$ signifies the standard deviation of gene j expression across the samples under group k. The use of the maximum of $\sigma[k, j]$ aims to circumvent any aberrant effects attributed to minor variations within a given group and considerable differences between separate groups. For the final transformation to be suitable for the leaky ReLU activation function, the standard scaled RLD values were adjusted such that rescaled $\mathrm{RLD} = \mathrm{SS} / (3.918\sigma_{\mathrm{SS}}) + 0.5$, which ensures that 95% of the values fall within the range [0,1].
4. For the TPR50 dataset, applying the same normalization as the previous 5xFAD dataset results in a distribution that is

significantly narrower with a longer tail compared to the standard Gaussian distribution. If this distribution is utilized for GAN training, the generated gene values frequently fall outside the range of the existing values. Consequently, we made a new modification to the normalization process. We applied the standard-distribution-like scaling operations as follows: rescaled RLD = SR/(3.918σ_{SR}) + 0.5, where SR is defined for a given sample (i), gene (j), and group (k) as follows: SR [i, j] = (RLD[i, j] − μ[j]) / σ_{IQR}[j], where μ[j] denotes the mean of gene j across all samples. Here, we utilized a complex denominator, σ_{IQR}[j], for each gene. We calculated σ_{GM}[j], the geometric mean between σ[j] (the standard deviation of gene j across all samples) and max(σ[k, j]) (the maximum standard deviation of gene j among the samples within group k). From the distribution of σ_{GM}[j], we conditionally selected σ_{IQR}[j] as follows:

$$\begin{aligned}
\sigma_{IQR}[j] &= \sigma_{GM}[j] \quad if \quad Q1 < \sigma_{GM}[j] < Q3, \\
\sigma_{IQR}[j] &= Q1 \quad if \quad \sigma_{GM}[j] \leq Q1, \\
\sigma_{IQR}[j] &= Q3 \quad if \quad \sigma_{GM}[j] \geq Q3
\end{aligned}$$

where $Q1$ represents the first quartile (25th percentile) of the distribution over σ_{GM}[j], while $Q3$ denotes the third quartile (75th percentile) of the distribution over σ_{GM}[j]. Through this intricate normalization procedure, we achieved a distribution that closely resembles the Gaussian normal distribution $N(\mu = 0.5, \sigma^2 = (1/3.918)^2)$. This is crucial for generating synthetic data that accurately simulates the real data with a high degree of precision.

5. GitHub repository: https://github.com/KBRI-Neuroinformatics/WGAN-for-RNASeq-analysis for 5xFAD and https://github.com/KBRI-Neuroinformatics/WGAN-for-toupathy for TPR50.

Acknowledgments

This research was supported by the KBRI basic research program through the Korea Brain Research Institute funded by the Ministry of Science and ICT (24-BR-02-04 [YK, MC]), the National Research Foundation of Korea (NRF) grant funded by the Korea government (MSIT) (2021R1A2C1003657 [MC]).

References

1. Scheltens P, De Strooper B, Kivipelto M, Holstege H, Chetelat G, Teunissen CE, Cummings J, van der Flier WM (2021) Alzheimer's disease. Lancet 397(10284):1577–1590
2. Gan L, Cookson MR, Petrucelli L, La Spada AR (2018) Converging pathways in neurodegeneration, from genetics to mechanisms. Nat Neurosci 21(10):1300–1309
3. Hou Y, Dan X, Babbar M, Wei Y, Hasselbalch SG, Croteau DL, Bohr VA (2019) Ageing as a risk factor for neurodegenerative disease. Nat Rev Neurol 15(10):565–581
4. Litke R, Garcharna LC, Jiwani S, Neugroschl J (2021) Modifiable risk factors in Alzheimer disease and related dementias: a review. Clin Ther 43(6):953–965
5. Myszczynska MA et al (2020) Applications of machine learning to diagnosis and treatment of neurodegenerative diseases. Nat Rev Neurol 16(8):440–456
6. Bennett DA, Buchman AS, Boyle PA, Barnes LL, Wilson RS, Schneider JA (2018) Religious orders study and rush memory and aging project. J Alzheimers Dis 64(s1):S161–S189
7. Hansson O (2021) Biomarkers for neurodegenerative diseases. Nat Med 27(6):954–963
8. Consortium GT (2020) The GTEx Consortium atlas of genetic regulatory effects across human tissues. Science 369(6509):1318–1330
9. Wang Y, Chen L (2022) DeepPerVar: a multimodal deep learning framework for functional interpretation of genetic variants in personal genome. Bioinformatics 38(24):5340–5351
10. Kingma DP, Welling M (2013) Auto-encoding variational bayes. arXiv preprint arXiv:13126114
11. Goodfellow I, Pouget-Abadie J, Mirza M, Xu B, Warde-Farley D, Ozair S, Courville A, Bengio Y (2020) Generative adversarial networks. Commun ACM 63(11):139–144
12. Vaswani A, Shazeer N, Parmar N, Uszkoreit J, Jones L, Gomez AN, Kaiser Ł, Polosukhin I (2017) Attention is all you need. Adv Neural Inf Proces Syst, 30
13. Ho J, Jain A, Abbeel P (2020) Denoising diffusion probabilistic models. Adv Neural Inf Proces Syst 33:6840–6851
14. Ghahramani A, Watt FM, Luscombe NM (2018) Generative adversarial networks simulate gene expression and predict perturbations in single cells. BioRxiv:262501
15. Park J, Kim H, Kim J, Cheon M (2020) A practical application of generative adversarial networks for RNA-seq analysis to predict the molecular progress of Alzheimer's disease. PLoS Comput Biol 16(7):e1008099
16. Kim H, Kim Y, Lee CY, Kim DG, Cheon M (2023) Investigation of early molecular alterations in tauopathy with generative adversarial networks. Sci Rep 13(1):732
17. Lee CD, Daggett A, Gu X, Jiang L-L, Langfelder P, Li X, Wang N, Zhao Y, Park CS, Cooper Y (2018) Elevated TREM2 gene dosage reprograms microglia responsivity and ameliorates pathological phenotypes in Alzheimer's disease models. Neuron 97(5): 1032–1048.e1035
18. Swarup V et al (2019) Identification of evolutionarily conserved gene networks mediating neurodegenerative dementia. Nat Med 25(1): 152
19. Leinonen R, Sugawara H, Shumway M, International Nucleotide Sequence Database C (2011) The sequence read archive. Nucleic Acids Res 39(Database issue):D19–D21
20. Bolger AM, Lohse M, Usadel B (2014) Trimmomatic: a flexible trimmer for Illumina sequence data. Bioinformatics 30(15): 2114–2120
21. Kim D, Langmead B, Salzberg SL (2015) HISAT: a fast spliced aligner with low memory requirements. Nat Methods 12(4):357
22. Anders S, Pyl PT, Huber W (2015) HTSeq—a Python framework to work with high-throughput sequencing data. Bioinformatics 31(2):166–169
23. Love MI, Huber W, Anders S (2014) Moderated estimation of fold change and dispersion for RNA-seq data with DESeq2. Genome Biol 15(12):550
24. Canepa E, Fossati S (2020) Impact of tau on neurovascular pathology in Alzheimer's disease. Front Neurol 11:573324
25. Wang W, Zhao F, Ma X, Perry G, Zhu X (2020) Mitochondria dysfunction in the pathogenesis of Alzheimer's disease: recent advances. Mol Neurodegener 15(1):30
26. Zhang B, Horvath S (2005) A general framework for weighted gene co-expression network analysis. Stat Appl Genet Mol Biol 4:Article17

Chapter 16

Protocol for Analyzing Epigenetic Regulation Mechanisms in Breast Cancer

Sri Lakshmi Bhavani Pagolu and Nita Parekh

Abstract

DNA methylation and gene expression are two critical aspects of the epigenetic landscape that contribute significantly to cancer pathogenesis. Analysis of aberrant genome-wide methylation patterns can provide insights into how these affect the cancer transcriptome and possible clinical implications for cancer diagnosis and treatment. The role of tumor suppressors and oncogenes is well known in tumorigenesis. Epigenetic alterations can significantly impact the expression and function of these critical genes, contributing to the initiation and progression of cancer. This protocol chapter presents a unified workflow to explore the role of DNA methylation in gene expression regulation in breast cancer by identifying differentially expressed genes whose promoter or gene body regions are differentially methylated using various *Bioconductor* packages in R environment. Functional enrichment analysis of these genes can help in understanding the mechanisms leading to tumorigenesis due to epigenetic alterations.

Key words DNA methylation, Gene expression, Epigenetics, Breast cancer, Promoter, Correlation, RnBeads, DESeq2

1 Introduction

DNA methylation is a well-established epigenetic mechanism that plays a pivotal role in regulating gene expression. Methylation of CpG islands in gene promoters is associated with gene silencing. This epigenetic modification prevents the binding of transcription factors and RNA polymerase, inhibiting gene expression. DNA methylation patterns can be tissue-specific and change during developmental phase, allowing for the fine-tuning of gene expression in different cell types. This enables the differentiation of cells with distinct functions in various tissues [1, 2]. DNA methylation can be influenced by environmental factors such as diet, stress, and exposure to toxins. These changes can lead to alterations in gene expression and contribute to disease susceptibility [3]. Aberrant DNA methylation patterns are associated with various diseases, including cancer. Recent studies in understanding the role of

Rajeev K. Azad (ed.), *Transcriptome Data Analysis*, Methods in Molecular Biology, vol. 2812,
https://doi.org/10.1007/978-1-0716-3886-6_16,

DNA methylation in various diseases have led to the development of epigenetic therapies and drugs targeting DNA methyltransferase inhibitors (DNMT inhibitors) are being explored for cancer treatment [4]. Thus, in-depth exploration of the role of DNA methylation in gene expression regulation in various biological contexts is necessary. The field of epigenetics continues to evolve, with ongoing research uncovering new insights into this important regulatory mechanism.

Advancements in sequencing and array-based technologies have revolutionized DNA methylation profiling, enabling researchers to unravel epigenetic mechanisms across various fields. In addition to its crucial role in cancer research, DNA methylation profiling has become indispensable in understanding diseases beyond cancer, such as neurodevelopmental disorders, cardiovascular diseases, and autoimmune conditions. Apart from providing insights into disease mechanisms, this has opened up new category of drugs. Sequencing-based techniques such as Whole-Genome Bisulfite Sequencing (WGBS) provide high-resolution, genome-wide methylation data by capturing methylated cytosines. On the other hand, array-based approaches such as Infinium Methylation Arrays offer cost-effective and high-throughput methylation profiling. Both methods involve bisulfite conversion, which chemically distinguishes methylated and unmethylated cytosines. The choice of technique depends on research goals, resources, and the scale of methylation analysis.

1.1 DNA Methylation in Cancer

Aberrant DNA methylation patterns are commonly observed in cancer and can lead to the activation of oncogenes or the silencing of tumor suppressor genes, contributing to tumor development and progression. DNA hypomethylation can lead to the overexpression of oncogenes, promoting uncontrolled cell growth and proliferation. For example, hypomethylation of RAS oncogene family members (e.g., KRAS) is associated with their increased expression and has been implicated in the development of various cancers, including pancreatic cancer [5]. DNA hypermethylation in the promoter regions of tumor suppressor genes is a hallmark of many cancers, resulting in their repression and loss of tumor-inhibiting functions. For example, the promoter of tumor suppressor PTEN, involved in cell cycle regulation and apoptosis, is frequently hypermethylated in various cancers, including prostate cancer [6]. DNA methylation can impact genes involved in cell cycle control and DNA repair. Aberrant methylation of these genes can lead to genomic instability and promote carcinogenesis. The BRCA1 gene, involved in DNA repair, is frequently silenced through DNA hypermethylation in breast and ovarian cancers, increasing the risk of these cancers [7]. Because of their role in tumorigenesis, DNA methylation patterns are being used as diagnostic and prognostic biomarkers in cancer. Recent advancements in genome-wide DNA methylation profiling have led to the discovery of novel biomarkers. For example, "EPICUP" test has been developed to help identify the tissue of origin in cancers of unknown primary [8]. Epigenetic therapies, including DNMT inhibitors and histone deacetylase inhibitors, are under investigation for their potential to reverse abnormal DNA methylation patterns and reactivate silenced genes in cancer treatment. Thus, understanding the role of DNA methylation in cancer would provide insight into disease mechanisms, facilitate the development of targeted therapies, and improve cancer diagnosis, prognosis and cancer management.

While promoter methylation is extensively studied, it's important to note that negative correlations with downstream gene expression are observed in ~70% of cases, while in ~20% of cases positive correlations are seen. Additionally, it's worth discussing the concept of gene body methylation, which involves methylation patterns within the coding region of genes and can have distinct regulatory roles in gene expression. Methylation within gene bodies can have both positive and negative effects on gene expression. While high gene body methylation generally correlates with active transcription, it can also serve to regulate alternative splicing. In breast cancer, gene body methylation of CD44 gene is associated with its alternative splicing, contributing to the generation of cancer stem cell-like phenotypes [9]. Intragenic methylation can influence alternative promoter usage, leading to the generation of different transcript variants in breast cancer. For example, methylation within the ESR1 gene is shown to result the production of distinct Erα isoforms with varying functions in breast cancer cells [10]. Thus, analysis of genome-wide DNA methylation profiling can provide a deeper understanding of epigenetic regulation in complex diseases such as cancer.

2 Materials

2.1 Data

For this protocol, DNA methylation data from Illumina Infinium HumanMethylation EPIC BeadChip (EPIC array) platform and gene expression data as raw sequence read counts from RNA sequencing (RNA-Seq) from AURORA US breast cancer cohort were downloaded from NCBI GEO (Acc: GSE212375) [11]. It contains normal tissue, primary tumor before and after treatment, and metastatic samples from 55 breast cancer patients belonging to African-American descent and Hispanic ethnicity. For this protocol, we considered 19 samples, 17 from pretreated primary tumor and 2 normal tissue patients to understand the role of methylation and gene expression.

2.2 System Requirements

Linux or Mac OS system is required to install the software. Computer requirement depends on the size of the dataset. For this tutorial, the analysis was run on a 64-bit computer with 8 GB of RAM. It may be noted that the commands to be executed on "terminal" are prefixed with "$", while ">" is used for the commands to be run in R environment and "#" is used for comments to describe the commands.

2.3 Tools

The workflow is carried out in the R environment (version 4.1.2) using several freely available Bioconductor packages. First step is to identify and install various R packages required for DNA methylation analysis. Before installing these packages, the user needs to initiate the R environment from the working directory (by typing the command R). Next, install BiocManager (version 3.14) and initiate it in R environment to install the required packages. The packages used for this protocol are "GEOquery" (version 2.62.1) for data retrieval from NCBI GEO [12], "RnBeads" (version 2.12.2) for identifying differentially methylated sites and regions [13], and "DESeq2" (version 1.34.0) for identifying differentially expressed genes [14]. A command-line toolset, "bedtools" which provides a wide-range of genomics analysis tasks, is used for mapping CpG sites to genic regions. Install these packages using the commands given below. It is important to note that when installing these packages, you will be prompted to accept the installation or update of all necessary dependencies. One may use the alternate versions of these packages; but make sure to assess their compatibility with existing dependencies. Ensuring that all packages function smoothly together is important for maintaining the reliability and reproducibility of the analysis.

We used various web-based resources for functional analysis of the genes. The Network of Cancer Genes (NCG), which contains manually curated genes with known or predicted role in cancer development and advancement, was used to identify oncogenes and tumor suppressor genes from the shortlisted genes [15]. To explore protein interactions and functional associations of the selected genes, the user may use STRING database, which is a database of known and predicted protein-protein interactions and can be accessed using STRINGdb package in R [16]. DAVID [17], g:Profiler [18], and EnrichR [19] are other popular resources which can be accessed via R packages gprofiler2, RDAVIDWebService, and enrichR, respectively, for gene set enrichment analysis. For prediction of prognostic markers, the user may perform survival analysis [20] using R package, "survival."

```
# in terminal
$ R
# Refer Notes 1,2
```

```
# in R install BiocManager
> if (!require("BiocManager", quietly = TRUE))
 install.packages("BiocManager")
 BiocManager::install(version = "3.14")

# install GEOquery
> BiocManager::install("GEOquery", version = "2.62.1")
# install RnBeads
> BiocManager::install("RnBeads", version = "2.12.2")
> BiocManager::install("RnBeads.hg19")

# install DESeq2
> BiocManager::install("DESeq2", version = "1.34.0")

# install bedtools on console
$ sudo apt update
$ sudo apt install bedtools
```

3 Methods

3.1 Data Download

To download DNA methylation and gene expression data files from NCBI GEO, the "GEOquery" package from Bioconductor is used. Begin with first loading the library and configuring download options. First create a directory, for example, "AURORA," and set it as working directory. Next, download methylation data (Acc ID: GSE212370) in idat file format and raw read count files for gene expression data (Acc ID: GSE209998) using the "getGEOSuppFiles" command. The metadata for both EPIC array and RNA-seq experiments are downloaded using the commands given below to build the annotation file.

```
#load GEOquery
> library(GEOquery)
> dir.create("~/Desktop/AURORA")
> setwd("~/Desktop/AURORA/")
> dir.create("~/Desktop/AURORA/METH")
> dir.create("~/Desktop/AURORA/DEG")

#set the download method to 'wget'
> options(timeout = max(300, getOption("timeout")))
> options(download.file.method.GEOquery = "wget")

# download the methylation files
> getGEOSuppFiles('GSE212370', makeDirectory = TRUE, baseDir
= getwd())
```

```
# download the expression files
> getGEOSuppFiles('GSE209998', makeDirectory = TRUE, baseDir
= getwd())

# download metadata from methylation and expression
> meth_data <- getGEO("GSE212370", GSEMatrix = TRUE)
> meth_metadata <- pData(meth_data[[1]])
> write.csv(meth_metadata, "meth_metadata.csv", row.names =
FALSE)

> exp_data <- getGEO("GSE209998", GSEMatrix = TRUE)
> exp_metadata <- pData(exp_data[[1]])
> write.csv(exp_metadata, "exp_metadata.csv", row.names =
FALSE)

# extract the files
$ cd ~/Desktop/AURORA/GSE212370/
$ tar -xvf GSE212370_RAW.tar
$ gzip -d *.gz

$ cd ~/Desktop/AURORA/GSE209998/
$ gzip -d *.gz
```

3.2 Data Preparation

The metadata files contain clinical information about patient samples such as sample type, tissue of origin, patient ID, sample ID, and treatments status. For this tutorial, using the sample IDs and sample types, we identified 17 pretreated primary tumor samples and 2 normal tissue samples for which both DNA methylation and gene expression files are available, as shown in Fig. 1. Next, we generated *"AUR.csv"* an annotation file in the working directory, *"~/Desktop/AURORA/,"* containing the information in Table 1 (*see* **Notes 3–7**). The microarray based Illumina Infinium HumanMethylationEPIC BeadChip (EPIC array) platform generates two primary datafiles for each sample. These files correspond to the "red" and "green" channels, that capture distinct color signals used in the microarray to measure DNA methylation levels (β values). The β value is calculated as the ratio of methylated signal intensity to the sum of methylated and unmethylated signal intensities, ranging from 0 (completely unmethylated) to 1 (completely methylated), providing a quantitative measure of DNA methylation levels, and is given by

$$\beta = M/(M + U + \alpha)$$

where, M represents the signal intensity of methylated cytosines, U represents the signal intensity of unmethylated cytosines, α is a small constant to avoid division by zero or very small values. Typically, α is set to a small positive number like 100. The

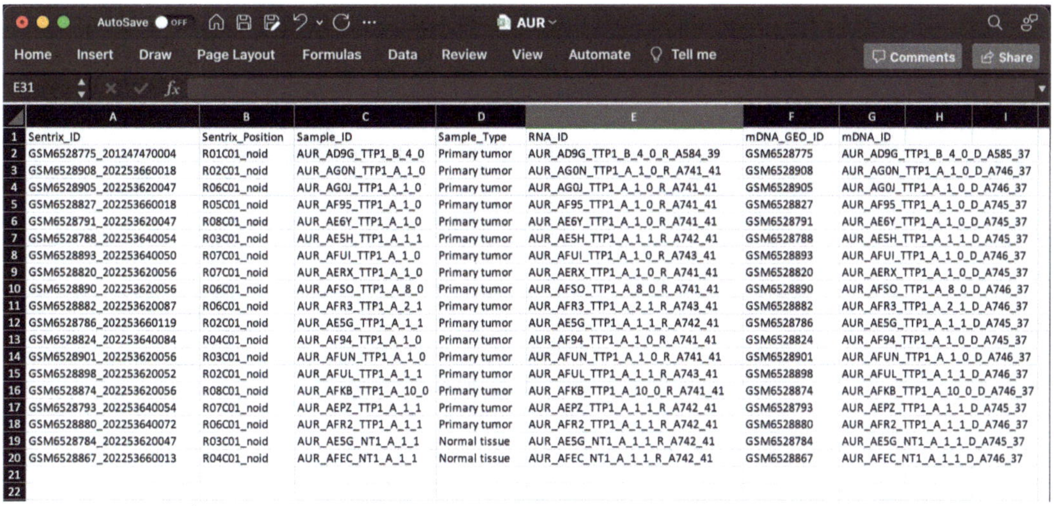

Sentrix_ID	Sentrix_Position	Sample_ID	Sample_Type	RNA_ID	mDNA_GEO_ID	mDNA_ID
GSM6528775_201247470004	R01C01_noid	AUR_AD9G_TTP1_B_4_0	Primary tumor	AUR_AD9G_TTP1_B_4_0_R_A584_39	GSM6528775	AUR_AD9G_TTP1_B_4_0_D_A585_37
GSM6528908_202253660018	R02C01_noid	AUR_AG0N_TTP1_A_1_0	Primary tumor	AUR_AG0N_TTP1_A_1_0_R_A741_41	GSM6528908	AUR_AG0N_TTP1_A_1_0_D_A746_37
GSM6528905_202253620047	R06C01_noid	AUR_AG0J_TTP1_A_1_0	Primary tumor	AUR_AG0J_TTP1_A_1_0_R_A741_41	GSM6528905	AUR_AG0J_TTP1_A_1_0_D_A746_37
GSM6528827_202253660018	R05C01_noid	AUR_AF95_TTP1_A_1_0	Primary tumor	AUR_AF95_TTP1_A_1_0_R_A741_41	GSM6528827	AUR_AF95_TTP1_A_1_0_D_A745_37
GSM6528791_202253620047	R08C01_noid	AUR_AE6Y_TTP1_A_1_0	Primary tumor	AUR_AE6Y_TTP1_A_1_0_R_A741_41	GSM6528791	AUR_AE6Y_TTP1_A_1_0_D_A745_37
GSM6528788_202253640054	R03C01_noid	AUR_AE5H_TTP1_A_1_1	Primary tumor	AUR_AE5H_TTP1_A_1_1_R_A742_41	GSM6528788	AUR_AE5H_TTP1_A_1_1_D_A745_37
GSM6528893_202253640050	R07C01_noid	AUR_AFUI_TTP1_A_1_0	Primary tumor	AUR_AFUI_TTP1_A_1_0_R_A743_41	GSM6528893	AUR_AFUI_TTP1_A_1_0_D_A746_37
GSM6528820_202253620056	R07C01_noid	AUR_AERX_TTP1_A_1_0	Primary tumor	AUR_AERX_TTP1_A_1_0_R_A741_41	GSM6528820	AUR_AERX_TTP1_A_1_0_D_A745_37
GSM6528890_202253620056	R06C01_noid	AUR_AFSO_TTP1_A_8_0	Primary tumor	AUR_AFSO_TTP1_A_8_0_R_A741_41	GSM6528890	AUR_AFSO_TTP1_A_8_0_D_A746_37
GSM6528882_202253620087	R06C01_noid	AUR_AFR3_TTP1_A_2_1	Primary tumor	AUR_AFR3_TTP1_A_2_1_R_A743_41	GSM6528882	AUR_AFR3_TTP1_A_2_1_D_A746_37
GSM6528786_202253660119	R02C01_noid	AUR_AE5G_TTP1_A_1_1	Primary tumor	AUR_AE5G_TTP1_A_1_1_R_A742_41	GSM6528786	AUR_AE5G_TTP1_A_1_1_D_A745_37
GSM6528824_202253640084	R04C01_noid	AUR_AF94_TTP1_A_1_0	Primary tumor	AUR_AF94_TTP1_A_1_0_R_A741_41	GSM6528824	AUR_AF94_TTP1_A_1_0_D_A745_37
GSM6528901_202253620056	R03C01_noid	AUR_AFUN_TTP1_A_1_0	Primary tumor	AUR_AFUN_TTP1_A_1_0_R_A741_41	GSM6528901	AUR_AFUN_TTP1_A_1_0_D_A746_37
GSM6528898_202253620052	R02C01_noid	AUR_AFUL_TTP1_A_1_1	Primary tumor	AUR_AFUL_TTP1_A_1_1_R_A743_41	GSM6528898	AUR_AFUL_TTP1_A_1_1_D_A746_37
GSM6528874_202253620056	R08C01_noid	AUR_AFKB_TTP1_A_10_0	Primary tumor	AUR_AFKB_TTP1_A_10_0_R_A741_41	GSM6528874	AUR_AFKB_TTP1_A_10_0_D_A746_37
GSM6528793_202253640054	R07C01_noid	AUR_AEPZ_TTP1_A_1_1	Primary tumor	AUR_AEPZ_TTP1_A_1_1_R_A742_41	GSM6528793	AUR_AEPZ_TTP1_A_1_1_D_A745_37
GSM6528880_202253640072	R06C01_noid	AUR_AFR2_TTP1_A_1_1	Primary tumor	AUR_AFR2_TTP1_A_1_1_R_A742_41	GSM6528880	AUR_AFR2_TTP1_A_1_1_D_A746_37
GSM6528784_202253620047	R03C01_noid	AUR_AE5G_NT1_A_1_1	Normal tissue	AUR_AE5G_NT1_A_1_1_R_A742_41	GSM6528784	AUR_AE5G_NT1_A_1_1_D_A745_37
GSM6528867_202253660013	R04C01_noid	AUR_AFEC_NT1_A_1_1	Normal tissue	AUR_AFEC_NT1_A_1_1_R_A742_41	GSM6528867	AUR_AFEC_NT1_A_1_1_D_A746_37

Fig. 1 Sample annotation file containing Sentrix ID, Sentrix Position, Sample ID, Sample Type, etc

Table 1
Contents of the input annotation file

Column name	Description
Sentrix_ID	From IDAT file names
Sentix_position	From IDAT file names
Sample_ID	Sample specific ID
Sample_Type	Primary tumor or normal tissue sample
RNA_ID	IDs for each patient's gene expression data in AURORA dataset
DNA_GEO_ID	GSM ID for each patient file in the dna methylation data
DNA_ID	IDs for each patient's methylation data in AURORA dataset

expression data file comprises raw read counts obtained from the RNA sequencing experiment. It is structured with gene identifiers in the first column, followed by the corresponding read counts for each gene across all the samples in consecutive columns.

3.3 DNA Methylation Analysis

For DNA methylation profiling, we first load the library for the Bioconductor package "RnBeads" and define the path for input files (i.e., idat files), annotation file, and the output folder where analysis reports will be generated and visualized, using the commands given below. RnBeads also provides a tailored analysis and graphical user interface (GUI) for performing the analysis (*see* **Note 8**).

```
#loading the library RnBeads
> library(RnBeads)
```

```
# path to the idat files
> idat.dir <- "~/Desktop/AURORA/GSE212370/"

# path to the annotation file
> sample.annotation <- "~/Desktop/AURORA/AUR.csv"

# path to generate the analysis and reports folder
> analysis.dir <- "~/Desktop/AURORA/"
> report.dir <- file.path(analysis.dir, "reports")

# generating the reports folder
> rnb.initialize.reports(report.dir)
[1] TRUE
> data.source <- c(idat.dir, sample.annotation)

# importing methylation data creating an RnB object
> result <- rnb.run.import(data.source=data.source, data.
type="infinium.idat.dir", dir.reports=report.dir)
```

Output:

```
2023-08-28 18:14:19 5.7 STATUS STARTED Loading Data

...

2023-08-28 18:20:04 7.0 STATUS Loaded 19 samples and 866895
sites
2023-08-28 18:20:05 7.0 INFO Output object is of type RnBea-
dRawSet
2023-08-28 18:20:05 7.0 STATUS COMPLETED Loading Data
```

Once the path has been defined, import the data from the .idat files. Next, create an R object and perform quality check to assess the data. The rnb.set object encapsulates essential details, including the sample count, probe descriptions, region information, significance status, and whether normalization and background correction procedures have been applied. In this context, "CpG" signifies a DNA sequence where a cytosine (C) is followed by guanine (G), "CpH" represents a sequence where cytosine is followed by any nucleotide except guanine, and "rs" denotes single nucleotide polymorphisms (SNPs). During preprocessing, sequences classified as "CpH" and "rs" are excluded or filtered out.

```
# creating an R object
> rnb.set <- result$rnb.set
> rnb.set
```

Output:

```
Object of class RnBeadRawSet
 19 samples
 866895 probes
 of which: 863904 CpG, 2932 CpH, and 59 rs
Region types:
 257075 regions of type tiling
 35214 regions of type genes
 45124 regions of type promoters
 26603 regions of type cpgislands
Intensity information is present
Detection p-values are present
Bead counts are present
Quality control information is present
Summary of normalization procedures:
 The methylation data was not normalized.
 No background correction was performed.
```

```
# Quality check
> rnb.run.qc(rnb.set, report.dir)
```

Output:

```
2023-08-28 18:47:44 7.0 STATUS STARTED Quality Control

...

2023-08-28 18:49:27 5.9 STATUS COMPLETED Visualizing SNP Probe
Data
2023-08-28 18:49:34 6.7 STATUS COMPLETED Quality Control
```

3.3.1 Data Preprocessing

Preprocessing of methylation data was carried out using RnBeads, to remove probes containing overlapping SNPs, context non-specific probes, cross-reactive probes, probes on sex chromosomes, low and high coverage probes, and low significant probes with $p \leq 0.05$. Probes containing more than 30% missing values were removed and for the remaining probes, imputation of missing values was performed by "knn" method using *impute* package. Various normalization techniques are available in RnBeads, namely, "SWAN," "PBC," "noob," and "Quantile". Here, beta mixture quantile normalization (BMIQ) was used for type-bias correction. These preprocessing tasks were performed using the commands given below.

```
# Set different options for carrying out preprocessing
> rnb.options(normalization.method = "bmiq", region.types=c
("sites", "promoters", "genes", "cpgislands"), identifiers.
column="RNA_ID", filtering.context.removal = c("CC","CAG","-
```

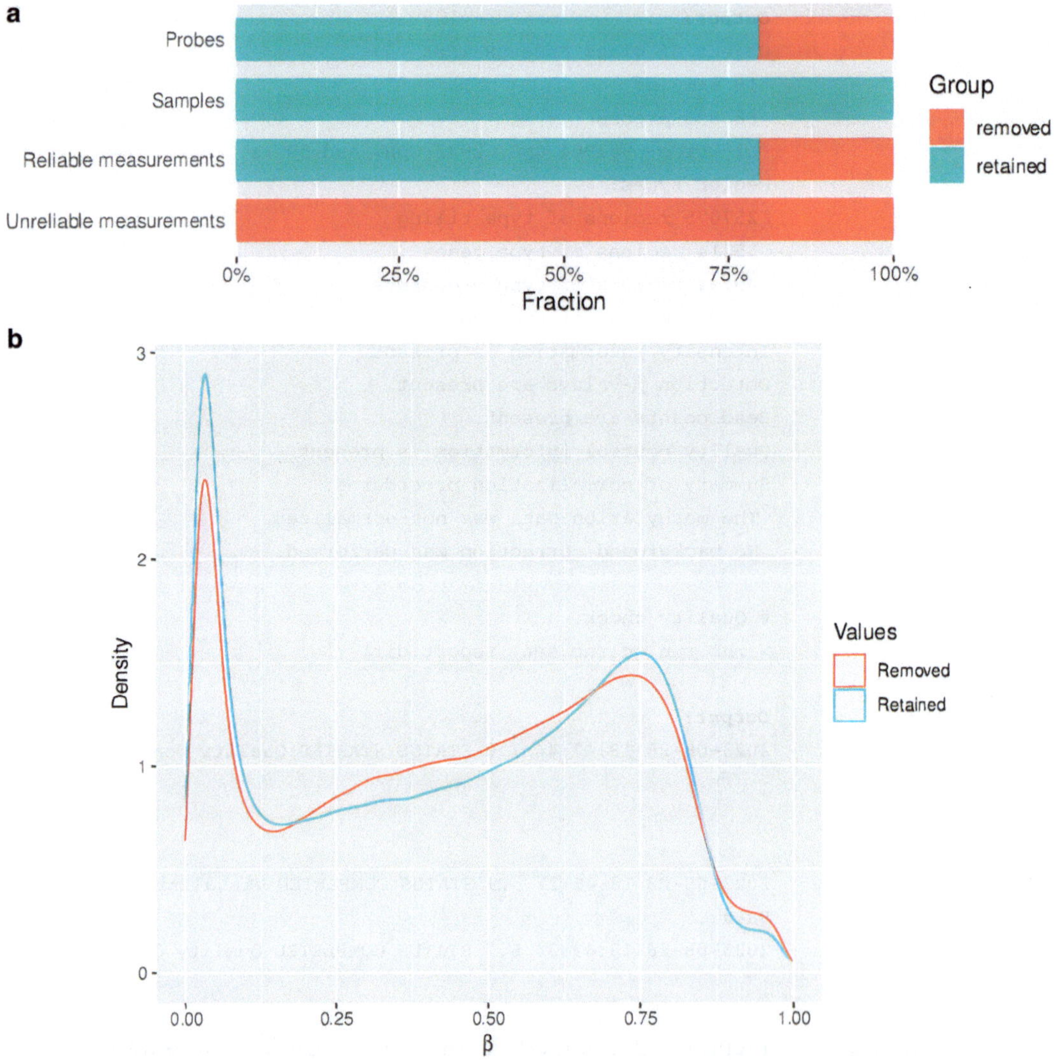

Fig. 2 (**a**, **c**) Percentage of sites and samples removed and retained after performing filtering **steps 1** and **2**. (**b**, **d**) Density distribution of sites removed and retained after filtering 1 and filtering 2

```
CAH", "CTG", "CTH", "Other"), filtering.snp = "any", filter-
ing.cross.reactive = TRUE, filtering.greedycut = TRUE, fil-
tering.sex.chromosomes.removal = TRUE, filtering.missing.
value.quantile = 0.3, filtering.low.coverage.masking = TRUE,
filtering.high.coverage.outliers = TRUE, imputation.method =
"knn")

# Copying the RnB object
> rnb.set.unfiltered <- rnb.set

# Running preprocessing step
> result <- rnb.run.preprocessing(rnb.set.unfiltered, dir.re-
ports=report.dir)
```

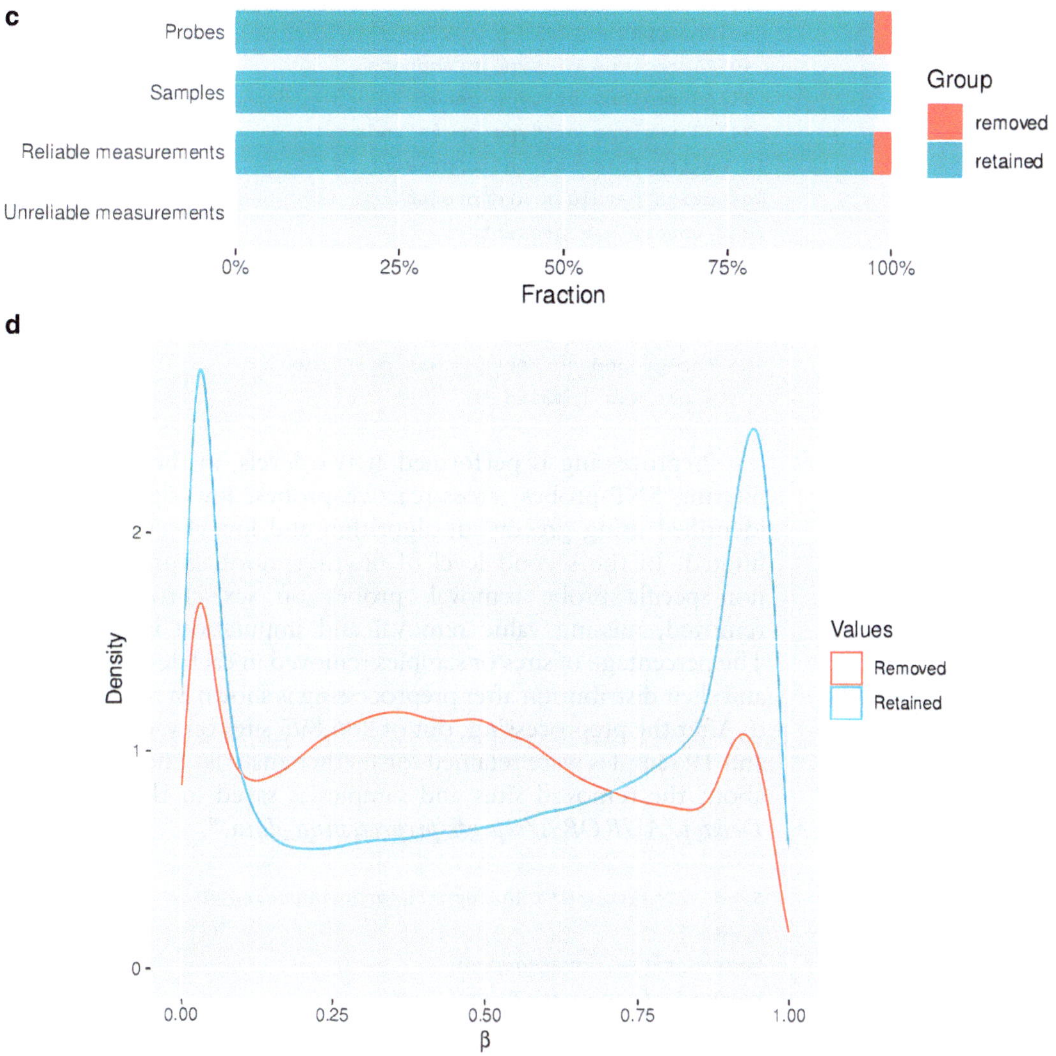

Fig. 2 (continued)

Output:

```
2023-08-28 19:06:16 6.5 STATUS STARTED Preprocessing

...

Coordinate system already present. Adding new coordinate
system, which will replace the existing one.
2023-08-28 19:26:07 9.7 STATUS COMPLETED Preprocessing

> rnb.set <- result$rnb.set
> rnb.set
Object of class RnBeadRawSet
 19 samples
 670569 probes
 of which: 670569 CpG, 0 CpH, and 0 rs
Region types:
 40571 regions of type promoters
 31148 regions of type genes
 24338 regions of type cpgislands
Intensity information is present
Detection p-values are present
Bead counts are present
Quality control information is present
Summary of normalization procedures:
 The methylation data was normalized with method bmiq.
 No background correction was performed.
 Data set was imputed.
```

Preprocessing is performed at two levels, in the first level of filtering SNP-probes, cross-reactive probes, low significant sites identified using greedy-cut algorithm and low-coverage sites are filtered. In the second level of filtering, normalization, context-non-specific probe removal, probes on sex chromosome are removed, missing value removal and imputation is performed. The percentage of sites or samples removed in each level of filtering and their distribution after preprocessing is shown in red in Fig. 2a–d. After the preprocessing, out of 866,895 sites only, 670,569 sites and 19 samples were retained for further analysis. The information about the removed sites and samples is saved in the folder *"~/Desktop/AURORA/reports/preprocessing_data."*

```
$ cd ~/Desktop/AURORA/reports/preprocessing_data
$ ls
masked_sites_coverage.csv
removed_sites_context.csv
removed_sites_cross_reactive.csv
removed_sites_greedycut.csv
removed_sites_na.csv
```

```
removed_sites_sex.csv
removed_sites_snp.csv
summary_1.csv
summary_2.csv
```

3.3.2 Differential Methylation Analysis

Differential methylation analysis involves identifying CpG sites (DMCs) and regions (DMRs) that are differentially methylated in primary tumor samples with respect to the normal tissue. First select the appropriate region type for which analysis needs to be performed and then select the columns for comparison (e.g., tumor vs. normal). Here, we selected "Sample_Type" column for differential methylation analysis. For identifying DMRs, we selected promoter regions, gene body regions, and CpG islands. Promoter regions are 2000 base pairs long and span from the TSS1500 (Transcription Start Site) to 500 base pairs into the gene region. The gene coordinates are derived from the hg19 reference genome. Various methods such as "limma," "wilcoxon," "ANOVA," "*t*-test," "minfi," and "svaseq" are available for identifying differentially methylated sites/regions. Here we used "limma" method for identifying DMCs and DMRs. In the case of pairwise data, we have the flexibility to specify columns. Additionally, one can also perform enrichment analysis using R packages such as "GOstats" and "LOLA."

```
# Specify the sample annotation table columns for which differential methylation is to be computed

#Setting options for differential methylation

> rnb.options(region.types=c("promoters", "genes", "cpgislands"), identifiers.column="RNA_ID", differential.site.test.method = "limma", differential.comparison.columns = "Sample_Type", differential.report.sites = TRUE)

# Running differential methylation

> rnb.run.differential(rnb.set, report.dir)
```

Output:

```
2023-08-28 20:21:43 2.9 STATUS STARTED Differential Methylation

...

2023-08-28 20:32:44 2.5 STATUS COMPLETED Differential Methylation
```

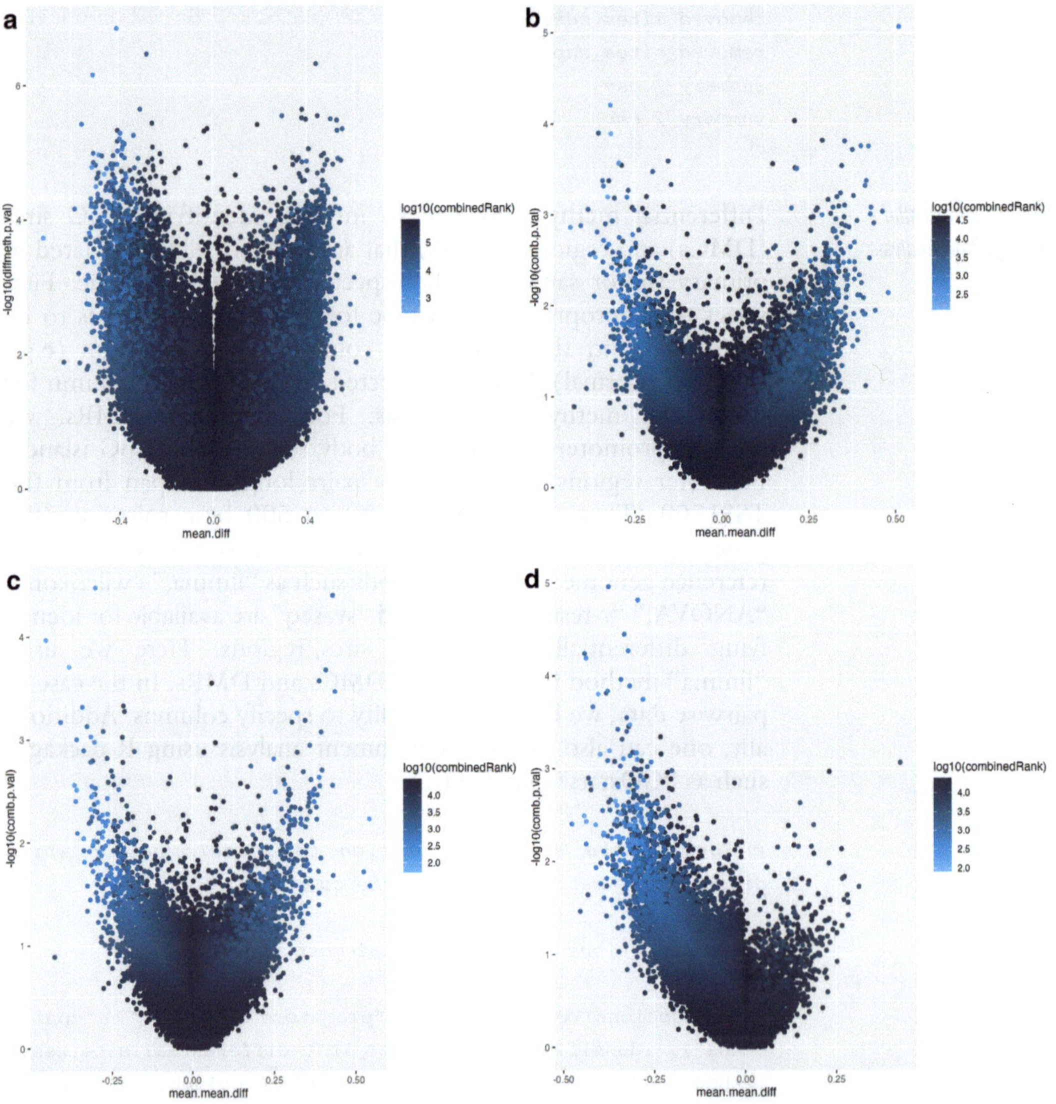

Fig. 3 Volcano plots showing Δβ (tumor vs. normal) on *X*-axis and log transformed *p*-value on *Y*-axis for (**a**) CpG sites, (**b**) promoter regions, (**c**) gene regions, and (**d**) CpG islands

Differential methylation analysis was performed on CpG sites, promoter regions, gene regions, and CpG islands. The volcano plots generated using the mean difference, Δβ, and *p*-value represent the distribution of differentially methylated CpG sites, promoter, gene, and CpG islands in Fig. 3a, b, c, and d, respectively. Separate files are generated for each case in the folder /reports/differential_methylation_data/.

```
# list of contents in the reports folder after the analysis
$ cd ~/Desktop/AURORA/reports/
$ ls
```

```
Output:
configuration
data_import_data
data_import_images
data_import_pdfs
data_import.html
differential_methylation_data
differential_methylation_images
differential_methylation_pdfs
differential_methylation.html
preprocessing_data
preprocessing_images
preprocessing_pdfs
preprocessing.html
quality_control_data
quality_control_images
quality_control_pdfs
quality_control.html

$ cd differential_methylation_data
$ ls

Output:
diffMethTable_region_cmp1_cpgislands.csv
diffMethTable_region_cmp1_genes.csv
diffMethTable_region_cmp1_promoters.csv
diffMethTable_site_cmp1.csv
differential_rnbDiffMeth
```

For identifying DMCs and DMRs, we used the cut-off criteria, $|\Delta\beta| \geq 0.2$ and p-value ≤ 0.05. The columns containing the CpG site ID (gene ID in case of promoter and gene body regions), mean beta-value of tumor (β_{tumor}), mean beta-value of normal (β_{normal}), mean.diff ($\Delta\beta$), and *p-value* are extracted.

```
# Refer point Note 9
# cut-off criteria for extracting differentially methylated
sites
# header line
$ awk -F "," '{if(NR==1) print $2","$7","$6","$8","$10}'
<diffMethTable_site_cmp1.csv > dmc.csv
$ awk -F "," '{if(($8 >= 0.2 || $8 <= -0.2) && $10 <= 0.05)
print $2","$7","$6","$8*-1","$10}' < diffMethTable_site_cmp1.
csv >> dmc.csv

# cut-off criteria for extracting differentially methylated
regions
```

```
# header line
$ awk -F "," '{if(NR==1) print $1","$8","$7","$9","$11}' <
diffMethTable_region_cmp1_promoters.csv > dmp.csv
$ awk -F "," '{if(($9 >= 0.2 || $9 <= -0.2) && $11 <= 0.05)
print $1","$8","$7","$9*-1","$11}' < diffMethTable_re-
gion_cmp1_promoters.csv >> dmp.csv

$ awk -F "," '{if(NR==1) print $1","$8","$7","$9","$11}' <
diffMethTable_region_cmp1_genes.csv > dmg.csv
$ awk -F "," '{if(($9 >= 0.2 || $9 <= -0.2) && $11 <= 0.05)
print $1","$8","$7","$9*-1","$11}' < diffMethTable_re-
gion_cmp1_genes.csv >> dmg.csv

$ awk -F "," '{if(NR==1) print $1","$6","$5","$7","$9}' <
diffMethTable_region_cmp1_cpgislands.csv > dmcg.csv
$ awk -F "," '{if(($7 >= 0.2 || $7 <= -0.2) && $9 <= 0.05)
print $1","$6","$5","$7*-1","$9}' < diffMethTable_re-
gion_cmp1_cpgislands.csv >> dmcg.csv

# No of differentially methylated sites and regions
$ wc -l dm*
```

Output:

```
42283 dmc.csv
1018 dmcg.csv
556 dmg.csv
1287 dmp.csv
```

Using the specified cut-off criteria, 42,283 CpG sites, 1018 CpG islands, 556 genes, and 1287 promoter regions that are differentially methylated are identified.

3.4 Gene Expression Analysis

Differential expression analysis using DESeq2 in R typically involves several steps. One can utilize the “read.table” or “read.csv” function to import raw read count data into R, ensuring that the data file is in appropriate format (e.g., tab-delimited, or comma-separated). The annotation file should contain sample-related information, including group assignments and sample IDs. Next, convert the data frame into a matrix using the as.matrix function. For data preparation, we consider only columns with relevant sample information, such as sample type, sample identifiers and ignore other columns. Finally, confirm that the column names of the count data correspond to the row names in the annotation data file (*see* **Note 10**). The data is then grouped into primary tumor and normal tissue.

```
# improting the data
> library(DESeq2)
```

```
> sampleData <- read.delim("~/Desktop/AURORA/AUR.csv", sep =
",")
> sampleData_v2 <- sampleData
> raw <- read.delim("~/Desktop/AURORA/GSE209998/
GSE209998_AUR_129_raw_counts.txt", sep = "\t")
> selected_columns <- c( colnames(raw)[1], sampleData$RNA_ID)
> rawCounts <- raw[,selected_columns]

# converting the dataframe into a matrix and data preparation
> geneID <- rawCounts$X
> head(geneID)
[1] "DDX11L1" "WASH7P" "MIR6859-3" "RP11-34P13.3" "MIR1302-9"
"FAM138A"

> sampleIndex <- grepl("AUR", colnames(rawCounts))
> sampleIndex
[1] FALSE TRUE TRUE TRUE TRUE TRUE TRUE TRUE TRUE TRUE TRUE
TRUE TRUE TRUE
[15] TRUE TRUE TRUE TRUE TRUE TRUE

> rawCounts <- as.matrix(round(rawCounts[,sampleIndex]),0)
> rownames(rawCounts) <- geneID
> head(rawCounts)
 AUR_AD9G_TTP1_B_4_0_R_A584_39 ........
DDX11L1 49
WASH7P 246
MIR6859-3 0
RP11-34P13.3 0
MIR1302-9 0
FAM138A 33
> rownames(sampleData) <- sampleData$RNA_ID
> keep <- c("Sample_Type" ,"Sample_ID")
> sampleData <- sampleData[,keep]
> colnames(sampleData) <- c("Sample_Type" ,"ID")
> head(sampleData)
 Sample_Type ID
AUR_AD9G_TTP1_B_4_0_R_A584_39 Primary tumor AUR_-
AD9G_TTP1_B_4_0
AUR_AG0N_TTP1_A_1_0_R_A741_41 Primary tumor AUR_AG0N_TT-
P1_A_1_0
AUR_AG0J_TTP1_A_1_0_R_A741_41 Primary tumor AUR_AG0J_TT-
P1_A_1_0
AUR_AF95_TTP1_A_1_0_R_A741_41 Primary tumor AUR_AF95_TT-
P1_A_1_0
AUR_AE6Y_TTP1_A_1_0_R_A741_41 Primary tumor AUR_AE6Y_TT-
P1_A_1_0
AUR_AE5H_TTP1_A_1_1_R_A742_41 Primary tumor AUR_AE5H_TT-
P1_A_1_1
```

```
> rawCounts <- rawCounts[,unique(rownames(sampleData))]
> all(colnames(rawCounts) == rownames(sampleData))
[1] TRUE

# grouping the data
> sampleData$Sample_Type <- factor(sampleData$Sample_Type,
levels=c("Primary tumor", "Normal tissue"))
```

3.4.1 Data Preprocessing

For gene expression analysis, data preprocessing is a crucial initial step and involves first creating a data object from the raw count matrix which is subsequently used for analysis. To enhance the quality of data and reduce noise, filtering is performed by setting a threshold for row sums, retaining only those genes for which minimum sum >10, to filter out genes with very low expression values. To account for variations in sequencing depth and composition biases, and to facilitate valid gene expression comparisons, normalization is essential. Several options are available for normalization in RNA-seq experiments, including DESeq2's Size Factor Normalization, Variance Stabilizing Transformation (VST), Regularized Log Transformation (rlog), and Normalization to a Control Group (DESeq2 for time-series data). Here, we use size factor normalization which reduces technical variation from sequencing depth. After normalization, the data is ready for downstream analysis, and the normalized dataset is saved for further investigation and interpretation. Out of 58,387 genes (transcripts), 43,741 are retained after preprocessing.

```
# Preprocessing
> deseq2Data <- DESeqDataSetFromMatrix(countData =rawCounts,
colData=sampleData, design= ~ Sample_Type)
> deseq2Data1 <- deseq2Data[rowSums(counts(deseq2Data)) >
10, ]
> dds<-estimateSizeFactors(deseq2Data1, type = c("ratio",
"poscounts", "iterate"), locfunc = stats:: median)
> NormCounts<-counts(dds, normalized = TRUE, replaced = FALSE)
> write.csv(NormCounts, "~/Desktop/AURORA/DEG/AUR_PT_Normalized_geneexp.csv")
> dim(deseq2Data)
[1] 58387 19

> dim(deseq2Data1)
[1] 43741 19
```

3.4.2 Differential Expression Analysis

For identifying differentially expressed genes between primary tumor and normal tissue samples, the package DESeq2 is used with following criteria: $|\log_2 FC| \geq 1$ (column 3) and *p-value* ≤ 0.05

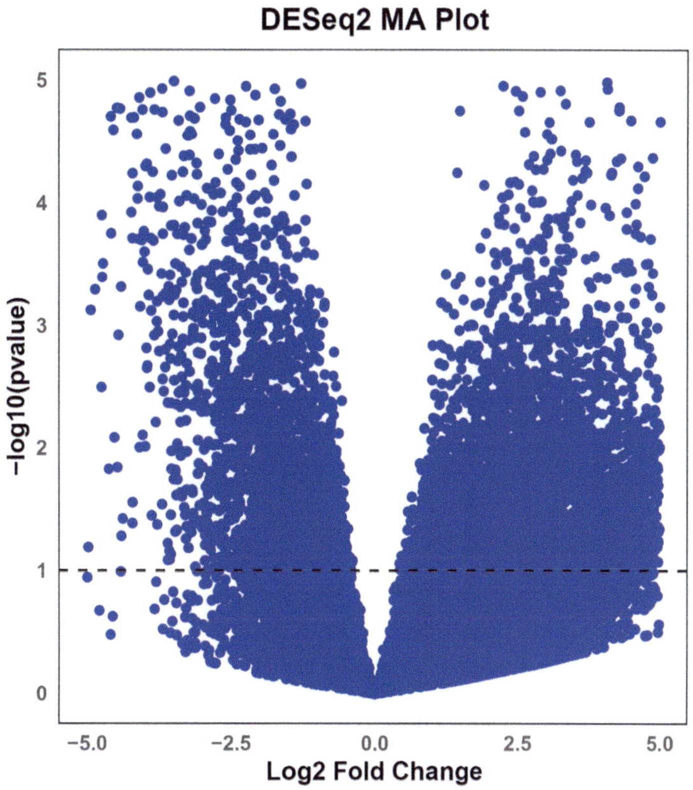

Fig. 4 The volcano plot obtained using ggplot2

(column 6). The log transformation minimizes the effects from outliers and helps prevent high-abundance genes from dominating downstream analysis due to their higher technical variability. In the previous step of designing the DESeq2 object, we incorporated sample type information to distinguish between primary tumor and normal tissue. To extract $\log_2$FC for "Primary tumor" vs. "Normal tissue", specify the order in the contrast command and save the results in "*AUR_PT_deseqres_PN.csv.*" This resulted in a total of 5012 differentially expressed genes (DEGs). Figure 4 shows the volcano plot obtained using ggplot2.

```
# DESeq2 analysis and saving the results
> deseq2Data2 <- DESeq(deseq2Data1)
> results <- results(deseq2Data2, contrast=c("Sample_Type",
"Primary tumor", "Normal tissue"))
> head(results)
log2 fold change (MLE): Sample_Type Primary tumor vs Normal
tissue
Wald test p-value: Sample_Type Primary tumor vs Normal tissue
DataFrame with 6 rows and 6 columns
 baseMean log2FoldChange lfcSE stat pvalue padj
```

```
 <numeric> <numeric> <numeric> <numeric> <numeric> <numeric>
DDX11L1 16.63846 1.2028774 1.528716 0.7868550 0.4313668
0.730871
WASH7P 513.89581 -0.0665788 0.528683 -0.1259331 0.8997849
0.967009
FAM138A 1.59674 -0.0430147 2.760409 -0.0155827 0.9875673 NA
OR4G4P 4.61570 4.8788319 2.777072 1.7568257 0.0789475 NA
OR4G11P 2.27563 1.5616941 2.864275 0.5452320 0.5855940 NA
RP11-34P13.7 134.73522 0.2377437 0.730705 0.3253620 0.7449071
0.904312

>write.csv(results,"~/Desktop/AURORA/DEG/AUR_PT_deseqres_PN.
csv")

$ cd ~/Desktop/AURORA/DEG
$ ls
```

Output:

```
AUR_PT_Normalized_geneexp.csv
AUR_PT_deseqres_PN.csv

# cut-off criteria for differentially expressed genes
$ awk -F "," '{if(($3 >= 1 || $3 <= -1) && $6 <= 0.05) print
$1","$3","$6}' < AUR_PT_deseqres_PN.csv > deg.csv
$ wc -l deg.csv
```

Output:

```
5102 deg.csv

#plotting the log2FC
> library(ggplot2)
> ma_plot <- ggplot(results, aes(x = log2FoldChange, y =
-log10(pvalue)))
+ geom_point(color = "blue")
# Set dot color to blue
+ geom_hline(yintercept = 1, linetype = "dashed") # Add a
horizontal line at y = 1
 + xlim(c(-5, 5)) # Adjust the x-axis limits as needed
 + ylim(c(0, 5)) # Set y-axis limits to 0 and 5
 + labs(x = "Log2 Fold Change", y = "-log10(pvalue)",
 title = "DESeq2 MA Plot") +
 theme_minimal() + # Plain background
  theme(panel.border = element_rect(color = "black", fill =
NA), # Borders
 panel.grid.major = element_blank(),
# Remove major gridlines
panel.grid.minor = element_blank(),
```

```
# Remove minor gridlines
 axis.text = element_text(face = "bold"), # Bold axis text
 axis.title = element_text(face = "bold"), # Bold axis titles
  plot.title = element_text(hjust = 0.5, face = "bold")) #
Center-aligned and bold title

> # Save the plot to a file (e.g., "MA_Plot.png")
> ggsave("~/Desktop/AURORA/MA_Plot.png", plot = ma_plot, width
= 6, height = 6, units = "in")
```

3.5 Mapping CpG Sites to Promoter and Genic Regions

To extract the annotations of CpG sites, promoter regions, gene regions, and CpG islands for the EPIC array from RnBeads, the following commands were used. Next, create a "*.bed*" file for each case by extracting the chromosome, coordinate start, end, and CpG site ID /gene ID. Then using "*bedtools intersect*" identify CpG sites overlapping to the promoter, gene body, and CpG islands.

```
> library(RnBeads)
#Annotations to .csv files
> anno.sites <- annotation(rnb.set.unfiltered, type="sites")
> anno.genes <- annotation(rnb.set.unfiltered, type="genes")
> anno.cpgislands <- annotation(rnb.set.unfiltered, type="-
cpgislands")
> anno.promoters <- annotation(rnb.set.unfiltered, type="pro-
moters")

> write.csv(anno.sites, "~/Desktop/AURORA/METH/anno_sites.
csv")
> write.csv(anno.genes, "~/Desktop/AURORA/METH/anno_genes.
csv")
> write.csv(anno.cpgislands, "~/Desktop/AURORA/METH/anno_cp-
gislands.csv")
> write.csv(anno.promoters, "~/Desktop/AURORA/METH/anno_pro-
moters.csv")

# pick the chrom, start, end and cpg site/ gene name into bed
files

$ cd ~/Desktop/AURORA/METH/
$ awk -F "," '{if(NR>1) print $2"\t"$3"\t"$4"\t"$5"\t"$1}' <
anno_sites.csv > sites.bed
$ awk -F "," '{if(NR>1)print
$2"\t"$3"\t"$4"\t"$5"\t"$1"\t"$6}' < anno_promoters.csv > pro-
moters.bed
$ awk -F "," '{if(NR>1)print
$2"\t"$3"\t"$4"\t"$5"\t"$1"\t"$6}' < anno_genes.csv > genes.
bed
```

```
$ awk -F "," '{if(NR>1)print $2"\t"$3"\t"$4"\t"$1}' < anno_cp-
gislands.csv > cpgislands.bed

# using bedtools intersect we map the CpG sites with the
promoter and gene regions

$ bedtools intersect -wo -a sites.bed -b promoters.bed >
CpG_prom.bed
$ bedtools intersect -wo -a sites.bed -b genes.bed > CpG_gene.
bed
$ bedtools intersect -wo -a sites.bed -b cpgislands.bed >
CpG_islands.bed
$ head -5 CpG_prom.bed
```

Output:

```
"chr1" 10525 10526 "+" "cg14817997" "chr1"10369 12368 "+"
"ENSG00000223972" "DDX11L1" 1
"chr1" 10848 10849 "-" "cg26928153" "chr1"10369 12368 "+"
"ENSG00000223972" "DDX11L1" 1
"chr1" 10850 10851 "-" "cg16269199" "chr1"10369 12368 "+"
"ENSG00000223972" "DDX11L1" 1
"chr1" 29407 29408 "+" "cg12045430" "chr1"29307 31306 "-"
"ENSG00000227232" "WASH7P" 1
"chr1" 29407 29408 "+" "cg12045430" "chr1"28054 30053 "+"
"ENSG00000243485" "MIR1302-10;MIR1302-11;MIR1302-2;MIR1302-9"
1
```

```
# pick column 5 (cpgsite) and column 11 (gene symbol)

$ echo "cpg,gene_symbol" > cpg_gene.csv
$ awk -F "\t" '{print $5 "," $11}' < CpG_gene.bed >> cpg_gene.
csv

$ echo "cpg,gene_symbol" > cpg_prom.csv
$ awk -F "\t" '{print $5 "," $11}' < CpG_prom.bed >> cpg_prom.
csv

$ echo "cpg,cpg_island" > cpg_island.csv
$ awk -F "\t" '{print $5 "," $9}' < CpG_islands.bed >>
cpg_island.csv

$ head cpg_gene.csv
```

Output:

```
cpg,gene_symbol
"cg13869341","WASH7P"
"cg14008030","WASH7P"
"cg12045430","WASH7P"
```

```
"cg20826792","WASH7P"
"cg21870274","OR4F5"
"cg03130891",NA
"cg04098293",NA
"cg16382250",NA
"cg24335620",NA
```

3.6 Correlation Analysis

3.6.1 File Preparation

In this step, we identify differentially methylated CpG sites (DMCs) in the promoter and gene body regions of differentially expressed genes (DEGs) using the files generated in the previous step, as shown below. This pairing information of CpG sites (DMCs) and associated genes (DEGs) is stored in the *aur_pair* file. Next, the methylation values of the sites and regions from the preprocessed RnB object are saved in an object for computing correlations.

```
#loading files
> library(dplyr)
> library(readr)
> dmc <- read.csv("~/Desktop/AURORA/reports/differential_-
methylation_data/dmc.csv")
> deg <- read.csv("~/Desktop/AURORA/DEG/deg.csv")
> cpg_prom <- read.csv("~/Desktop/AURORA/METH/cpg_prom.csv")
> cpg_gene <- read.csv("~/Desktop/AURORA/METH/cpg_gene.csv")

# filtering dmcs and degs
> aur_cpg_prom <- cpg_prom %>% filter(cpg %in% dmc$cgid &
gene_symbol %in% deg$X)
> aur_cpg_gene <- cpg_gene %>% filter(cpg %in% dmc$cgid &
gene_symbol %in% deg$X)

#pairing dmcs and degs
> aur_pair <- rbind(aur_cpg_gene, aur_cpg_prom)
> aur_pair <- unique(aur_pair)
> dim(aur_pair)
[1] 6564 2

> length(unique(aur_pair$cpg))
[1] 6156

> length(unique(aur_pair$gene_symbol))
[1] 1650

> head(aur_pair)
 cpg gene_symbol
1 cg01511231 TTLL10-AS1
2 cg03278726 TTLL10-AS1
3 cg15417666 TTLL10-AS1
```

```
4 cg26020826 LINC01134
5 cg27527922 RPL22
6 cg26935168 RPL22

# methylation values of sites, cpgislands, promoter regions
and gene regions are saved to .csv files.
> mm_sites <- meth(rnb.set,row.names=TRUE)
> write.csv(mm_sites, "~/Desktop/AURORA/METH/meth_sites.csv")
> mm_cpgislands <- meth(rnb.set, type="cpgislands", row.name-
s=TRUE)
> mm_promoters <- meth(rnb.set, type="promoters", row.name-
s=TRUE)
> mm_genes <- meth(rnb.set, type="genes", row.names=TRUE)
```

3.6.2 Calculating Correlations Between Methylation and Gene Expression

Object *mm_sites* contain the normalized β values of all the CpG sites and *NormCounts* contains the normalized expression values of all the genes. Using *mm_sites* and *NormCounts*, we create R objects containing the β values (*meth_t_aur*) and expression values (*gene_t_aur*) of only those sites and genes that are present in *aur_pair*. Next, compute Pearson's correlation coefficients using the command *cor.test*(), between methylation and gene expression values. First, extract β values from *meth_t_aur*, gene expression from *gene_t_aur* for each pair in *aur_pair*. Next, compute the correlation and save the correlation coefficient (r) and *p-value* in *correlation_results*. Then filter the significantly correlated pairs using the cut-off criteria $|r| \geq 0.5$ and *p-value* ≤ 0.05 and save in *filtered_results*.

```
# reading the normalized beta values and read counts
> meth_aur <- mm_sites[rownames(mm_sites) %in% aur_pair$cpg,]
> meth_t_aur <- t(meth_aur)
> dim(meth_t_aur)
[1] 19 6156

> gene_aur <- NormCounts[rownames(NormCounts) %in% aur_pair
$gene_symbol,]
> gene_t_aur <- t(gene_aur)
> dim(gene_t_aur)
[1] 19 1650

# Initialize an empty data frame to store the correlation and
p-value results
> correlation_results <- data.frame(
 Gene = character(0),
 CpGSite = character(0),
 Correlation = numeric(0),
 PValue = numeric(0))
```

```
# Loop through each pair to calculate correlations and p-values
> for (i in 1:nrow(aur_pair)) {
 gene <- aur_pair$gene_symbol[i]
 cpg_site <- aur_pair$cpg[i]

 # Extract gene expression and methylation data for the current
pair
 gene_expression <- gene_t_aur[, gene]
 methylation_values <- meth_t_aur[, cpg_site]

 # Calculate the correlation
 correlation_test <- cor.test(gene_expression, methylation_-
values)

 # Add the result to the correlation_results data frame
correlation_results <- rbind(correlation_results,
 data.frame(Gene = gene, CpGSite = cpg_site, Correlation =
correlation_test$estimate,PValue = correlation_test$p.value))
}
> write.csv(correlation_results, "~/Desktop/AURORA/correla-
tions.csv")

# Filter the correlation_results data frame
> filtered_results <- correlation_results %>% filter(abs(Cor-
relation) >= 0.5, PValue <= 0.05)
> write.csv(filtered_results, "~/Desktop/AURORA/filtered_cor-
relations.csv")
> dim(filtered_results)
[1] 1784 4

> length(unique(filtered_results$CpGSite))
[1] 1727

> length(unique(filtered_results$Gene))
[1] 713
```

A total of 6156 DMCs and 1650 DEGs, corresponding to unique 6564 CpG-gene pairs, were sent as input for correlation analysis. For the specified criteria ($|r| \geq 0.5$ and *p-value* ≤ 0.05), 1784 highly significant correlated CpG-gene pairs were identified from 1727 DMCs and 713 DEGs.

3.6.3 Extracting Most Correlated CpG Site for Each Gene

To identify the most influential methylation marker associated with gene expression, only the most highly correlated CpG site was selected for each gene, resulting in 713 CpG-gene pairs. An analysis of these differentially methylated and highly correlated CpG sites with differentially expressed genes would provide an insight into the regulatory mechanisms of these 713 genes.

```
# Group the filtered results by Gene
> library(dplyr)
> grouped_results <- filtered_results %>% group_by(Gene)
# Find the highest absolute correlation and its corresponding
CpG site for each gene
> max_correlation_results <- grouped_results %>% slice(which.
max(abs(Correlation)))
> dim(max_correlation_results)
[1] 713 4
```

3.7 Analysis of Shortlisted CpG Sites (Genes)

Correlation between methylation status of these CpG sites and the expression of the associated genes is performed using the following commands. In Fig. 5 is depicted the number of CpG sites that are Hyper and Hypo methylated in the promoter and gene body regions. Of the 713 CpG sites, 135 sites are in the promoter region and 578 in the gene body region. As expected, in the promoter region, majority of the CpG sites exhibit negative correlation with downstream gene expression (~78%); however, a significant number of the promoter CpG sites (~22%) do exhibit positive correlation. In the case of methylated CpG sites within the gene body region, the number of CpG sites positively (~43%) and negatively (~56%) correlated with the expression of associated gene is comparable.

```
# distribution of CpG sites
> max_correlation_results$CpG_Gene_Pair <- paste(max_correla-
tion_results$CpGSite, max_correlation_results$Gene, sep = '-
')
> aur_cpg_prom$CpG_Gene_Pair <- paste(aur_cpg_prom$cpg,
aur_cpg_prom$gene_symbol, sep = '-')
> max_correlation_results$Region <- ifelse(max_correlation_-
results$CpG_Gene_Pair %in% aur_cpg_prom$CpG_Gene_Pair, 'Pro-
moter', 'Gene')
> write.csv(max_correlation_results, "~/Desktop/AURORA/H-cor.
csv")

> distribution <- table(max_correlation_results$Region, sign
(max_correlation_results$Correlation))
> print(distribution)

 -1 1
 Gene 326 252
 Promoter 105 30
```

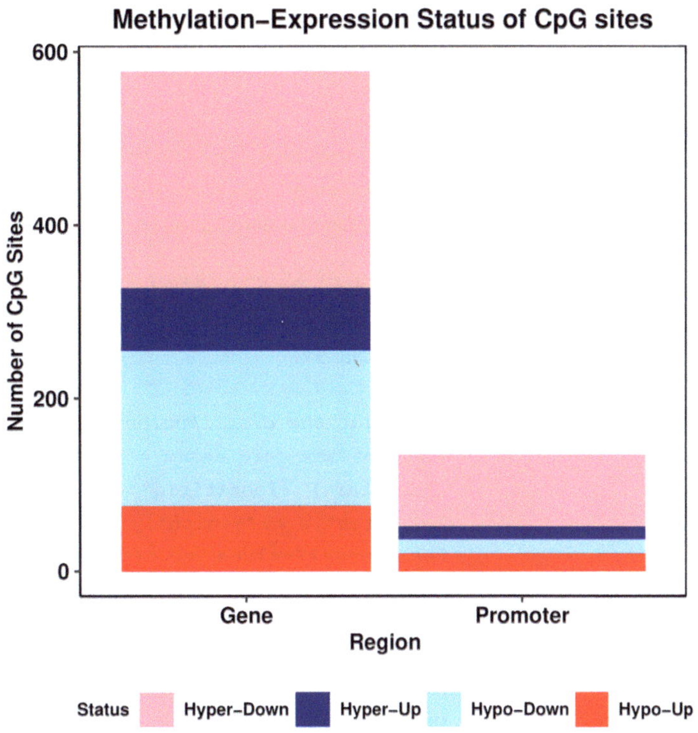

Fig. 5 Distribution of methylation and expression status of 713 CpG sites and associated genes, Hyper-Down (Pink) and Hypo-Up (Red), correspond to the negative correlations. Hypo-Down (light blue) and Hyper-Up (dark blue) correspond to the positive correlations

Correlation analysis of methylation status of these 713 CpG sites with expression of the associated genes revealed following four categories: 334 hypermethylated and downregulated, 87 hypermethylated and upregulated, 195 hypomethylated and downregulated, 97 hypomethylated and upregulated, as shown in Fig. 5. In most cases, oncogenes tend to exhibit hypomethylation coupled with upregulated expression, while tumor suppressor genes often show hypermethylation accompanied by downregulated expression in individuals with tumors. Following are the command to plot the methylation and expression status of the CpG sites across gene body and the promoter regions.

```
# distribution of hypo and hypermethylated genes
> max_cor <- max_correlation_results
> max_cor <- merge(max_cor, dmc, by.x = "CpGSite", by.y =
"cgid", all.x = TRUE)
> max_cor <- merge(max_cor, deg, by.x = "Gene", by.y = "X",
all.x = TRUE)
> max_cor$meth <- ifelse(max_cor$mean.diff > 0, 'Hyper',
'Hypo')
> max_cor$exp <- ifelse(max_cor$log2FoldChange > 0, 'Up',
```

```
'Down')
> max_cor$Status <- paste(max_cor$meth, max_cor$exp, sep = "-
")

> write.csv(max_cor, "~/Desktop/AURORA/HHUD.csv")
> contingency_table <- table(max_cor$meth, max_cor$exp)
> print(contingency_table)

 Down Up
 Hyper 334 87
 Hypo 195 97

# to plot the distribution of the CpG sites
> ggplot(max_cor, aes(x = Region, fill = Status)) +
 geom_bar() +labs(title = "Methylation-Expression Status of
CpG sites",x = "Region",y = "Number of CpG Sites") + scale_-
fill_manual(values = c("Hypo-Up" = "red", "Hypo-Down" =
"lightblue", "Hyper-Up" = "darkblue","Hyper-Down" = "light-
pink")) + theme_classic() +theme(panel.border = element_rect
(color = "black", fill = NA),axis.title = element_text(size =
10, face = "bold"), axis.text = element_text(size = 10, face =
"bold", color = "black"), plot.title = element_text(size =
12, face = "bold", hjust = 0.5), legend.position = "bottom",
legend.justification = "center", legend.title = element_text
(face = "bold", size = 8),legend.text = element_text(face =
"bold", size = 8))
```

3.8 Functional Analysis

To find if any of these 713 genes are known cancer genes, we searched these genes in the Network Cancer Genes database [15]. We identified 32 genes reported to be associated with any cancer, 15 of which are oncogenes and 17 tumor suppressors. To explore protein-protein interactions (PPI) and pathways involved, we analyzed these 32 genes in StringDB [16], and the PPI network is depicted in Fig. 6. Pathway enrichment analysis revealed these genes participating in key pathways such as PI3K-Akt, MAPK, ErbB, Estrogen signaling, and their involvement in diverse cancer types. This analysis provides insights into epigenetic regulation of these affected pathways in breast cancer. A similar analysis of the remaining genes can be carried out to identify novel genes and pathways involved in breast cancer initiation and progression. Other important resources for functional enrichment of genes and pathways that can be used are RDAVIDWebService [17], gprofiler2 [18], and enrichR [19]. Additionally, if survival data of the patients is available, one can perform survival analysis by computing COX proportional hazard ratios using "survival" package [20], to prioritize genes that may have an impact on patient survival outcomes. This would aid in the discovery of potential biomarkers and prognostic markers for breast cancer.

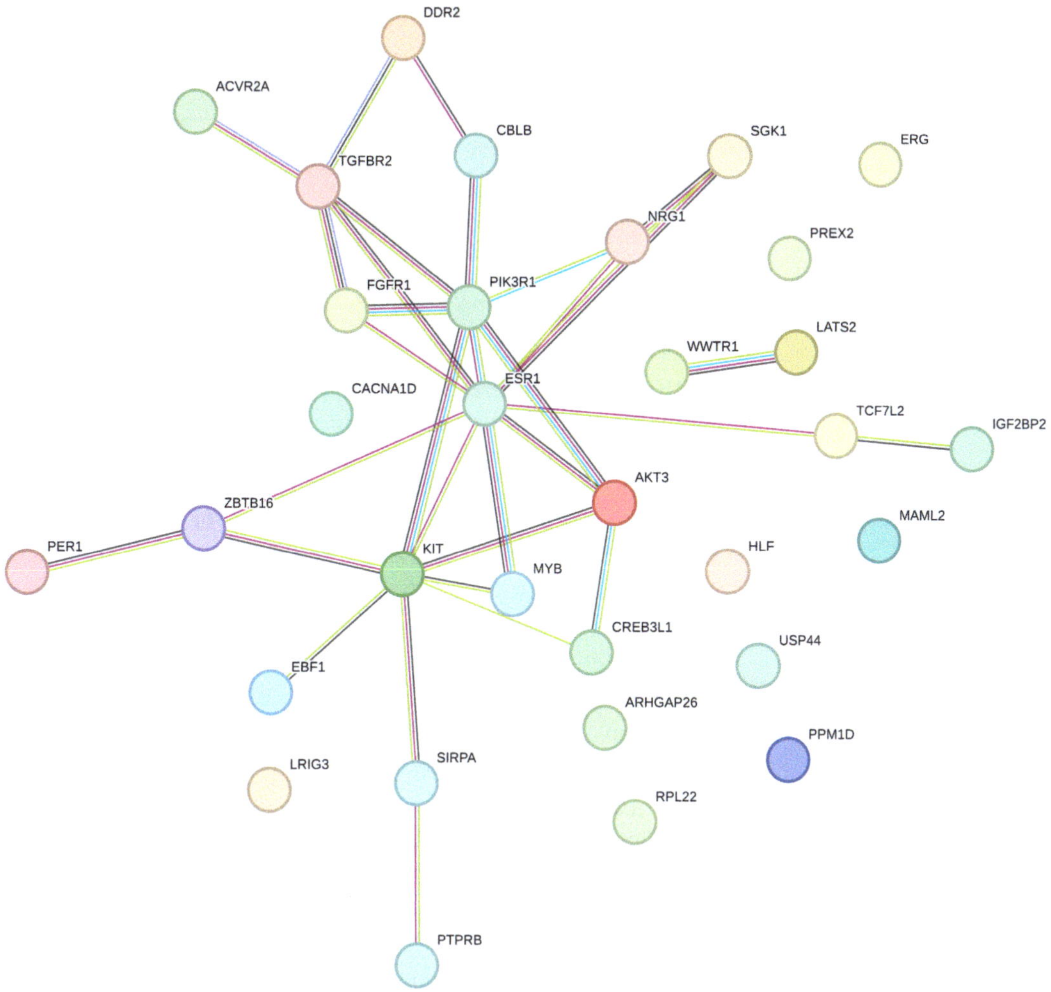

Fig. 6 Protein-protein interaction network involving the 32 oncogenes and tumor suppressors using STRINGDB

4 Notes

1. Users are encouraged to review the prerequisites for "RnBeads" package and ensure that all dependencies are installed, preferably using the latest versions. In case of any errors arising during either the installation or usage of the package, it is recommended to refer to the error messages for specific guidance regarding which versions or dependencies require attention.
2. If a different version of R is already installed, one can proceed to install all the required packages without providing the version information. This will install the compatible versions of the packages for that version of R.

3. For preparing the annotation file, we require the sample type (whether tumor or normal), sample identifiers (for each sample), mapping identifiers (to map the expression data to the methylation data). The supplementary material of AURORA study contains sample type and sample identifier information [11]. For example, "AUR-AF98-TTP1-A-1-0-D-A745-37" is the methylation data identifier present in metadata file for methylation data. The same samples expression data identifier is "AUR-AF98-TTP1-A-1-0-R-A741-41." We can see that the first part "AF98-TTP1-A-1-0" of both the identifiers are the same.
4. As "RnBeads" requires "Sentrix ID" and "Sentrix Positions" in the annotation file for each sample, we generate them using the file name (present in the metadata file). For example, for GSM ID "GSM6528775," there are two raw intensity files "GSM6528775_201247470004_R01C01_noid_Red.idat" and "GSM6528775_201247470004_R01C01_noid_green.idat." The first part "GSM6528775_201247470004" is "Sentrix ID" and the second part "R01C01_noid" is "Sentrix Position" are concatenated in the idat file name (refer Fig. 1).
5. When using Excel on MacOS to create and save a .csv file, make sure to save the file in the "CSV UTF-8 (comma-delimited) (.csv)" format for proper encoding and compatibility.
6. While using the annotation file, it's crucial to ensure that the identifiers used for methylation analysis and differential analysis are consistent. In this context, we have utilized "RNA_ID" as the common identifier for both datasets, as the expression file is structured with "RNA_IDs" as its column names.
7. In R, when creating an R object, any hyphens "-" in the column names are automatically replaced with periods "." by default. To prevent any potential issues, it's advisable to replace all hyphens "-" in the annotation file and in column names of the gene expression file "GSE209998_AUR_129_raw_counts.txt" with underscores "_" before processing.
8. RnBeads provides users with the flexibility to conduct custom analyses, encompassing step-by-step procedures for filtering, normalization, exploratory analysis, and differential analysis. You can find detailed instructions in the reference manual. Moreover, RnBeads offers a graphical user interface (GUI) to facilitate these tasks. To activate the GUI, simply execute the following commands in R.

```
> library(RnBeads)
> rnb.run.dj()
```

9. When calculating differential methylation, the program by default considers the groups in alphabetical order. For instance, in the comparison of "primary tumor vs. normal tissue," the program calculates the difference between "normal tissue" and "primary tumor." To obtain the desired file with "primary tumor vs. normal tissue" comparison, modify the results by multiplying the "mean.diff" or "mean.mean.diff" column by −1.
10. Please make sure to record the gene identifiers present in the raw read count file, whether they are in the form of gene symbols or Ensembl IDs. When performing mapping in Subheading 3.5, use column 10 for Ensembl IDs and column 11 for gene symbols. Maintaining this consistency is essential for ensuring clarity and accuracy throughout the analysis process.

References

1. Meissner A, Mikkelsen TS, Gu H et al (2008) Genome-scale DNA methylation maps of pluripotent and differentiated cells. Nature 454: 766–770
2. Smith ZD, Chan MM, Humm KC et al (2014) DNA methylation dynamics of the human preimplantation embryo. Nature 511:611–615
3. Waterland RA, Jirtle RL (2003) Transposable elements: targets for early nutritional effects on epigenetic gene regulation. Mol Cell Biol 23: 5293–5300
4. Issa JPJ (2007) DNA methylation as a therapeutic target in cancer. Clin Cancer Res 13: 1634–1637
5. Bournet B, Muscari F, Buscail C et al (2016) KRAS G12D mutation subtype is a prognostic factor for advanced pancreatic adenocarcinoma. Clin Transl Gastroenterol 7:157
6. Bhat AV, Palanichamy Kala M, Rao VK et al (2019) Epigenetic regulation of the PTEN–AKT–RAC1 Axis by G9a is critical for tumor growth in alveolar rhabdomyosarcoma. Cancer Res 79:2232–2243
7. Farolfi A, Gurioli G, Fugazzola P et al (2019) Immune system and DNA repair defects in ovarian cancer: implications for locoregional approaches. Int J Mol Sci 20:2569
8. Moran S, Martínez-Cardús A, Sayols S et al (2016) Epigenetic profiling to classify cancer of unknown primary: a multicentre, retrospective analysis. Lancet Oncol 17:1386–1395
9. Hurtado A, Holmes KA, Geistlinger TR et al (2008) Regulation of ERBB2 by oestrogen receptor-PAX2 determines response to tamoxifen. Nature 456:663–666
10. Rusidzé M, Adlanmérini M, Chantalat E et al (2021) Estrogen receptor-α signaling in post-natal mammary development and breast cancers. Cell Mol Life Sci 78:5681–5705
11. Garcia-Recio S, Hinoue T, Wheeler GL et al (2022) Multiomics in primary and metastatic breast tumors from the AURORA US network finds microenvironment and epigenetic drivers of metastasis. Nat Cancer 4:128–147
12. Davis S, Meltzer PS (2007) GEOquery: a bridge between the Gene Expression Omnibus (GEO) and BioConductor. Bioinformatics 23: 1846–1847
13. Müller F, Scherer M, Assenov Y et al (2019) RnBeads 2.0: comprehensive analysis of DNA methylation data. Genome Biol 20:55
14. Love MI, Huber W, Anders S (2014) Moderated estimation of fold change and dispersion for RNA-seq data with DESeq2. Genome Biol 15:550
15. Acha-Sagredo A, Malanchi I, Ciccarelli FD (2021) Clone competition as a mechanism to reduce tumor formation. Dev Cell 56:3307–3308
16. Szklarczyk D, Kirsch R, Koutrouli M et al (2023) The STRING database in 2023: protein–protein association networks and functional enrichment analyses for any sequenced genome of interest. Nucleic Acids Res 47:607–613
17. Sherman BT, Hao M, Qiu J et al (2022) DAVID: a web server for functional enrichment analysis and functional annotation of gene lists. Nucleic Acids Res 50:216–221

18. Raudvere U, Kolberg L, Kuzmin I et al (2019) g:Profiler: a web server for functional enrichment analysis and conversions of gene lists. Nucleic Acids Res 47:191–198
19. Chen EY, Tan CM, Kou Y et al (2013) Enrichr: interactive and collaborative HTML5 gene list enrichment analysis tool. BMC Bioinformatics 14:128
20. Therneau TM, Grambsch PM (eds) (2000) Modeling survival data: extending the cox model. Springer, Berlin, New York

Chapter 17

Identification of Virus-Derived Small Interfering RNAs (vsiRNAs) from Infected sRNA-Seq Samples

A. T. Vivek and Shailesh Kumar

Abstract

Plants have developed sophisticated defense mechanisms to combat viral infections, prominently utilizing Dicer-like enzymes (DCL) for generating virus-derived small interfering RNAs (vsiRNAs) through RNA interference (RNAi). This intrinsic mechanism effectively impedes virus replication. Exploiting their potential, vsiRNAs have become a major focus area for comprehensive viral investigations in plants, integrating both bioinformatics and experimental strategies. This chapter introduces an up-to-date computational workflow optimized for identifying and comprehensively annotating vsiRNAs with the utilization of small RNA sequencing (sRNA-seq) data collected from virus-infected plants. The workflow detailed in this chapter centers on known plant-targeting viruses, providing step-by-step guidance to enhance vsiRNA analysis, ultimately advancing the comprehension of plant-virus interactions.

Key words Plant-virus interactions, RNA interference, Virus-derived small interfering RNAs, Small RNA sequencing

1 Introduction

The RNA interference (RNAi) mechanism is a fundamental defense mechanism against viruses, producing vsiRNAs that encompass the genetic material of RNA and DNA viruses [1–3]. These vsiRNAs are 20–24 nt small RNAs that the host generates as a defense mechanism against viral infections. First, vsiRNAs are pivotal in comprehending the molecular-level interactions between host cells and viruses. They participate in silencing viral genes and curtailing viral replication, offering insights into the dynamic host-virus relationship [4]. Second, the RNAi pathway responsible for generating vsiRNAs is a conserved defense mechanism across various organisms [4–7]. Analyzing vsiRNAs provides a window into the host's immune response and its counter-strategies against viral infections. Furthermore, vsiRNA analysis aids in identifying specific segments of viral genomes that are targeted by the host defense

Rajeev K. Azad (ed.), *Transcriptome Data Analysis*, Methods in Molecular Biology, vol. 2812,
https://doi.org/10.1007/978-1-0716-3886-6_17,

machinery [1, 6]. This knowledge sheds light on which parts of the viral genome are vulnerable to degradation through RNAi. The study of vsiRNAs also contributes to our understanding of the co-evolution between viruses and their hosts. By evading or suppressing host RNAi responses, viruses can establish persistent infections. Analyzing vsiRNAs helps uncover the mechanisms by which viruses adapt to counteract host defenses. Additionally, certain vsiRNAs have the potential to serve as biomarkers for specific viral infections. Detecting unique vsiRNA profiles associated with different viruses or infection stages could greatly enhance rapid and accurate diagnostic methods [8, 9]. Comparative studies of vsiRNA profiles across various viral infections or different infection stages reveal shared and distinct features of host-virus interactions. This broader understanding contributes toward knowledge of the intricate mechanisms underlying viral infections. Finally, beyond its virological implications, studying vsiRNAs enriches our understanding of RNA biology, gene regulation, and the complex interplay between hosts and pathogens. Thus, analyzing vsiRNAs extracted from infected small RNA samples not only illuminates host defense mechanisms and viral adaptations but also offers potential therapeutic avenues. This step is integral to unraveling the intricate molecular interplay between viruses and plant host cells.

Analyzing different deep sequencing studies of small RNAs in naturally or experimentally infected host plants demonstrated that all 26 groups of land plant viruses generate small RNAs that assemble into sequences representing sections or entire genomes of viruses or viroids [4]. These findings underscore that the build-up of small silencing RNAs derived from viruses is a prevalent aspect of an effective immune reaction to viral infections in various plant hosts [3, 6]. With the application of deep sequencing and bioinformatics analysis of the small RNA population, we can achieve two important outcomes, (1) detection of vsiRNAs, (2) accurate reconstruct of viral genomes [8, 9]. The detection of vsiRNAs from sRNA-seq samples is a comprehensive process that seamlessly integrates bioinformatics analysis with experimental validation. Primarily, the focus shifts to characterizing the distribution of vsiRNAs along the viral genome. This analysis not only reveals specific regions targeted by the host's defense machinery but also offers a nuanced understanding of the dynamic interplay between the virus and the host. This approach also facilitates the comprehensive reconstruction of the entire virome in cases of mixed infections [8]. Considering the mentioned aspects, we propose a basic analytical pipeline for profiling vsiRNAs in virus-infected plants. Our goal is not an exhaustive list of specialized software or detailed library preparation methods. Instead, we offer a simple guide for analyzing sRNA-seq data using free and in-house bioinformatics resources.

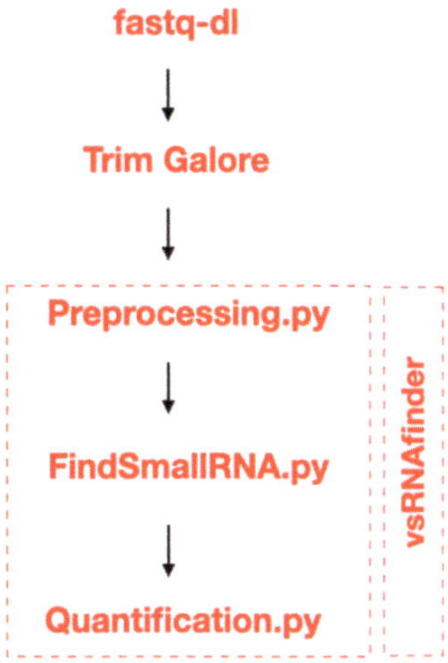

Fig. 1 Illustration of a straightforward workflow for vsiRNA identification and annotation

In this chapter, we emphasize the implementation of the vsRNAfinder [10] and a few other bioinformatic tools for the identification and annotation of vsiRNAs from virus-infected plant sRNA-seq samples. The proposed vsRNAfinder workflow comprises three major multistep pipelines. At first, raw data undergo preprocessing, which involves adapter trimming and size filtering. The second step is dedicated to characterizing and identifying vsiRNAs, and it includes filtering sRNAs based on the abundance of boundary sites, length, and significance of sRNAs. If the target viral genome is available, sRNA-seq reads can be mapped onto the viral genome to trace the origin of the vsRNAs. The final step 3 facilitates the quantification of vsiRNAs in each chromosome of the viral genome. The steps involved in this workflow are outlined sequentially in Fig. 1.

2 Materials

2.1 Hardware Requirements

- Operating System: 32-bit/64-bit Linux.
- Available RAM: Minimum of 4 GB.
- Free disk space varies with the number of samples and sequencing depth.

2.2 Software Requirements

Software Packages	Bowtie (v1.3.0), samtools (v1.9), bedtools (v2.29.2), RNAfold (v2.4.18), fastq-dl (2.0.4), Trim Galore (v0.6.10),
R Dependencies	Base on R 3.6.1, pracma 2.3.3
Python Dependencies	Base python 3.6.8, Numpy 1.19.2, Pandas 1.1.5, Scipy 1.5.2, Pysam 0.15.3

2.3 Installation Documentation

(i) *Download vsRNAfinder with Git from GitHub*:

Retrieve the vsRNAfinder repository from GitHub by executing the following command:

```
git clone https://github.com/ZenaCai/vsRNAfinder.git
```

(ii) *Create a New Conda Environment*:

Set up a new Conda environment named "vsRNAfinder" and activate it with the following commands:

```
conda create -n vsRNAfinder python=3.6.8
conda activate vsRNAfinder
```

(iii) *Install Required Packages*:

Install necessary packages using the Conda package manager with the following command:

```
conda install -c bioconda bowtie=1.3.0 samtools=1.9 bed-
tools=2.29.2 viennarna=2.4.18
```

(iv) *Install R and Python Dependencies*:

Install R and Python dependencies using Conda with the following command:

```
conda install r-base=3.6.1 \
    -c conda-forge r-pracma=2.3.3 \
    numpy=1.19.2 pandas=1.1.5 scipy=1.5.2 \
    -c bioconda pysam=0.15.3 fastq-dl trimgalore
```

By following these steps, the user will successfully set up the vsRNAfinder environment with all the required packages and dependencies for running the tools effectively.

3 Methods

The aim of this analysis is to extract small RNA reads from the original sequencing data, as well as to classify and describe vsiRNA reads. This process commences by identifying and trimming sRNA reads from the raw sequences upon detecting 3′ adaptor sequences.

Following this, any low-quality, contaminated, or extraneous sequence reads are eliminated through filtering. The sample utilized in this example is retrieved from NCBI SRA belongs to *Solanum lycopersicum* infected with Tomato mosaic virus (TMV). In this chapter, the analyses were conducted on Ubuntu 20.04.5 LTS. However, it is important to note that other operating systems like Windows or MacOS X can also be used, though installation and execution procedures may vary. Furthermore, the execution times for the analyses are subject to the computer's capacity. For our computations, we utilized a desktop with an Intel Xeon Silver 4210 CPU @ 2.20 GHz (40 cores), 64-bit architecture, and 125 GB of memory.

3.1 Obtaining sRNA-Seq Data from NCBI SRA and Genome TMV

- The example data can be obtained from the Short Read Archive (SRA) under the accession number SRR6872534. To directly download it, type

```
fastq-dl --accession SRR6872534 --provider SRA
```

- Get the reference genome of TMV using NCBI Reference Sequence: NC_002692.1 (https://www.ncbi.nlm.nih.gov/nuccore/NC_002692.1?report=fasta)

3.2 Adapter Trimming

Adapter trimming serves as a targeted solution to the removal of unwanted adapter sequences, a common occurrence during the sRNA library preparation process. This step becomes especially crucial in the context of vsiRNA identification. To attain a dataset of high quality, it is imperative to trim adaptor sequences from the raw data, a practice that guarantees the acquisition of clean reads. The significance of this lies in the accurate detection of genuine vsiRNAs.

For instance, in the provided dataset, the trim_galore tool was utilized for adapter trimming to achieve trimming of adapters and removal of low-quality bases. An example command used is below:

```
trim_galore -a AGATCGGAAGAG SRR6872534.fastq -o
SRR6872534_trimmed.fq
```

The summary below shows the details of the raw data filtered:

- Total reads processed: 22,227,962
- Reads with adapters: 22,124,813 (99.5%)
- Reads written (passing filters): 22,227,962 (100.0%)
- Total basepairs processed: 1,111,398,100 bp
- Quality-trimmed: 8,420,715 bp (0.8%)
- Total written (filtered): 507,363,349 bp (45.7%)

3.3 Preprocessing

The following command utilizes the "Preprocessing.py" script from the vsRNAfinder workflow to preprocess a fastq file for subsequent analysis:

```
python Preprocessing.py --cleanfq SRR6872534_trimmed.fq --
genome NC_002692.1.fna --outdir test_SRR6872534 --prefix
SRR6872534 --threads 10
```

This command performs preprocessing using the "Preprocessing.py" script. It processes the input fastq file named "SRR6872534_trimmed.fq" and references the genome file "NC_002692.1.fna." The data is processed using ten threads. Output files will be saved in the directory "test_SRR6872534," with filenames prefixed by "SRR6872534." This step is an integral part of the comprehensive vsRNAfinder workflow, ensuring data preparation for downstream analyses. The script generates index files for the TMV genome, such as "NC_002692.1.fna.bowtie.1.ebwt," "NC_002692.1.fna.bowtie.2.ebwt," and others. Alignment files ("SRR6872534.bam" and "SRR6872534.bam") are generated using the Bowtie tool, along with separate files for mapped and unmapped reads, namely "SRR6872534.unmap.fastq" and "SRR6872534.map.fastq." The genomic positions of aligned reads are further transformed using the BAM file, producing "SRR6872534.sort.bed." Additionally, a mapping summary of small RNA reads is documented in "SRR6872534.mapInfo.txt." Moreover, the script calculates and stores the depth of aligned reads on both the positive and negative strands of the genome. These depth results are saved as "SRR6872534.sort.positive.bga.txt" and "SRR6872534.sort.negative.bga.txt." This comprehensive process contributes to preparing the data for subsequent analyses.

3.4 Identifying vsiRNAs from TMV Infected *Solanum lycopersicum* sRNA-Seq Sample

The FindSmallRNA.py script automates a multistep analysis process for identifying significant vsiRNA regions in a genome. The script takes various command-line arguments to customize the analysis parameters. Below is an explanation of the provided command and its parameters:

```
python FindSmallRNA.py --data test_SRR6872534 --genomecovfile
SRR6872534.sort.positive.bga.txt --chromosome NC_002692.1 --
outdir test_SRR6872534 --len 20,24 --strand positive --read-
file SRR6872534.sort.bed --mapInfor SRR6872534.mapInfo.txt --
threads 10
```

This command enables to conduct of an analysis on sRNA-seq data for a specific objective. The command is structured as follows:

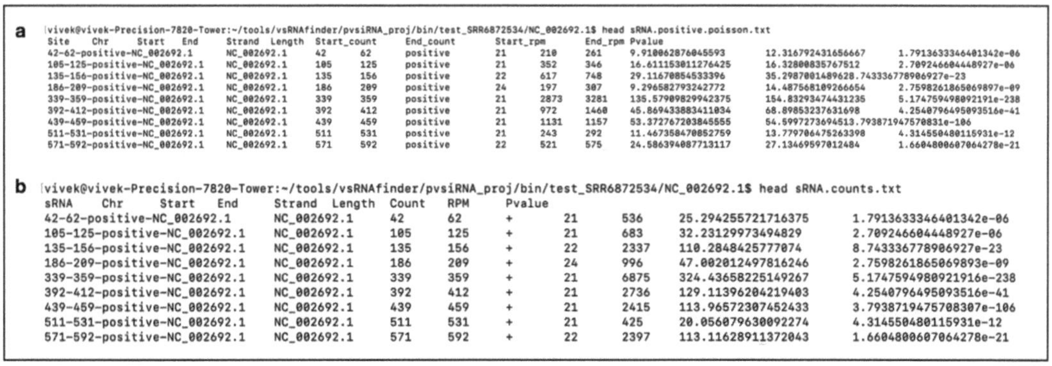

a

```
vivek@vivek-Precision-7820-Tower:~/tools/vsRNAfinder/pvsiRNA_proj/bin/test_SRR6872534/NC_002692.1$ head sRNA.positive.poisson.txt
Site    Chr    Start    End    Strand  Length  Start_count    End_count    Start_rpm    End_rpm Pvalue
42-62-positive-NC_002692.1    NC_002692.1    42    62    positive    21    210    261    9.91006287604559З    12.316792431656667    1.7913633346401342e-06
105-125-positive-NC_002692.1    NC_002692.1    105    125    positive    21    352    346    16.611153011276425    16.32800835767512    2.709246604448927e-06
135-156-positive-NC_002692.1    NC_002692.1    135    156    positive    22    617    748    29.11670854533396    35.2987001489628.743336778906927e-23
186-209-positive-NC_002692.1    NC_002692.1    186    209    positive    24    197    307    9.296582793242772    14.487568109266654    2.7598261865069897e-09
339-359-positive-NC_002692.1    NC_002692.1    339    359    positive    21    2873    3281    135.57909829942375    154.8329347443123 5    5.174759498092191e-238
392-412-positive-NC_002692.1    NC_002692.1    392    412    positive    21    972    1460    45.869433883411034    68.89853237631698    4.254079649509351 6e-41
439-459-positive-NC_002692.1    NC_002692.1    439    459    positive    21    1131    1157    53.37276720384555 5    54.5997273694513.79387194757083 1e-106
511-531-positive-NC_002692.1    NC_002692.1    511    531    positive    21    243    292    11.467358470852759    13.779706475263398    4.314550480115931e-12
571-592-positive-NC_002692.1    NC_002692.1    571    592    positive    22    521    575    24.586394087713117    27.13469597012484    1.6604800607064278e-21
```

b

```
vivek@vivek-Precision-7820-Tower:~/tools/vsRNAfinder/pvsiRNA_proj/bin/test_SRR6872534/NC_002692.1$ head sRNA.counts.txt
sRNA    Chr    Start    End    Strand  Length  Count    RPM    Pvalue
42-62-positive-NC_002692.1    NC_002692.1    42    62    +    21    536    25.294255721716375    1.7913633346401342e-06
105-125-positive-NC_002692.1    NC_002692.1    105    125    +    21    683    32.23129973494829    2.709246604448927e-06
135-156-positive-NC_002692.1    NC_002692.1    135    156    +    22    2337    110.2848425777074    8.743336778906927e-23
186-209-positive-NC_002692.1    NC_002692.1    186    209    +    24    996    47.002012497816246    2.7598261865069893e-09
339-359-positive-NC_002692.1    NC_002692.1    339    359    +    21    6875    324.43658225149267    5.1747594980921916e-238
392-412-positive-NC_002692.1    NC_002692.1    392    412    +    21    2736    129.11396204219403    4.2540796495093516e-41
439-459-positive-NC_002692.1    NC_002692.1    439    459    +    21    2415    113.96572307452433    3.7938719475708307e-106
511-531-positive-NC_002692.1    NC_002692.1    511    531    +    21    425    20.056079630092274    4.314550480115931e-12
571-592-positive-NC_002692.1    NC_002692.1    571    592    +    22    2397    113.11628911372043    1.6604800607064278e-21
```

Fig. 2 Snapshots of output files for vsiRNA analysis. (**a**) Initial lines in "sRNA.positive.poisson.txt" and (**b**) "sRNA.counts.txt." Columns detail genomic position, chromosome, coordinates, orientation, size, read distribution, normalized abundance, and statistical significance (*p*-value). "Sequence" displays the nucleotide sequence, "Count" reveals total mapped reads, and "RPM" normalizes overall vsiRNA abundance

--data test_SRR6872534: Specifies the location of the analysis data.

--genomecovfile SRR6872534.sort.positive.bga.txt': Points to the genome coverage file containing coverage information for the positive strand.

--chromosome NC_002692.1: Defines the chromosome (or accession) under analysis as NC_002692.1.

--outdir test_SRR6872534': Sets the output directory for storing analysis results.

--len 20,24: Specifies the length range for the small RNA reads, ranging from 20 to 24 nucleotides.

--strand positive': Indicates the analysis focuses on the positive strand of the genome.

--readfile SRR6872534.sort.bed': Specifies the BED file containing sequence alignment information.

--mapInfor SRR6872534.mapInfo.txt': Provides the bowtie mapping information file.

--threads 10: Specifies the number of threads (parallel processes) for the analysis.

This command is part of a broader analysis process that identifies small RNA sequences meeting specific criteria within the provided data. The script analyzes both positive and negative strands of the genome. A similar command is used to analyze negative strand small RNAs:

```
python FindSmallRNA.py --data test_SRR6872534 --genomecovfile
SRR6872534.sort.negative.bga.txt --chromosome NC_002692.1 --
outdir test_SRR6872534 --len 20,24 --strand negative --read-
file SRR6872534.sort.bed --mapInfor SRR6872534.mapInfo.txt --
```

```
threads 10
```

In the output directory "test_SRR6872534," files named "sRNA.negative.poisson.txt" and "sRNA.positive.poisson.txt" will be generated. These files display site and locus information along with strand details and significance values (p-values), providing insights into significance of vsiRNAs and locus from which it is generated as displayed in Fig. 2a.

3.5 Quantifying the Abundance of vsiRNAs Across TMV Genome

This step focuses on quantifying the abundance of vsiRNAs all along the TMV genome. The process is automated using the Python script "Quantification.py," to perform the quantification of sRNA abundance across different regions of the TMV genome. Here is an explanation of the command and its associated parameters:

```
python Quantification.py --threads 10 --data test_SRR6872534/
NC_002692.1 --bam test_SRR6872534/SRR6872534.sort.bam --ma-
pInfor test_SRR6872534/SRR6872534.mapInfo.txt
```

--threads 10: Specifies the number of threads (parallel processes) to be used for analysis, set to 10 for efficient computation.

--data test_SRR6872534/NC_002692.1: Points to the output path of the preceding FindSmallRNA step, specific to the NC_002692.1 chromosome of TMV.

--bam test_SRR6872534/SRR6872534.sort.bam: Specifies the BAM file containing the aligned sRNA sequence data.

--mapInfor test_SRR6872534/SRR6872534.mapInfo.txt: Provides the bowtie mapInfor file, which encapsulates mapping information from the alignment process.

In total, 85 vsiRNAs are identified from the positive strand and 89 from the negative strand. Quantified vsiRNA data, normalized in RPM, is saved in the "sRNA.counts.txt" output file within the designated directory for the NC_002692.1 chromosome (Fig. 2b).

References

1. Ding SW, Voinnet O (2007) Antiviral immunity directed by small RNAs. Cell 130:413–426
2. Guo Z, Li Y, Ding SW (2019) Small RNA-based antimicrobial immunity. Nat Rev Immunol 19:31–44
3. Llave C (2010) Virus-derived small interfering RNAs at the core of plant-virus interactions. Trends Plant Sci 15:701–707
4. Pooggin MM (2018) Small RNA-omics for plant virus identification, virome reconstruction, and antiviral defense characterization. Front Microbiol 9. https://doi.org/10.3389/fmicb.2018.02779
5. Li L, Andika IB, Xu Y et al (2017) Differential characteristics of viral siRNAs between leaves and roots of wheat plants naturally infected with wheat yellow mosaic virus, a soil-borne

virus. Front Microbiol 8:284468. https://doi.org/10.3389/FMICB.2017.01802/BIBTEX

6. Wu Q, Luo Y, Lu R et al (2010) Virus discovery by deep sequencing and assembly of virus-derived small silencing RNAs. Proc Natl Acad Sci USA 107:1606–1611. https://doi.org/10.1073/pnas.0911353107
7. Donaire L, Wang Y, Gonzalez-Ibeas D et al (2009) Deep-sequencing of plant viral small RNAs reveals effective and widespread targeting of viral genomes. Virology 392:203–214. https://doi.org/10.1016/j.virol.2009.07.005
8. Massart S, Chiumenti M, De Jonghe K et al (2019) Virus detection by high-throughput sequencing of small RNAs: Large-scale performance testing of sequence analysis strategies. Phytopathology 109:488–497. https://doi.org/10.1094/PHYTO-02-18-0067-R/ASSET/IMAGES/LARGE/PHYTO-02-18-0067-R_F4.JPEG
9. Vivek AT, Zahra S, Kumar S (2020) From current knowledge to best practice: a primer on viral diagnostics using deep sequencing of virus-derived small interfering RNAs (vsiRNAs) in infected plants. Methods 183:30–37. https://doi.org/10.1016/J.YMETH.2019.10.009
10. Cai Z, Fu P, Qiu Y et al (2022) vsRNAfinder: a novel method for identifying high-confidence viral small RNAs from small RNA-Seq data. Brief Bioinform 23. https://doi.org/10.1093/BIB/BBAC496

Chapter 18

Incorporating Sequence-Dependent DNA Shape and Dynamics into Transcriptome Data Analysis

Manisha Kalsan, Almas Jabeen, and Shandar Ahmad

Abstract

Differentially expressed genes in a cellular context may be co-regulated by the same transcription factor. However, in the absence of a concurrent transcription factor binding data, such interactions are difficult to detect, especially at the single cell expression level. Motif enrichments in such genes can be used to gain insight into differential expressions caused by the shared upstream TFs. However, it is now established that many genes are co-regulated by the same TF due to a shared DNA shape or sequence-dependent conformational dynamics instead of sequence motif. In this work, we demonstrate how, starting from a gene expression data, such DNA shape and dynamics signatures can be potentially detected using publicly available tools, including *DynaSeq*, developed in our group for predicting the sequence-dependent components of these DNA shape features.

Key words Gene expression, Transcription factor, Sequence motifs, DNA shape, DNA conformational dynamics

1 Introduction

Expression of specific genes in a given cellular environment, stages of development and in response to environmental stimuli such as infection or diseased conditions are all mediated through a process of selectivity at the transcriptional level [1–4]. Thus, the decoding of transcriptional specificity and cellular attributes of gene expression for each gene and their genome-scale patterns are essential to the understanding of cellular functions [5]. During transcription, the DNA-binding proteins called transcription factors (TFs) interact with genomic DNA to subsequently initiate and regulate the expression of genes resulting in different types of RNAs, namely, mRNA, miRNA, and lncRNAs [6]. With the availability of large-scale data sets on transcriptional events, data-driven technologies have become critical in the identification of condition-specific and single-cell level of heterogeneity in transcription. Chromatin

Rajeev K. Azad (ed.), *Transcriptome Data Analysis*, Methods in Molecular Biology, vol. 2812,
https://doi.org/10.1007/978-1-0716-3886-6_18,

immunoprecipitation followed by sequencing (ChIP-seq), DNAse-Seq, ATAC-Seq, and several other techniques have been successful in providing insights into genomic events related to TF binding [7–10]. Most of these techniques work on bulk biological samples and thereby do not capture the cellular heterogeneity, which is now the state-of-the-art in the case of gene expression data analysis. Emerging technologies for gene expression data analysis such as RNA-seq, Small RNA-seq, Targeted RNA-seq, Long-read RNA sequencing, and Single-Cell RNA-seq (scRNA-seq), on the other, hand provide insights into gene expression patterns at high resolutions [11–16]. However, the mechanistic basis of gene expression at the transcriptional level is not fully revealed by RNA-Seq data alone and that is where directly estimating the TF binding to gene expression will be helpful. If a link between RNA-Seq experiments (both at single cell and bulk samples) and TF binding could be established, insights into many biological phenotypes will be obtained. In a genomic context, there are numerous copies of the binding sites of a TF. Binding specificity of TFs incorporates an interplay of multiple mechanisms that contribute to the final selection of a subset of sites which will eventually be bound by them [17]. Intuitively, TF binding can be understood at the level of target sequences by building a consensus among the promoters of co-regulated genes. However, it is now known that sequence-level consensus of TF binding is incomplete and *DNA shape* and *dynamics* must be considered for identifying consensus targets of a TF. In this work, we propose a protocol that can be used to estimate an upstream TF binding to DNA from the gene expression data and to develop hypotheses of their co-regulation at the genomic scale. In this process, we also explain the working of our tool *DynaSeq* which can be used to analyze whole genome intrinsic propensities of DNA conformational dynamics of DNA sequences using a dictionary-based approach. The protocols presened here will therefore also be useful in investigating consensus patterns between a large data set of sequences beyond sequence motifs, particularly for genomic DNA.

2 Role of Sequence and Sequence-Dependent DNA Shape in TF-DNA Interactions

The binding between TFs and DNA has been traditionally characterized by studying the sequences of the target genomic locations of these proteins and is analyzed in the form of *sequence motifs*. These target motifs of proteins are formalized as positional probabilities of occurrence of each nucleotide base in the target sites and represented as a position weight matrix (PWM) or position frequency matrix (PFM) [18, 19]. The widely accepted usage of PWMs to characterize the binding sites of TFs assumes DNA sequence to be the only factor directing these interactions, thereby not considering that features beyond the sequence also contribute to the specificity of these interactions [20, 21]. In the three-dimensional space, where these interactions take place between a TF and its genomic target DNA locations, two major mechanisms have been identified and widely studied, namely, ***direct or sequence readout*** and ***indirect or shape readout*** [22]. Sequence readout refers to the direct physical recognition of specific DNA sequences by TFs through direct contacts with the DNA bases whereas shape readout refers to the recognition through the overall structural properties and deformability of the DNA helix, which does indirectly depend on the base sequence but does not have a one to one relationship with it due to the complex relationship between sequence and double helical structures. In shape readout, the protein interacts with the DNA backbone, sugar-phosphate groups, or other structural features of the DNA rather than directly contacting the DNA bases and is apparently less sequence-dependent compared to direct readout [23, 24]. Nonetheless, DNA shape is all that is accessible to an approaching TF before it reaches the target motifs as bases are located inside a less accessible interior of the double helix. Understanding the recognition of targets by TFs will therefore be severely limited without considering these interactions. Several recent studies have focused on elucidating the interactions between TFs and DNA by studying the shape readout and DNA shape profiles of genomic target sites [25–30].

DNA shape refers to the three-dimensional structure and conformational characteristics of DNA molecules. While DNA is commonly represented as a double helix, it is a flexible and dynamic molecule that can adopt different shapes and conformations as influenced by the factors such as sequence composition, base stacking interactions, base-pair hydrogen bonding, and the presence of TFs, and modifications such as methylation and acetylation (not being considered in this work) [31, 32]. The structural properties of DNA can influence protein binding and recognition, subsequently affecting processes such as transcription, DNA repair, and chromatin organization [33, 34].

To study DNA structural conformations, researchers have used various experimental approaches including X-ray crystallography, electron microscopy, nuclear magnetic resonance (NMR) spectroscopy, and cryo-electron microscopy (cryo-EM) [35–38]. Although these methods offer in vivo experimental advantages, these are laborious, expensive, and time-consuming. Further, the DNA shape is driven both by the intrinsic sequence propensities as well as the genomic context of long range folding and epigenetics forces. Sequence-dependent interactions of protein-DNA recognition cannot be separated easily from the indirect and epigenetic factors using a purely crystallographic study of a single binding event. Hence, computational approaches employing statistical mechanics, molecular dynamics simulations, and bioinformatics algorithms have been developed and extensively used to model DNA structure and to predict shape features based on DNA sequence information. Most of the computational approaches have characterized this level of DNA shape in terms of base pairs, step *parameters* and *groove widths*. There are multiple resources available that provide functionalities to analyze, visualize, and interpret this DNA shape information in the form of multiple aspects of DNA *conformational parameters*. To name a few, *Nucleic Acid Database* (NDB) [39, 40] *x3DNA* [41–43], *DNAshape* [44, 45], *and DynaSeq* [28] are some of computational tools that can aid in the analysis of DNA conformational *parameters* for an existing DNA structure or their sequence-dependent predictions. For example, NDB provides a database of atomic coordinates of experimentally determined structures of DNA molecules and their assemblies accessible online at http://ndbserver.rutgers.edu/. NDB is planned to be transformed into a new *Nucleic Acid Knowledgebase (NAKB)* preserving all the functionalities of NDB including reports, statistics, visualization tools, software resources, and community web and resources at https://nakb.org/ [46]. *x3DNA* is one of the most widely used tools for the analysis of DNA structural properties, when the atomic coordinates have already been solved using X-ray crystallography or other methods. It also allows analysis, and visualization of known structures and building atomic coordinates for model DNA topologies (e.g., A, B, or Z-DNA structures). This tool is freely available online at http://web.x3dna.org/ and offline as a package of multiple executable programs. It is also available for download as a standalone program to run locally on *Unix/Linux*-like environments. *DNAshape* is another publicly available tool that can be used as an online tool or an R package (*DNAshapeR*) to generate genome-wide DNA shape profiles based on structural information derived from Monte Carlo simulations incorporating a total of 13 conformational parameters into the analysis. *DynaSeq* (http://www.sciwhylab.org/dynaseq/) was developed by us to address the limitations of *DNAshape* by providing a more comprehensive set of predictions for sequence-dependent DNA shape properties including a dictionary of 13 DNA shape parameters and their 65-global equal frequency bin population distributions in a *conformational ensemble*. Instead of the Monte Carlo approach of *DNAshape*, *DynaSeq* was based on the Molecular Dynamics (MD) simulations of an exhaustive set of DNA tetramers. The DNA shape dynamics are available in the form of shape ensemble as five equal global frequency bins for each pentamer where each bin of a pentamer represents its occupancies in the global equal frequency bins of a parameter which spans the observed range of that parameter in the simulations. The shape ensemble helps to evaluate the preference of a pentamer for a specific range of a conformational parameter (more details in the coming section).

3 Analysis of Sequence-Dependent DNA Shape Profiles Using *DynaSeq*

As stated above, a simple description of DNA shape and subtle changes in topologies can be characterized by assuming an ideal DNA structure and treating planar nucleic acid bases inside a double helix as rectangular stacking subunits. Complementary bases in a pair of stacking subunits (corresponding to a base position) and a pair of successively stacked base pairs can take any *deformed* geometries slightly deviating from their ideal positional arrangements. These deformations may further be summarized as *translational* and *rotational* changes in stacking or pairing along the helical z-axis or in the x-y plane normal to it (Table 1). A base pair here refers to the pairing of two complementary nucleotide bases from opposite strands held together by hydrogen bonds whereas a base step refers to the combination of two adjacent nucleotide bases and the associated sugar-phosphate backbone in the DNA double helix (Fig. 1). The DNA shape conformations have been defined with the help of rotations and translations across x-, y-, axis in base plane and the helical z-axis. The base pairs and base steps have been thus formalized as 12 conformational parameters (6 base pair and 6 base step), namely, *Shear, Stretch, Stagger, Buckle, Prop-Tw, Opening, Shift, Slide, Rise, Tilt, Roll, Twist,* and a 13th parameter namely the *minor groove width* at a given nucleotide position (Table 1 and Fig. 2).

Actual elastic deformations that the DNA undergoes in a cellular environment cannot be directly measured using current structure-determination methods. However, specific sequences prefer a specific set of conformational parameter values, as observed in crystallographic data and molecular simulations. For example, the average value for *shear* in the central position of CGACC (0.133 radians) is much higher than that of corresponding sequence TCAGG (0.033 radians) and TCAGT (0.022 radians) as predicted by *DynaSeq,* suggesting that stacking of Adenine base by CG and CC on either sides gives it a natural *shear* that can be exploited by a TF with complementary local structures. Sequence-specific predictions of these conformational parameters and their conformational ensembles can be made for any 5-mer window of nucleic acid sequence using an *SVM*-based model inbuilt into *DynaSeq* [28]. *DynaSeq* was originally trained on the trajectory data of all-atom MD simulations on an exhaustive set of tetrameric DNA sequences flanked by GCGC adapters resulting in 12-mers with unique central tetramers. The MD simulated structures corresponding to different snapshots of the simulation trajectory were used to generate measures of the 13 conformational parameters. These conformations have been formalized as two types of shape features: static shape and shape ensemble. The static features are representations of averages of conformations in those MD

Table 1
Summary of rotational and translational parameters defined as DNA conformational parameters: Base pair and base step parameters which make a total of 12 conformational parameters

S. no	Base pair/base step	Translational/rotational	Along axis	Parameter
1	Base pair	Translational	X-axis	Shear
2			Y-axis	Stretch
3			Z-axis	Stagger
4		Rotational	X-axis	Buckle
5			Y-axis	Propeller
6			Z-axis	Opening
7	Base step	Translational	X-axis	Shift
8			Y-axis	Slide
9			Z-axis	Rise
10		Rotational	X-axis	Tilt
11			Y-axis	Roll
12			Z-axis	Twist
13	–	–	–	Minor groove width

Minor groove width forms the 13th conformational parameter

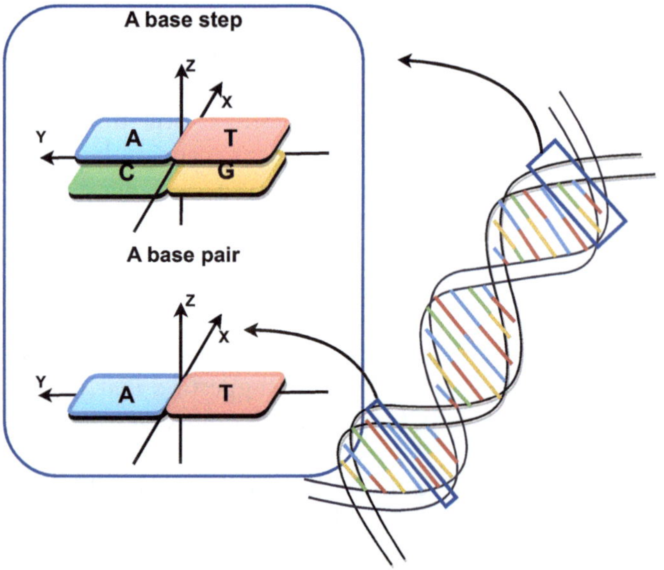

Fig. 1 Definition of a base pair and a base step for shape readout calculations

simulation snapshots or predictions based on them. Ensemble parameters have been created by first defining *global ensemble bins* of each conformational parameter by pooling values for the parameter at all positions of all tetramers and then creating five equal frequency bins. These global ensemble bins were used to calculate the differences at each base in each 12-mer to create their occupancy

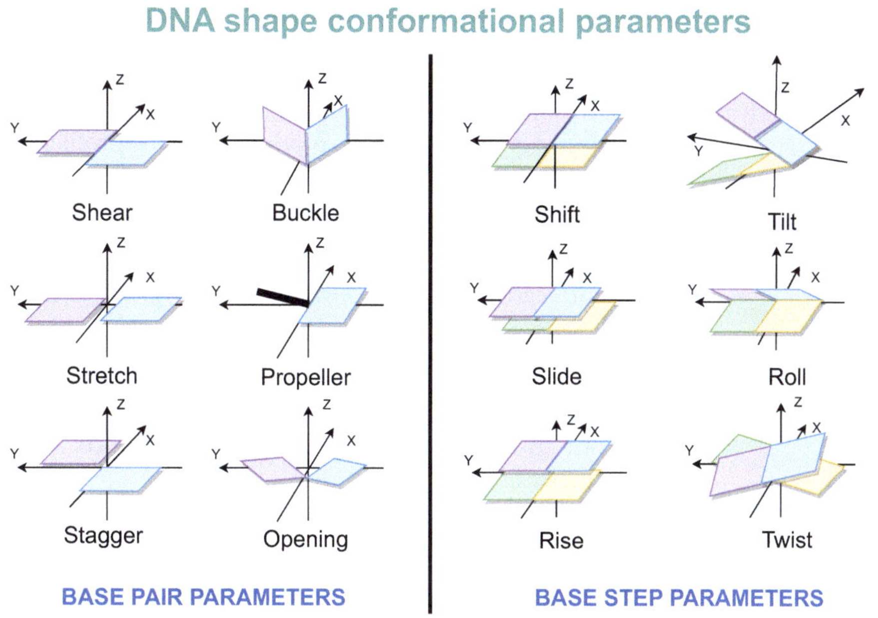

Fig. 2 DNA shape conformational parameters as *translations* and *rotations* along axes defined as specific base pair and base step parameters

profiles in reference to these *global ensemble bins*. The static shape helps to understand how a DNA sequence would behave structurally on an average whereas the shape *ensemble* incorporates the dynamic aspect of DNA structures in the form of divergence of a parameter from its average shape. Bin-wise occupancies of DNA, derived using MD simulations may better model *conformational selection* approaches to molecular interactions [47–49]. These features of conformational dynamics are not captured by their averages and *DynaSeq* tool is the only tool that provides such extensive predictions to the best of our knowledge.

In order to extend the motif-based analysis of genomic sequences regulated by the same TFs to the *DNA shape* and dynamics, we need to first collect a set of potential targets and align them for investigating shape consensus among them. Genome-wide set of sequences come from the promoters of genes *co-expressed* with a TF or more generally likely to be *co-regulated*. When a direct TF binding data is available, the target sites of TFs are inferred from the *peak-calling* algorithms on high quality unique alignment maps from the high-throughput sequencing techniques producing the in vivo TF-DNA interactions such as ChIP-seq. The binding specificities of TFs are explored by taking a reference position in these binding sites and extending them upstream and downstream to generate the length of genomic sites suitable for insights into the DNA shape properties. For comparison of the specificities of sequences from TF binding and nonbinding or low-affinity sets of sequences, a length as small as 10–20 bp has the ability to explain the differences [44]. But, if the binding sites of proteins from the same family are compared where TFs have similar sequential preferences or similar binding motifs, the analysis may also require exploring the shape profiles of flanking regions.

Starting from a data set of gene expression profiles, a DNA shape based analysis can be carried out in the following steps.

1. Identification of *differentially expressed genes* (DEGs) in two cellular conditions.
2. Filtering DEGs with and without potentially shared upstream TFs and collection of their promoter sequences.
3. Motif discovery and enrichment in both sets of promoter sequences.
4. Sequence and shape based alignment and analysis of selected promoters sequence sets.
5. Identification of potential new TF targets.
6. Refining results and further analysis.

We provide tools to carry out these analyses in our GitHub repository and explain each of these steps in the methods section below, by way of taking a real case-study and explaining the protocol in the most general manner for wider application by potential users.

4 Implementation and Case Study

4.1 Systems and Software Requirements

- OS: Linux/macOS
- RNA-seq NGS pipeline
  - RNA-seq *fastq* files
  - Reference genome
  - *FastQC* [50] (Quality check for dataset)
  - *Trimmomatic* [51] (Adapter removal and trimming of low-quality bases)
  - *Bowtie2* [52]/*BWA* [53]/*HISAT2* [54]/*STAR* [55]/ *TopHat2* [56] aligners (either tool for read mapping/alignment to reference genome)
  - *StringTie* [57]/*Cufflinks* [58]/*featureCounts* [59] (Transcript assembly and mapped read quantification)
- R/RStudio [60] and its libraries
  - DESeq2 [61] (differential gene expression analysis)
  - enrichR [62] (enrichment for gene sets)

  - biomaRt [63] (mapping between gene identifiers and extraction of metadata)
  - Other required bioconductor packages (e.g., *BSgenome, TxDb.Hsapiens.UCSC.hg38.knownGene, BSgenome.Hsapiens.UCSC.hg38, GenomicFeatures*)
- Genomic data for genes (*UCSC genome browser*) [64]
- Bedtools [65] (for handling of genomic coordinates data and extraction of sequence or other features)
- *HOMER* [66] (Motif discovery)
- *DynaSeq* dictionaries (https://github.com/Sciwhylab/DynaSeq) [28] (DNA shape profile analysis) (This is the main in-house tool used in this protocol, while all others are publicly available third-party softwares)

4.2 Detailed Methods: A Case Study

As stated above, the DNA-shape based analysis of RNASeq data will involve the steps outlined in Fig. 3. Programming codes for running these examples and files described in the following sections refer to our GitHub repository hosted at the URL: https://github.com/Sciwhylab/DynaSeq_transcriptomics. The proposed pipeline is demonstrated here as applied on a selected dataset (explained later) with the following steps.

- Identification of DEGs
- Identification of genes with shared upstream TFs
- Consensus motif prediction in the set of promoter sequences
- Shape analysis of selected promoters

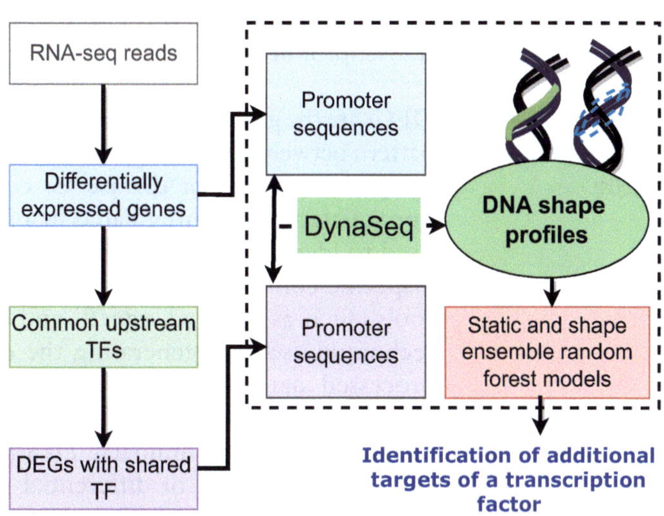

Fig. 3 The proposed pipeline makes use of the DEGs subsetted into two: regulated by a common TF and remaining DEGs to extract their corresponding promoter sequences for use in DNA shape profile analysis

- Identification of potential new TF targets
- Refining results and further analysis

4.2.1 Dataset Description

In this work, we demonstrate the proposed pipeline on a gene expression dataset generated by the study entitled "*Mutant FOXL2C134W Hijacks SMAD4 and SMAD2/3 to Drive Adult Granulosa Cell Tumors*" by Weis-Banke et al. [67] (the choice of the case study is rather arbitrary and based on our ongoing involvement in the study of this system elsewhere). The Weis-Banke et al.'s work has focused on FOXL2 which is a forkhead TF essential for proper reproductive function in females. C134W mutation is characteristic of adult granulosa cell tumors of the ovary. Weis-Banke et al. [67] characterized the binding of this mutant with a novel hybrid DNA motif inducing expression of nearby genes characteristic of stemness and epithelial-to-mesenchymal transition. The processed RNA-seq datasets are available at the *Gene Expression Omnibus* (GEO) with the accession ID GSE138496. For the analysis of RNA-seq data, there are multiple steps that need to be performed before differential gene expression analysis. As stated, the available pipelines and tools for the analysis of transcriptomics data is now well established and is beyond the scope of this chapter, the elaboration of an RNA-seq pipeline starting from quality control and mapping is omitted here and we focus on the incorporation of DNA shape profiles analysis in transcriptomics and start our analysis from differential gene expression (a reader interested in RNA-Seq analysis pipelines may refer to numerous literature on the subject for example, there is a comprehensive tutorial provided by Galaxy Training! from Galaxy Project available at https://training.galaxyproject.org/training-material/topics/transcriptomics/).

4.2.2 Identification of Differentially Expressed Genes (DEGs)

DEGs are the genes which show significant changes in their expression pattern between different group of *samples* (e.g., wild-type vs. mutant sample conditions or disease vs. control). These genes may be upregulated or downregulated in a specific sample set and can be used to identify genes vital for a biological process or pathway or a specific condition. The DEGs can be predicted using multiple tools, such as *DESeq2*, *edgeR*, and *limma*, depending upon the technique used for generating the expression data. The *DESeq2* processed datasets from Weis-Banke et al. are available at the GEO with GSE138496 and were utilized for further protocol in this case study. In summary, *DESeq2* is a tool available as an R package for analysis of differential gene expression in RNA-seq datasets by providing statistical methods to assess the change in expression of genes in sample vs. control conditions. It takes raw counts data for genes per sample conditions and generates a list of DEGs between the sample and control conditions along with their gene expression level and significance values across the samples. When using *DESeq2* as a bioconductor package in R, we require a specific object called `DESeqDataset` containing `rowRanges`, `colData`, and `assay` which help to assemble multiple levels of information required for such analyses in one object.

- `rowRanges` correspond to genomic ranges
- `colData` refers to a matrix containing metadata about the sample names, treatment conditions, run numbers, etc.
- `assay` refers to any data corresponding to genes per sample such as the counts data.

For more details about this object or the application of *DESeq2*, multiple tutorials are available with one available at

http://bioconductor.org/packages/devel/bioc/vignettes/DESeq2/inst/doc/DESeq2.html#differential-expression-analysis

Briefly, the prepared `DESeqdataset` is subjected to the differential analysis pipeline on the raw counts using the function `DESeq` which estimates the size factors and dispersion values and uses these to fit a generalized linear model for estimation of fold changes. The result of `DESeq2` can be saved in a tabular format and the details in each column are as follows.

- *baseMean* is the average of the normalized count values, divided by size factors, taken over all samples in the `DESeqDataSet`.
- *log2FoldChange* is an effect size estimate which indicates the level of change of gene expression due to treatment vs. control samples reported on log2 levels.
- *lfcSE* represents a standard error estimate for the log2 fold change estimate which tries to ensure that the change indicated by the *log2FoldChange* is certain and there is enough evidence to reject the null hypothesis which suggests no significant difference between control samples and treated samples except than cause by experimental differences.
- *Pvalue* is the probability of observing a fold change estimate as strong or even more, produced in the results to be observed in a null hypothesis.
- *Padj* is an indicator of false discovery rate.

The file "GSE138496_Galaxy758_DESeq2_result_file_on_-MUT_TGF_vs_WT_TGF_.txt" contains the results for the DGE analysis between WT FOXL2 and MUT FOXL2 when both are stimulated with TGF and was used to implement the proposed pipeline, but this pipeline can be applied to the choice of conditions for the study.

In gene expression datasets, there are genes which go through a high degree of change between samples but most of the genes have no significant difference between samples and a large number are not even expressed at all such as housekeeping genes and genes specific to another cell-type, respectively. The negligibly expressed genes will not have count data and hence will result in a lot of NA's in the *DESeq2* output. Therefore, we removed rows with NA and selected genes with significant change using a cutoff of 0.05 or 0.01 on p-adjusted value.

```
data1<-de_data[complete.cases(de_data), ]
data_significant<-data1[as.numeric(data1$V7) < 0.05,]
```

Once the DEGs are found, the specific genes which are upregulated or downregulated are selected by using a threshold on *log2FoldChange* values. The genes with *log2FoldChange* value less than -1 have less than half of the expression in the sample or treated condition as compared to the control condition. The genes with a higher expression of more than twice the expression in control have *log2FoldChange* values of more than 1. A different threshold value is used to detect genes going through higher levels of changes. For simplicity of the steps in the current analysis, we used all the genes without separating the upregulated and downregulated genes. However, in practice, one may explore by analyzing the two gene sets separately. Based on this criterion, 5014 genes (adjusted p-value <0.05) were selected for further analysis.

4.2.3 Identification of Genes with Shared Upstream TFs

The list of DEGs identified above was used to predict the TFs which potentially regulate these genes. We used the `enrichr` function from the `enrichR` library to generate gene enrichment results against a chosen database. For enrichment of TFs whose binding sites are enriched in the promoter regions of genes, the database was set to *Transfac* and *Jaspar* PWMs which searched the PWMs from these databases in the promoter regions of the specified genes.

```
enrichment_result<-as.data.frame(enrichr(data_significant
[,1],"TRANSFAC_and_JASPAR_PWMs"))
```

To select the significant enrichment results, we filtered the results using a threshold on adjusted p-values.

```
padj_sig<-enrichment_result[enrichment_result$TRANSFAC_and_-
JASPAR_PWMs.Adjusted.P.value <= 0.05,]
```

This gave a list of TFs predicted to have significant enrichment values in the regions upstream of these genes. In the result, the columns correspond to *Term name, Overlap, P.value, Adjusted.P. value, Old.P.value, Old.Adjusted.P.value, Odds.Ratio, Combined.*

Score, and *Genes regulated*. We used the first TF SP1 for further analysis. Out of the 5014 genes we selected above, a total of 4126 genes were found to be enriched by at least one of the 36 TFs (found by these results) in their upstream regions with a 0.05 significance level. Out of these, we demonstrate the analysis on one TF, namely, SP1 which was found to be enriched in promoter regions of 802 genes. We call these 802 genes as *TF regulated* and the rest of the DEGs as *others* hereafter.

```
hgnc_symbols <- unlist(strsplit(enrichment_result[1,9],";"))
tf_regulated_genes<-data_significant[grep(paste(hgnc_symbols,
collapse = "|"),data_significant$V1),1]
remaining_genes<-data_significant[-grep(paste(hgnc_symbols,
collapse = "|"),data_significant$V1),1]
```

Our data contains gene names in the form of *HGNC* symbols, so we converted them to Entrez gene IDs by using the `biomaRt` tool. If not already installed, *biomaRt* can be installed by the following command.

```
BiocManager::install(c("biomaRt"))
```

We defined the mart to be used according to our data (*Homo Sapiens*). The list of all available marts can be found using the `listMarts()` or `listEnsembl()`.

```
library(biomaRt)
ensembl <- useMart(biomart = "ENSEMBL_MART_ENSEMBL", dataset
= "hsapiens_gene_ensembl")
```

To use a specific dataset from a mart, we need to specify it by choosing from the list of datasets from that mart. List of datasets can be assessed using.

```
listDatasets(mart).
listDatasets(mart = ensembl)
```

The list of available attributes for the specified mart can be assessed using `listAttributes(mart)`. We used the list of genes enriched by the TF SP1 to extract the metadata for genes.

```
mid_info<-getBM(attributes = c("hgnc_symbol", "ensembl_gen-
e_id_version",
 "ensembl_transcript_id_version",
 "Entrezgene_id", "strand", "transcription_start_site"),
 filters = "hgnc_symbol",
 values = list_genes,
 mart = ensembl))
```

Tabular output generated here contains the *HGNC* symbols, *ensembl* gene IDs with version, *ensembl* transcript ID with version, *entrez* gene ID, strand, and transcription start site (TSS) for the list of genes provided to `values` by filtering using `filter = "hgnc_symbol"` because we had our list of genes as HGNC symbols.

For generating the promoter sequences for a list of DEGs, we can approach in multiple ways e.g. using bioconductor libraries such as `GenomicFeatures`, `BSgenome`, `BSgenome.Hsapiens.UCSC.hg38`, `TxDb.Hsapiens.UCSC.hg38.knownGene` to call functions which facilitate the extraction of the genomic sequences. Some of these libraries are specific to organisms therefore, we use `BSgenome.Hsapiens.UCSC.hg38`, `TxDb.Hsapiens.UCSC.hg38.knownGene` which correspond to data for *Homo Sapiens* with the genome assembly hg38. This method requires a grouped transcripts `GRangesList` object prepared by `transcriptsBy` containing the transcripts grouped by genes, exons, or cds. This `GRangesList` object is used as an input to `getPromoterSeq` to extract the corresponding sequence by defining the length of upstream and downstream flank. The second approach uses linux-based tools such as *bedtools* which have specific functions dedicated for the same task but have to be called through an operating system terminal. The preparation of input data to be fed in both these approaches is different and multiple tutorials addressed to their usage can be referred to. We used the UCSC table browser to generate a file containing details for all genes for hg38. This file served as a reference for the generation of promoter sequences using the function `flank` from *bedtools.*

We used the unique *entrez* gene IDs extracted from *biomaRt* to first generate transcript IDs of one TSS per gene by selecting the lowest TSS coordinate (most upstream) if the gene is on positive strand and the largest coordinate (most downstream) if the gene is on negative strand. These transcript IDs were then used to extract corresponding locations and exon counts from genes.bed.

The genomic coordinates can often be represented as part of scaffolds which are contiguous sequences part of the genome assembly but not assigned to specific chromosomes or contigs. These scaffolds represent the sequences from unplaced or alternative portions of chromosomes which vary from assembly to assembly. We have focused our analysis on genomic sequences from standard chromosomes only. Hence, we removed the sequences from these scaffolds.

All the steps starting from *HGNC* symbols of DEGs enriched by a TF as generated by `enrichr` enrichment results, till generation of their promoter sequences have been made into an R function by the name `extract_unique_tss` in the file `generate_TSS_coded.R`. This script was specifically written to carry out the steps described till now using the input files provided by Weis-Banke et al. and also available at the repository page under `input`. For customization, the functions available in `generate_TSS.R` can be found on the same page.

The bed files containing TSS coordinates were used as input to `flank` from *bedtools* to convert these coordinates into genomic coordinates corresponding to gene promoter sequences by defining a flank as length to be called a promoter for a gene TSS. For example, if we define promoter as TSS -/+2 kb, we would need to specify this as an argument to `flank`. This can also be executed in a linux terminal directly or by using `system()` from inside the R. We executed this in the linux terminal by command.

```
bedtools flank -i genes_coord.bed -g hg38.chrom.sizes -l 20 -r
0 -s > test_flank.bed
```

The `flank` from *bedtools* is used when the flanking regions for a list of genomic features are to be generated as a bed file. Here, we will generate coordinates for promoters of the genes (features) stored in a bed file. The definition of a promoter mostly is user specific with researchers using either 1000, 2000, or 5000 bp (as an argument -l (left)) upstream of TSS of a gene dependent on their study. In the above command, `genes_coord.bed` is the bed file with coordinates of TSS generated in the steps above and should be replaced by the name of the corresponding file as input (–i). The -g takes the value of the file name containing the chromosome sizes of the hg38 assembly. For simplicity, we are going to extract sequences of 20 bp length but it can be changed according to the analysis requirements. If the downstream sequences of a specific length are also needed, the -r (right) can be set to the length, but we do not use any downstream sequence here hence it is set to 0 here. After `flank`, we use `bedtools getfasta`.

```
bedtools getfasta -fi hg38.fa -bed genes_flank.bed -fo genes_-
flank.fasta
```

The `bedtools getfasta` is used to extract the genomic sequences for a list of genomic coordinates contained in a bed file. Here, the genome is set by `-fi`, and `-bed` takes the value as name of a bed file containing promoter coordinates generated above to extract sequences to a file by name given to `-fo` and hence the values to attributes are to be set accordingly.

4.2.4 Sequence Motifs in Candidate Targets

Before motif search, we make sure that there are no duplicate sequence names in the *fasta* file. Motif discovery algorithms identify short, conserved DNA or protein motifs within input sequences which could represent functional elements, such as TFBSs, protein interaction domains, or regulatory elements. There are multiple motif discovery algorithms for DNA sequences widely used by researchers including *MEME* (Multiple Em for Motif Elicitation) [68] which uses *Expectation-Maximization* (EM) algorithm to identify motifs, *DREME* (Discriminative Regular Expression Motif Elicitation) [69] which employs a modified EM algorithm and produces position-specific scoring matrices (PSSMs) to represent the discovered motifs, *HOMER* (Hypergeometric Optimization of Motif EnRichment) [66] searches for overrepresented motifs in sets of DNA sequences, either in a given list or within the genome. *GLAM2* (Gapped Local Alignment of Motifs) [70] available as a part of *MEME* suite uses a gapped alignment method to identify multiple motifs with the presence of gaps (insertions or deletions) in the aligned sequences. Other motif discovery tools include *MEME-ChIP* [71], *YMF* (Yeast Motif Finder) [72], *RSAT* (Regulatory Sequence Analysis Tools) [73] to name a few.

HOMER can be used to predict the enriched motifs and for *motif annotation*, *motif comparison* and *clustering*. *HOMER* uses perl scripts to perform these analyses. The motif discovery and enrichment is executed by `findMotifsGenome.pl` which can be used on an input set of sequences or the whole genome. The basic syntax to use is as follows.

```
findMotifsGenome.pl <peak/BED file> <genome> <output direc-
tory> -size # [options]
```

Here, size refers to the length of input sequences. Please note that we have provided the BED file instead of the sequence file because `findMotifsGenome.pl` takes BED file coordinates along with reference genome information which uses the installed genome *hg38* inside *HOMER*. Motif discovery results in text files containing information about motif PWM and motif logo which is representation of a motif with information content per position in bits. The details of the results from this step are not included in this work.

4.2.5 Shape Analysis of Selected Promoters

We generated the DNA shape profiles of promoter sequences using a sliding window approach on 5-mers using *DynaSeq* which is a tool available as dictionaries of shape profiles for DNA sequences. These dictionaries contain representations of static shape and shape ensemble of an exhaustive set of pentamers in the form of the conformational parameters, available at the github repository mentioned above.

The script `dna_shape_analysis.R` generates the DNA shape profiles in a tabular format with each row corresponding to a sliding window position over a sequence and columns representing conformational parameters. Note that as *DynaSeq* uses a pentamer sequence information to generate conformations at a single position, the sequence information of 5 bp is converted to 1 bp (central) of structural information and hence for a sequence of length k, structural parameters for k-4 positions will be generated with the absence of parameter values for 2 bp flanks from each end of a sequence. For example, if we take a 20 bp sequence, shape profiles of only the third to 18th positions of the sequence will be generated. The output data generated will be a matrix of `(n*(len-4))*78` dimensions where `n` is the number of sequences in each dataset and `len` refers to the sequence length. The number 78 comes from the combination of 13 static and 13x5 conformational ensemble bin-wise populations, making it a total of 78 columns. The columns 1 through 13 represent the static values of a conformational parameters whereas 14 through 78 contain shape ensemble values for a sequence. The static values represent the average measure of the parameter whereas shape ensemble values represent the proportional occupancies of a nucleotide sequence in the five equal frequency bins constituting the global ensemble bins of that parameter. The script will also plot the shape profiles for these two gene promoter sequence sets and write the plots to PDF files named `plots_promoters_shape.pdf`, `plots_promoters_shape_boxplots.pdf`, and `ensemble_boxplots_5-bin.pdf`. Note that, as our original genomic sequences were 20 bp upstream of the genes TSS, we have used negative values of positions corresponding to 16 bp upstream of TSSs. This will differ if the positions are positive in reference to gene TSS or if, for example, regions flanking the summits of peaks in a ChIP-seq data are analyzed in which case both negative (upstream of summit) and positive (downstream of summit) positions are present.

The DNA shape profiles can be visualized and further refined as suitable for a given study. For example 5-bin ensemble values of sequence windows may be too voluminous. So, we can also represent the conformational ensemble in 3-bins instead of 5-bins by pooling together the occupancies of central bins. In this work, a smoothing with a window of size 3 has been applied to visualizations, and hence, the visualized shape profiles have 14 values per parameter as 1 bp from each end position will be removed while smoothing. These plots helped us to visualize the average patterns of the promoter regions of *TF regulated* and *others*. The line plots look as shown in Fig. 4 while the boxplots of the average values of all positions in both of these sets as shown in Fig. 5.

We observed that in the data set being analyzed, the profiles for the parameters such as *Stretch* are different for both sets of promoter sequences (Fig. 4). The plots show that the *opening* measures of promoter sequences of *others* have a higher value than the promoter sequences of *TF regulated* genes (Fig. 6).

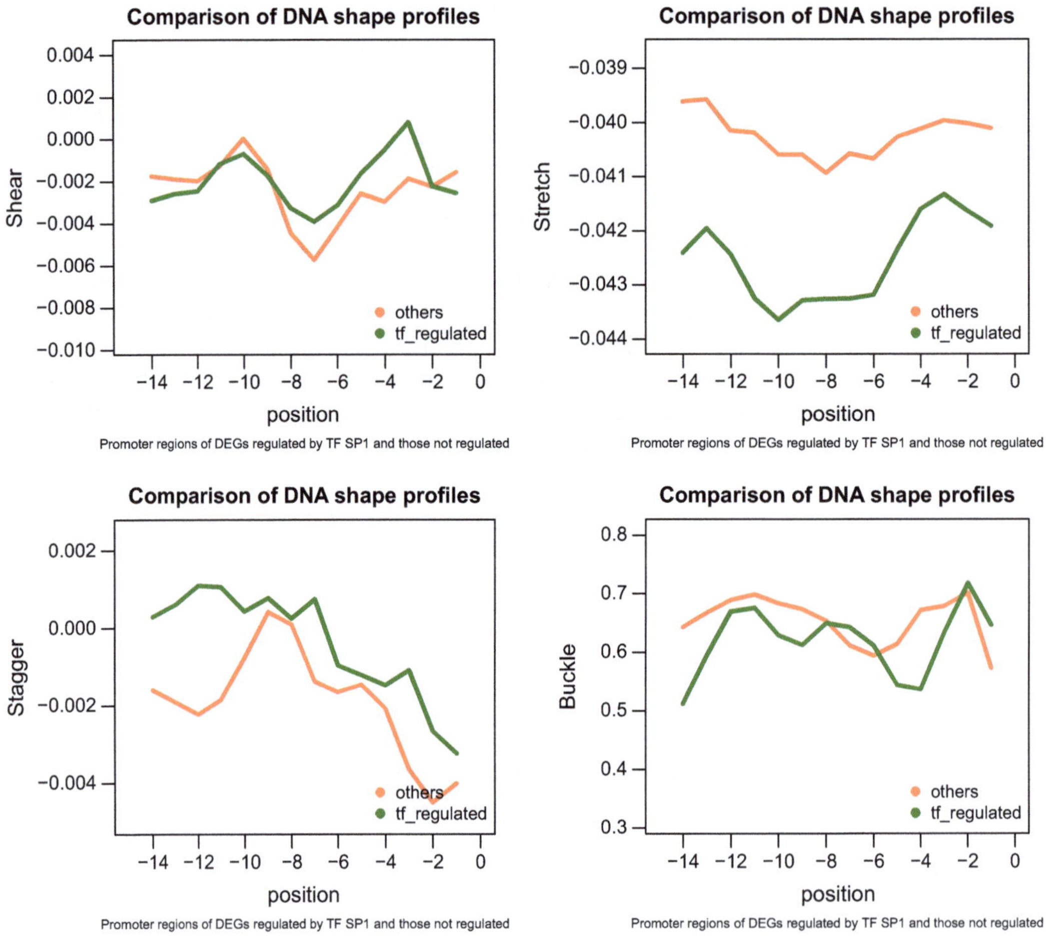

Fig. 4 The average value of each set of promoter sequences was plotted in a position wise manner. The green and red represent DNA shape profiles for promoter sequences from *TF regulated and others*

Upon looking at the shape ensembles of both sets of sequences, we can find the preferences of a set of sequences for a range of a specific parameter. For example, having higher occupancy of one set of sites in bin 1 or bin 2 indicates preference of those sequences for lower range of a parameter, whereas higher occupancy in bin 5 and bin 4 indicate preference toward higher values of the parameter.

4.2.6 Identification of Potential Novel TF Targets

The DNA shape profiles of the promoter regions of both sets of genes were used to build *random forest* classifiers using static shape and shape ensemble. The purpose of modeling the shape and dynamic features of these promoter sequences is identification of additional gene candidates which could be regulated by the TF of interest. Finding these gene candidates would underline the importance of static and shape ensemble in the mechanism of interactions between TFs and their targets and may help to elucidate other biological processes by incorporating additional evidence.

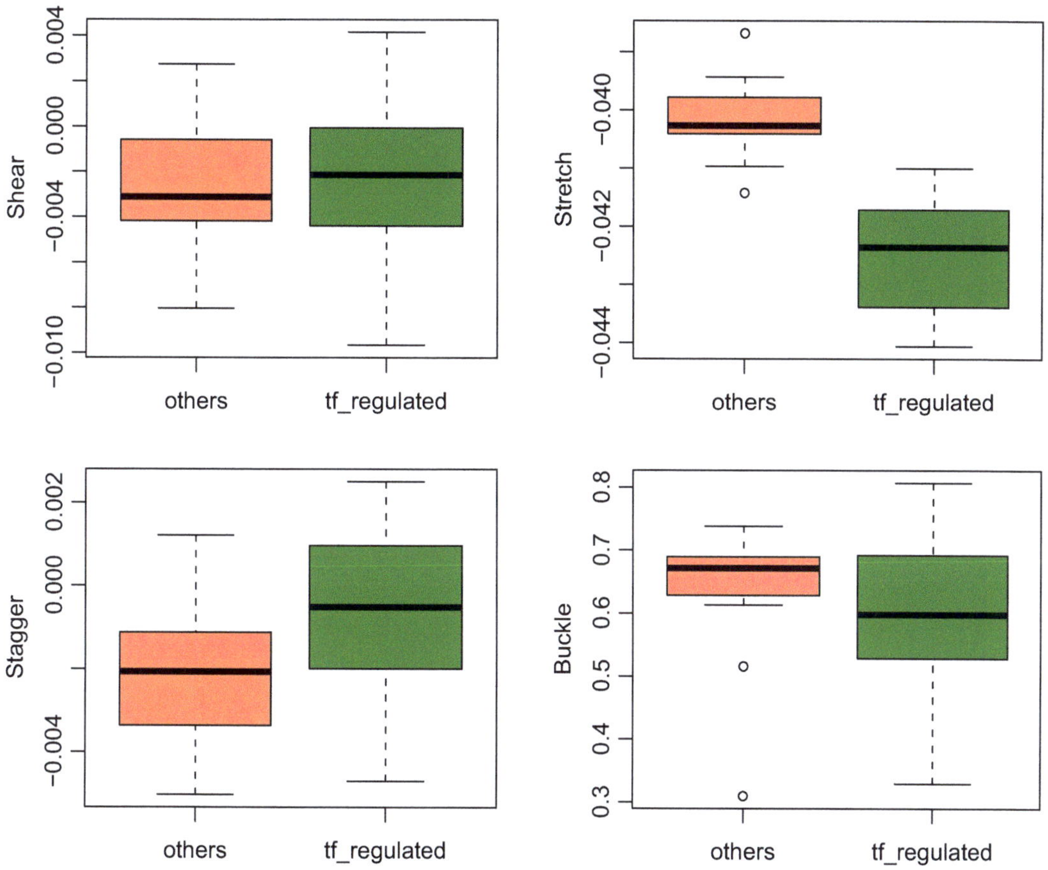

Fig. 5 The position wise average value of conformational parameters show differences in the range of a parameter preferred by both sets of promoter sequences. By looking at the plots, the promoter sequences of the *TF regulated* set have a specific preference compared to promoter sequences from *others* set for the same parameter

For a *random forest* model, we used the position-wise static shape parameters and bin-wise occupancies to prepare sequence wise vectors of features for static and ensemble shape, respectively, for both sets of promoters for identification of possible genes with promoters preferring shape patterns of the other group of promoters. These feature vectors were labeled as belonging to a class *TF regulated* or *others* and these labeled feature datasets were subsequently pooled together to form a single dataset for static and ensemble each. This data was divided into training and test sets. The training set is the subset which serves as an input to capture the relationship between the independent variable (features) and the dependent variable (label) using a machine learning model (example *random forest* here). In our analysis, features are measures of DNA conformational parameters and the label is the class of a promoter, that is, *TF regulated* or *others.* The test data serves as the set of instances (here promoter regions shape profiles) to

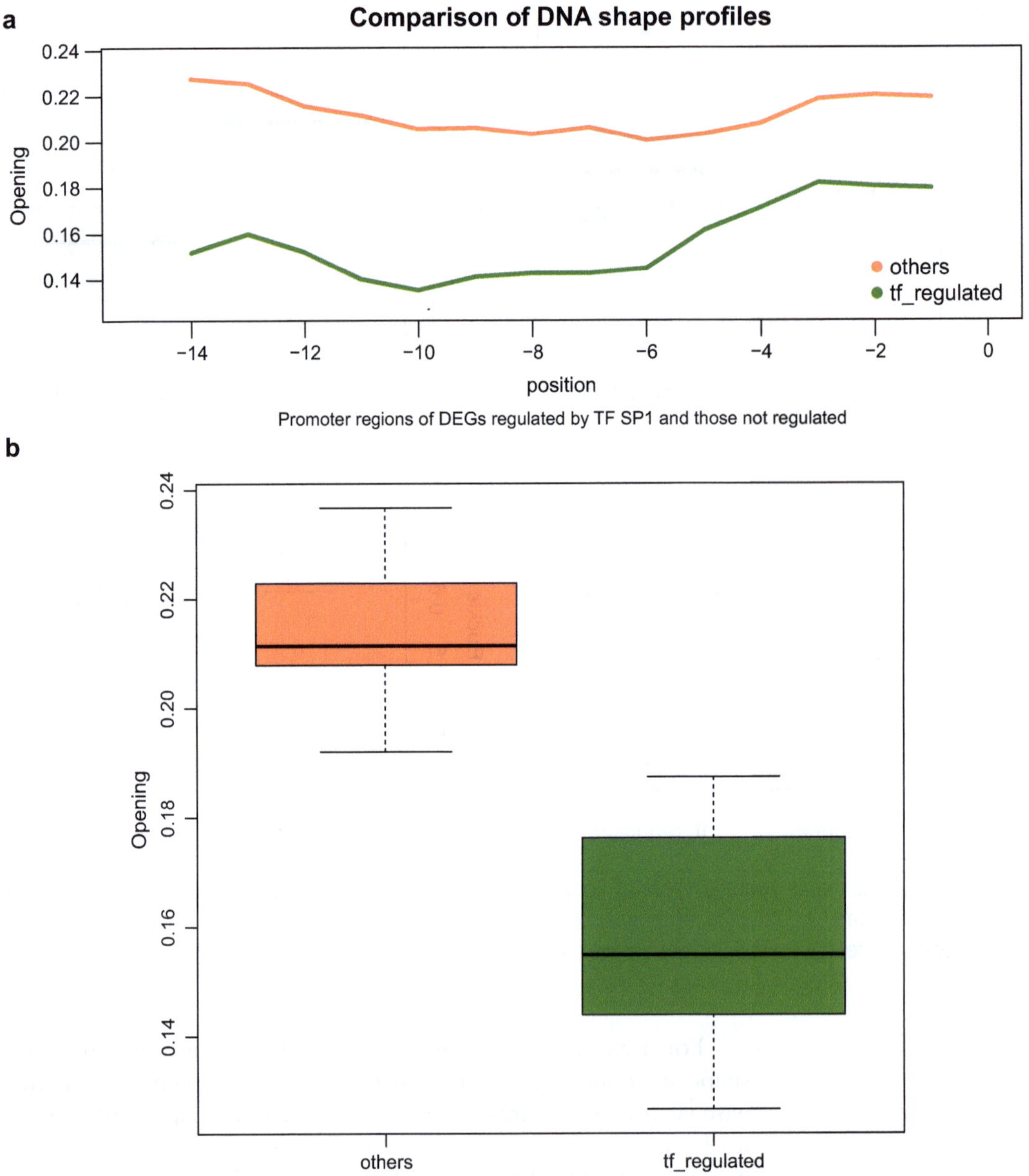

Fig. 6 The differential shape preference of promoters. (**a**) The average profiles of static shape for the *opening* is different between the *TF regulated* and *others* set of promoter sequences at position level. (**b**) In a set-wise manner, the distributions of average values per position between both sets differ where the *others* set prefers a higher opening than the *TF regulated* set

evaluate the prediction accuracy of the model as the instances in the test data are new to the model which is trained on the training data. The training to test ratio was kept as 0.7 and a test accuracy was calculated for each *random forest* model. For reproducibility, we use a value as a seed before starting the steps of the modeling of data. It ensures that if the same steps are run again, the results are the same which is taken care of by the seed. It can be set using `set.seed()` in R. We trained a *random forest* model of the labeled data training set using the function `randomForest()` from the library *randomForest* in R. The scripts available at the github repository contain the commands for steps mentioned above. The test accuracies of both the *random forest* models were calculated and all the DEG promoters which were mispredicted to be from a different class than its label were identified.

For the models using static shape and shape ensemble, test accuracies of 80.2% and 80.7%, respectively, were achieved. When we used these models to identify potential candidates which were regulated by the TF SP1, we looked for the gene promoters which were labeled as others but were predicted to be TF regulated by one or both of these models. We found 8 and 1 gene(s) to be additional targets of the TF by static and dynamic shape, respectively (Tables 2 and 3). The static shape profile distribution of these genes and the

Table 2
Number of genes explained or not, by the random forest models trained using the static and shape ensemble profiles of the promoter regions

		Original class	Predicted class	
			TF regulated	Others
static	TF regulated	775	541	234
	Others	3204	8	3196
ensemble	TF regulated	775	545	230
	Others	3204	1	3203

Original class column corresponds to the total number of genes in the *TF regulated* and *others* sets

Table 3
List of genes with promoter shape profiles mispredicted to belong to class *TF regulated* instead of their original class *others* (not regulated by TF SP1)

Mispredicted using static shape (8)	Mispredicted using shape ensemble (1)
KITLG	STMN3
ABHD4	
ANO8	
LETM1	
ME1	
GEM	
ST6GALNAC4	
RPGR	

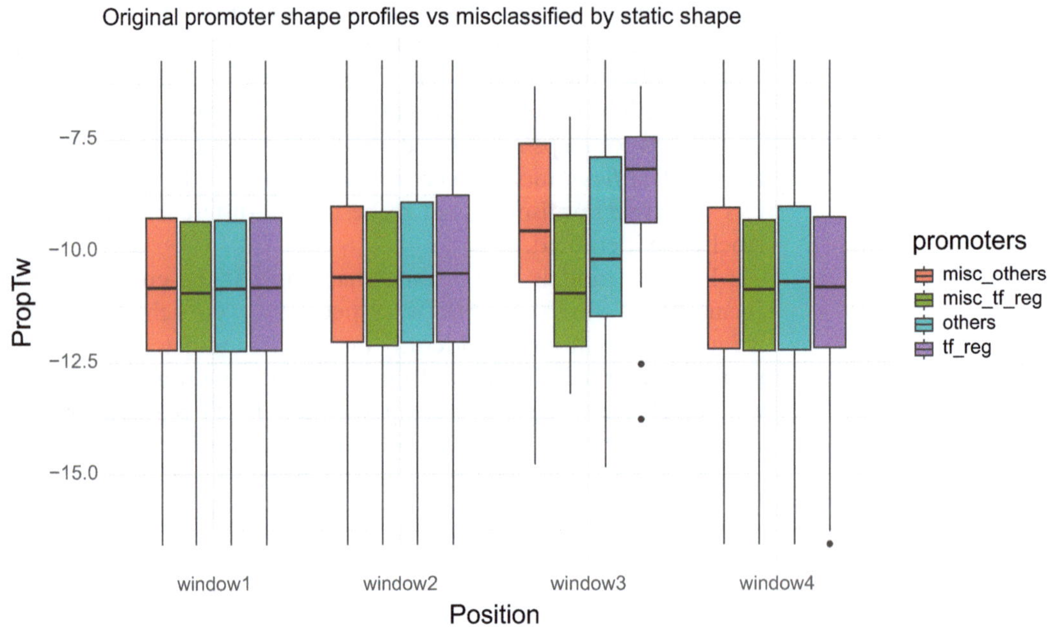

Fig. 7 The static shape profiles of *TF regulated*, *others*, *misclassified TF regulated*, and *misclassified others* for the conformational parameter *Propeller Twist*. Windows refer to pooled positions of the promoters where window1 through window4 correspond to 3–6, 7–10, 11–14, and 15–18 positions in the 20 bp input DNA sequences. We noticed a higher difference between the sets of promoter regions at the window3 in comparison to other windows

TF regulated and *others* set were visualized using boxplots for positional windows using a window size of 4 (Fig. 7). The visualized static shape profiles show the profiles for *TF regulated*, *others*, misclassified *TF regulated* (predicted to be *others*), and misclassified *others* (predicted to be *TF regulated*). We observed that window3 (positions 11–14 of the 20 bp sequences) had higher differences between all sets if compared with all other positional windows for all conformational parameters indicating the importance of these positions. As only one gene was found using shape ensemble, the ensemble distributions have not been shown here.

The eight genes predicted to be potential candidates for TF SP1 were KITLG, ABHD4, ANO8, LETM1, ME1, GEM, ST6GALNAC4, and RPGR using static shape and STMN3 was predicted using shape ensemble. The binding of SP1 to the promoter of STMN3 has been found to be facilitated by Transformation/transcription domain-associated protein (Trrap) for the maintenance of microtubule dynamics important in neurons [74]. When we mined the potential targets of SP1 more extensively beyond motif search, all these candidates were found to be listed in Harmonizone SP1 gene set dataset for ENCODE Transcription Factor Targets [75]. Other research works have also highlighted the links between some of these potential gene targets and SP1

[76, 77]. This strongly supported our approach as even though these targets were absent in the enrichment result of SP1 for Transfac and Jaspar PWMs, our analysis could pick up some of the ChIP-Seq outcomes without using ChIP-Seq data. Further discussion on these findings is out of the scope of this protocol and will be presented separately. Here we show how we can construct hypotheses of potentially regulated non-canonical targets of a TF de novo by comparing the shape signatures contained in their sequences.

These results highlight the importance of static shape and shape ensemble, respectively, other than sequence, in the recognition of target genes by a TF and underline the intrinsic preferences of target genomic regions of TF not directly explainable by their sequence properties. The difference in the list of genes captured as potential targets by using static and ensemble indicates different modes of recognition for a TF.

4.2.7 Refining Results and Further Analysis

There are multiple directions which this work can further proceed according to the aim of the study. These directions can be used to explain how the structural aspects and functional mechanisms of the potential gene targets align with the TF of interest and can also be further corroborated by supportive research studies available, if any.

To gain insights into the associated biological processes, *gene ontology* enrichment analysis of the gene candidates can be predicted using multiple tools such as *enrichR* used above for predicting genes regulated by TFs in *Transfac* and *Jaspar* database.

If the structure of TF is available, a comparative specificity analysis of the target promoter regions can be performed using our tool *Readout* available at http://sciwhylab.jnu.ac.in/servers/readout/ [24].

The findings can be further corroborated by using datasets already available such as ChIP-seq or by experimental studies such as hybridization or interaction assays such as Electrophoretic Mobility Shift Assay (EMSA), Yeast One-Hybrid Assay, and proximity ligation assays.

Note: The gene candidates found as potential targets of TF SP1 in this work have been generated using DNA shape profiles of short upstream regions (20 bp) of the genes for the purpose of demonstration. With a standard length of promoter regions (2 kb or 5 kb), the results might differ and help better explain the links in biological interactions or find some novel mechanisms.

4.3 Code Availability

The codes provided at https://github.com/Sciwhylab/DynaSeq_transcriptomics are available as five files:

1. `generate_TSS_coded.R`
2. `dna_shape_analysis.R`

3. `visualization.R`
4. `shape_models.R`
5. `analyze_misclassifications.R`

The `generate_TSS_coded.R` file will read the file of DESeq2 result named `GSE138496_Galaxy758_DESeq2_result_file_on_MUT_TGF_vs_WT_TGF_.txt` and create unique TSS coordinates for each gene in gene set corresponding to those regulated by the first TF to be observed in the enrichment results and those that are not regulated by this TF. This code will generate two bed files corresponding to *TF regulated* and *others* sets of genes and their genomic coordinates. For customizable functions, refer to `generate_TSS.R`. After this, the user would be required to use `bedtools flank` and `bedtools getfasta` on those files to extract genomic features into *fasta* files. These *fasta* files will be read by the `dna_shape_analysis.R` script to generate the shape profiles for each set of sequences. These files can be used to plot the profiles using `visualization.R`. The DNA shape profiles can be converted to vectorized features and modeled using `shape_models.R` which will generate the files corresponding to misclassified promoters which can be used to visualize their shape profiles along with the original sets of promoters using `analyse_misclassifications.R`.

References

1. Wray GA (2003) Transcriptional regulation and the evolution of development. Int J Dev Biol 47(7–8):675–684
2. Ohler U, Wassarman DA (2010) Promoting developmental transcription. Development 137(1):15–26. https://doi.org/10.1242/dev.035493
3. Gilmour KC, Reich NC (1995) Signal transduction and activation of gene transcription by interferons. Gene Expr 5(1):1–18
4. Lee TI, Young RA (2013) Transcriptional regulation and its misregulation in disease. Cell 152(6):1237–1251. https://doi.org/10.1016/j.cell.2013.02.014
5. Farnham PJ (2009) Insights from genomic profiling of transcription factors. Nat Rev Genet 10(9):605–616. https://doi.org/10.1038/nrg2636
6. Garvie CW, Wolberger C (2001) Recognition of specific DNA sequences. Mol Cell 8(5): 937–946. https://doi.org/10.1016/s1097-2765(01)00392-6
7. Johnson DS, Mortazavi A, Myers RM et al (2007) Genome-wide mapping of in vivo protein-DNA interactions. Science 316(5830):1497–1502. https://doi.org/10.1126/science.1141319
8. Mikkelsen TS, Ku M, Jaffe DB et al (2007) Genome-wide maps of chromatin state in pluripotent and lineage-committed cells. Nature 448(7153):553–560. https://doi.org/10.1038/nature06008
9. Visel A, Blow MJ, Li Z et al (2009) ChIP-seq accurately predicts tissue-specific activity of enhancers. Nature 457(7231):854–858. https://doi.org/10.1038/nature07730
10. Boyle AP, Davis S, Shulha HP et al (2008) High-resolution mapping and characterization of open chromatin across the genome. Cell 132(2):311–322. https://doi.org/10.1016/j.cell.2007.12.014
11. Wang Z, Gerstein M, Snyder M (2009) RNA-Seq: a revolutionary tool for transcriptomics. Nat Rev Genet 10(1):57–63. https://doi.org/10.1038/nrg2484
12. Tang F, Barbacioru C, Wang Y et al (2009) mRNA-Seq whole-transcriptome analysis of a

single cell. Nat Methods 6(5):377–382. https://doi.org/10.1038/nmeth.1315

13. Hafner M, Landgraf P, Ludwig J et al (2008) Identification of microRNAs and other small regulatory RNAs using cDNA library sequencing. Methods 44(1):3–12. https://doi.org/10.1016/j.ymeth.2007.09.009
14. Martin DP, Miya J, Reeser JW et al (2016) Targeted RNA sequencing assay to characterize gene expression and genomic alterations. J Vis Exp 114:54090. https://doi.org/10.3791/54090
15. Sharon D, Tilgner H, Grubert F (2013) A single-molecule long-read survey of the human transcriptome. Nat Biotechnol 31(11):1009–1014. https://doi.org/10.1038/nbt.2705
16. Chen G, Ning B, Shi T (2019) Single-cell RNA-seq technologies and related computational data analysis. Front Genet 10:317. https://doi.org/10.3389/fgene.2019.00317
17. Slattery M, Zhou T, Yang L et al (2014) Absence of a simple code: how transcription factors read the genome. Trends Biochem Sci 39(9):381–399. https://doi.org/10.1016/j.tibs.2014.07.002
18. Stormo GD (2000) DNA binding sites: representation and discovery. Bioinformatics 16(1):16–23. https://doi.org/10.1093/bioinformatics/16.1.16
19. Stormo GD (2013) Modeling the specificity of protein-DNA interactions. Quant Biol 1(2):115–130. https://doi.org/10.1007/s40484-013-0012-4
20. von Hippel PH (1994) Protein-DNA recognition: new perspectives and underlying themes. Science 263(5148):769–770. https://doi.org/10.1126/science.8303292
21. Inukai S, Kock KH, Bulyk ML (2017) Transcription factor-DNA binding: beyond binding site motifs. Curr Opin Genet Dev 43:110–119. https://doi.org/10.1016/j.gde.2017.02.007
22. Bareket-Samish A, Cohen I, Haran TE (1998) Direct versus indirect readout in the interaction of the trp repressor with non-canonical binding sites. J Mol Biol 277(5):1071–1080. https://doi.org/10.1006/jmbi.1998.1638
23. Zhang Y, Xi Z, Hegde RS et al (2004) Predicting indirect readout effects in protein-DNA interactions. Proc Natl Acad Sci USA 101(22):8337–8341. https://doi.org/10.1073/pnas.0402319101
24. Ahmad S, Kono H, Araúzo-Bravo MJ et al (2006) ReadOut: structure-based calculation of direct and indirect readout energies and specificities for protein-DNA recognition. Nucleic Acids Res 34:W124–W127. https://doi.org/10.1093/nar/gkl104
25. Tiwari P, Tripathi LP, Nishikawa-Matsumura T et al (2013) Prediction and experimental validation of a putative non-consensus binding site for transcription factor STAT3 in serum amyloid A gene promoter. Biochim Biophys Acta 1830(6):3650–3655. https://doi.org/10.1016/j.bbagen.2013.01.024
26. Zhou T, Shen N, Yang L et al (2015) Quantitative modeling of transcription factor binding specificities using DNA shape. Proc Natl Acad Sci USA 112(15):4654–4659. https://doi.org/10.1073/pnas.1422023112
27. Mathelier A, Xin B, Chiu TP et al (2016) DNA shape features improve transcription factor binding site predictions in vivo. Cell Syst 3(3):278–286.e4. https://doi.org/10.1016/j.cels.2016.07.001
28. Andrabi M, Hutchins AP, Miranda-Saavedra D et al (2017) Predicting conformational ensembles and genome-wide transcription factor binding sites from DNA sequences. Sci Rep 7(1):4071. https://doi.org/10.1038/s41598-017-03199-6
29. Sielemann J, Wulf D, Schmidt R et al (2021) Local DNA shape is a general principle of transcription factor binding specificity in Arabidopsis thaliana. Nat Commun 12(1):6549. https://doi.org/10.1038/s41467-021-26819-2
30. Zhang Y, Liu Y, Wang Z et al (2022) Uncovering the relationship between tissue-specific TF-DNA binding and chromatin features through a transformer-based model. Genes (Basel) 13(11):1952. https://doi.org/10.3390/genes13111952
31. Schleif R (1992) DNA looping. Annu Rev Biochem 61:199–223. https://doi.org/10.1146/annurev.bi.61.070192.001215
32. Mathew-Fenn RS, Das R, Harbury PA (2008) Remeasuring the double helix. Science 322(5900):446–449. https://doi.org/10.1126/science.1158881
33. Cloutier TE, Widom J (2005) DNA twisting flexibility and the formation of sharply looped protein-DNA complexes. Proc Natl Acad Sci USA 102(10):3645–3650. https://doi.org/10.1073/pnas.0409059102
34. Travers A (2013) Dynamic DNA underpins chromosome dynamics. Biophys J 105(10):2235–2237. https://doi.org/10.1016/j.bpj.2013.10.011
35. Panne D, Maniatis T, Harrison SC (2004) Crystal structure of ATF-2/c-Jun and IRF-3 bound to the interferon-beta enhancer.

EMBO J 23(22):4384–4393. https://doi.org/10.1038/sj.emboj.7600453

36. Crowther RA, Klug A (1975) Structural analysis of macromolecular assemblies by image reconstruction from electron micrographs. Annu Rev Biochem 44:161–182. https://doi.org/10.1146/annurev.bi.44.070175.001113
37. Lamerichs RM, Boelens R, van der Marel GA et al (1989) H NMR study of a complex between the lac repressor headpiece and a 22 base pair symmetric lac operator. Biochemistry 28(7):2985–2991. https://doi.org/10.1021/bi00433a037
38. Guo TW, Bartesaghi A, Yang H et al (2017) Cryo-EM structures reveal mechanism and inhibition of DNA targeting by a CRISPR-Cas surveillance complex. Cell 171(2): 414–426.e12. https://doi.org/10.1016/j.cell.2017.09.006
39. Coimbatore Narayanan B, Westbrook J, Ghosh S et al (2014) The nucleic acid database: new features and capabilities. Nucleic Acids Res 42 (Database issue):D114–D122. https://doi.org/10.1093/nar/gkt980
40. Berman HM, Gelbin A, Clowney L et al (1996) The nucleic acid database: present and future. J Res Natl Inst Stand Technol 101(3):243–257. https://doi.org/10.6028/jres.101.026
41. Lu XJ, Olson WK (2003) 3DNA: a software package for the analysis, rebuilding and visualization of three-dimensional nucleic acid structures. Nucleic Acids Res 31(17):5108–5121. https://doi.org/10.1093/nar/gkg680
42. Lu XJ, Olson WK (2008) 3DNA: a versatile, integrated software system for the analysis, rebuilding and visualization of three-dimensional nucleic-acid structures. Nat Protoc 3:1213–1227. https://doi.org/10.1038/nprot.2008.104
43. Zheng G, Lu XJ, Olson WK (2009) Web 3DNA—a web server for the analysis, reconstruction, and visualization of three-dimensional nucleic-acid structures. Nucleic Acids Res 37(Web Server issue):W240–W246. https://doi.org/10.1093/nar/gkp358
44. Zhou T, Yang L, Lu Y et al (2013) DNAshape: a method for the high-throughput prediction of DNA structural features on a genomic scale. Nucleic Acids Res 41(Web Server issue):W56–W62. https://doi.org/10.1093/nar/gkt437
45. Li J, Sagendorf JM, Chiu TP et al (2017) Expanding the repertoire of DNA shape features for genome-scale studies of transcription factor binding. Nucleic Acids Res 45(22): 12877–12887. https://doi.org/10.1093/nar/gkx1145
46. Berman HM, Lawson CL, Schneider B (2022) Developing community resources for nucleic acid structures. Life (Basel) 12(4):540. https://doi.org/10.3390/life12040540
47. Boehr DD, Nussinov R, Wright PE (2009) The role of dynamic conformational ensembles in biomolecular recognition. Nat Chem Biol 5(11):789–796
48. Fenwick RB, Esteban-Martín S, Salvatella X (2011) Understanding biomolecular motion, recognition, and allostery by use of conformational ensembles. Eur Biophys J 40:1339–1355
49. Battistini F, Hospital A, Buitrago D et al (2019) How B-DNA dynamics decipher sequence-selective protein recognition. J Mol Biol 431(19):3845–3859
50. Andrew S. (2010) FastQC: a quality control tool for high throughput sequence data. Babraham Bioinformatics. https://www.bioinformatics.babraham.ac.uk/projects/fastqc/
51. Bolger AM, Lohse M, Usadel B (2014) Trimmomatic: a flexible trimmer for Illumina sequence data. Bioinformatics 30(15): 2114–2120. https://doi.org/10.1093/bioinformatics/btu170
52. Langmead B, Salzberg SL (2012) Fast gapped-read alignment with Bowtie 2. Nat Methods 9(4):357–359. https://doi.org/10.1038/nmeth.1923
53. Li H, Durbin R (2009) Fast and accurate short read alignment with Burrows-Wheeler transform. Bioinformatics 25(14):1754–1760. https://doi.org/10.1093/bioinformatics/btp324
54. Kim D, Paggi JM, Park C et al (2019) Graph-based genome alignment and genotyping with HISAT2 and HISAT-genotype. Nat Biotechnol 37(8):907–915
55. Dobin A, Davis CA, Schlesinger F et al (2013) STAR: ultrafast universal RNA-seq aligner. Bioinformatics 29(1):15–21. https://doi.org/10.1093/bioinformatics/bts635
56. Kim D, Pertea G, Trapnell C et al (2013) TopHat2: accurate alignment of transcriptomes in the presence of insertions, deletions and gene fusions. Genome Biol 14(4):R36. https://doi.org/10.1186/gb-2013-14-4-r36
57. Pertea M, Pertea GM, Antonescu CM et al (2015) StringTie enables improved reconstruction of a transcriptome from RNA-seq reads. Nat Biotechnol 33(3):290–295. https://doi.org/10.1038/nbt.3122
58. Trapnell C, Williams BA, Pertea G et al (2010) Transcript assembly and quantification by RNA-Seq reveals unannotated transcripts and isoform switching during cell differentiation.

Nat Biotechnol 28(5):511–515. https://doi.org/10.1038/nbt.1621

59. Liao Y, Smyth GK, Shi W (2014) featureCounts: an efficient general purpose program for assigning sequence reads to genomic features. Bioinformatics 30(7):923–930. https://doi.org/10.1093/bioinformatics/btt656
60. R Core Team (2019) R: a language and environment for statistical computing. R Foundation for Statistical Computing, Vienna. https://www.R-project.org/
61. Love MI, Huber W, Anders S (2014) Moderated estimation of fold change and dispersion for RNA-seq data with DESeq2. Genome Biol 15(12):550. https://doi.org/10.1186/s13059-014-0550-8
62. Kuleshov MV, Jones MR, Rouillard AD et al (2016) Enrichr: a comprehensive gene set enrichment analysis web server 2016 update. Nucleic Acids Res 44(W1):W90–W97. https://doi.org/10.1093/nar/gkw377
63. Durinck S, Spellman PT, Birney E et al (2009) Mapping identifiers for the integration of genomic datasets with the R/Bioconductor package biomaRt. Nat Protoc 4(8):1184–1191. https://doi.org/10.1038/nprot.2009.97
64. Karolchik D, Hinrichs AS, Furey TS et al (2004) The UCSC Table Browser data retrieval tool. Nucleic Acids Res 32(Database issue):D493–D496. https://doi.org/10.1093/nar/gkh103
65. Quinlan AR, Hall IM (2010) BEDTools: a flexible suite of utilities for comparing genomic features. Bioinformatics 26(6):841–842. https://doi.org/10.1093/bioinformatics/btq033
66. Heinz S, Benner C, Spann N et al (2010) Simple combinations of lineage-determining transcription factors prime cis-regulatory elements required for macrophage and B cell identities. Mol Cell 38(4):576–589
67. Weis-Banke SE, Lerdrup M, Kleine-Kohlbrecher D et al (2020) Mutant FOXL2C134W hijacks SMAD4 and SMAD2/3 to drive adult granulosa cell tumors. Cancer Res 80(17):3466–3479. https://doi.org/10.1158/0008-5472.CAN-20-0259
68. Bailey TL, Gribskov M (1998) Combining evidence using p-values: application to sequence homology searches. Bioinformatics (Oxford, England) 14(1):48–54
69. Bailey TL (2011) DREME: motif discovery in transcription factor ChIP-seq data. Bioinformatics 27(12):1653–1659
70. Frith MC, Saunders NF, Kobe B et al (2008) Discovering sequence motifs with arbitrary insertions and deletions. PLoS Comput Biol 4(5):e1000071
71. Machanick P, Bailey TL (2011) MEME-ChIP: motif analysis of large DNA datasets. Bioinformatics 27(12):1696–1697
72. Sinha S, Tompa M (2003) YMF: a program for discovery of novel transcription factor binding sites by statistical overrepresentation. Nucleic Acids Res 31(13):3586–3588
73. Santana-Garcia W, Castro-Mondragon JA, Padilla-Gálvez M et al (2022) RSAT 2022: regulatory sequence analysis tools. Nucleic Acids Res 50(W1):W670–W676. https://doi.org/10.1093/nar/gkac312
74. Tapias A, Lázaro D, Yin BK et al (2021) HAT cofactor TRRAP modulates microtubule dynamics via SP1 signaling to prevent neurodegeneration. elife 10:e61531
75. Rouillard AD, Gundersen GW, Fernandez NF et al (2016) The harmonizome: a collection of processed datasets gathered to serve and mine knowledge about genes and proteins. Database 2016:baw100
76. Shu X, Simpson JR, Hart AW et al (2012) Functional characterization of the human RPGR proximal promoter. Invest Ophthalmol Vis Sci 53(7):3951–3958
77. Kang NY, Park YD, Choi HJ et al (2004) Regulatory elements involved in transcription of the human NeuAcα2, 3Galβ1, 3GalNAcα2, 6-sialyltransferase (hST6GalNAc IV) gene. Mol Cell (Springer Science & Business Media BV) 18(2):157–162

Chapter 19

Utilizing RNA-seq Data to Infer Bacterial Transcription Termination Sites and Validate Predictions

Swati Gupta and Debnath Pal

Abstract

The transcription termination process is an important part of the gene expression process in the cell. It has been studied extensively, but many aspects of the mechanism are not well understood. The widespread availability of experimental RNA-seq data from high-throughput experiments provides a unique opportunity to infer the end of the transcription units genome wide. This data is available for both Rho-dependent and Rho-independent termination pathways that drive transcription termination in bacteria. Our book chapter gives an overview of the current knowledge of Rho-independent transcription termination mechanisms and the prediction approaches currently deployed to infer the termination sites. Thereafter, we describe our method that uses cluster hairpins to detect Rho-independent transcription termination sites. These clusters are a group of hairpins that lies at <15 bp from each other and are together capable of enforcing the termination process. The idea of a group of hairpins being extensively used for transcription termination is new, and results show that at least 52% of the total cases are of this type, while in the remaining cases, a single strong hairpin is capable of driving transcription termination. The reads derived from the RNA-seq data for corresponding bacteria have been used to validate the predicted sites. The predictions that match these RNA-seq derived sites have higher confidence, and we find almost 98% of the predicted sites, including alternate termination sites, to match the RNA-seq data. We discuss the features of predicted hairpins in detail for a better understanding of the Rho-independent transcription termination mechanism in bacteria. We also explain how users can use the tools developed by us to do transcription terminator predictions and design their experiments through genome-level visualization of the transcription termination sites from the precomputed INTERPIN database.

Key words Transcription termination, Bacteria, RNA-seq, Hairpins, Prediction method

1 Introduction

Transcription is the process of the formation of mRNA from the DNA strand in prokaryotes as well as eukaryotes. It occurs in three stages: initiation, elongation, and termination. Transcription starts with the binding of an enzyme, called RNA polymerase (RNAP), to specific locations on the DNA sequence, called the promoter, in the process called transcription initiation. Initiation is followed by elongation and finally, termination. RNA synthesis continues till

Rajeev K. Azad (ed.), *Transcriptome Data Analysis*, Methods in Molecular Biology, vol. 2812,
https://doi.org/10.1007/978-1-0716-3886-6_19,

the termination region, where the polymerase pauses and dissociates from the DNA, with the release of the newly synthesized mRNA chain. Termination leads to the recycling of RNAP, prevents interference from antisense reads, and controls inappropriate expression of downstream genes [1]. The pressing question is to understand how termination occurs and how the site of termination is determined. The transcription conditions can determine the kinetics and the efficiency of the termination signal.

Termination is principally regulated by two mechanisms—Rho-independent (or intrinsic) termination and Rho-dependent termination. The process of Rho-dependent termination requires the binding of the Rho protein to specific sites called RUT sites (or Rho utilization sites) [2]. Here, a Rho hexamer binds to the RNA and translocates on the RNA, close behind the polymerase. On encountering a pause, the Rho complex latches onto the polymerase and stops its movement, leading to dissociation with the release of the transcript [1, 3, 4]. On the other hand, intrinsic transcription termination (ITT) does not require any external factors. Here, a GC-rich site of complementary inverted repeats folds onto itself and forms a stem–loop structure called the hairpin. These secondary RNA structures stall the movement of RNAP and cause it to fall off from the DNA strand, after which the polymerase is reused by the cell for the next transcription cycle. The presence of hairpins in mRNA 3′ end prevents the transcript from terminating via Rho-mediated termination since it does not allow Rho to pull out nascent RNA from the hybrid [5].

Given the importance of transcription termination as a regulatory process, it is essential to find termination sites in order to better understand the mechanism and to study factors influencing it.

1.1 Parts of the Intrinsic Transcription Termination Machinery

RNA Polymerase (RNAP) and hairpins are the two main parts of the ITT machinery. The hairpins are secondary structure folding of a single strand of RNA that forms base pairs (bp) with itself and a structure that comprises stem and loop (*see* Fig. 1). A stem is formed of complementary bases, and its length typically ranges from 2 to 15 bp. The unpaired nucleotides (nt) on top of the stem in the hairpin structures form loops. A minimum of three residues are required to form a loop, due to steric clashes.

The RNAP transitions from an elongating to paused to the terminated state during transcription termination. It has different subunits—two alpha (α), one beta (β), one beta′ (β'), and one sigma (σ) subunit [6, 7]. The formation (or nucleation) of hairpin inside the RNAP leads to the weakening of RNA:DNA hybrid bonds, which combined with other conformational changes in RNAP subunit structures lead to transcription termination.

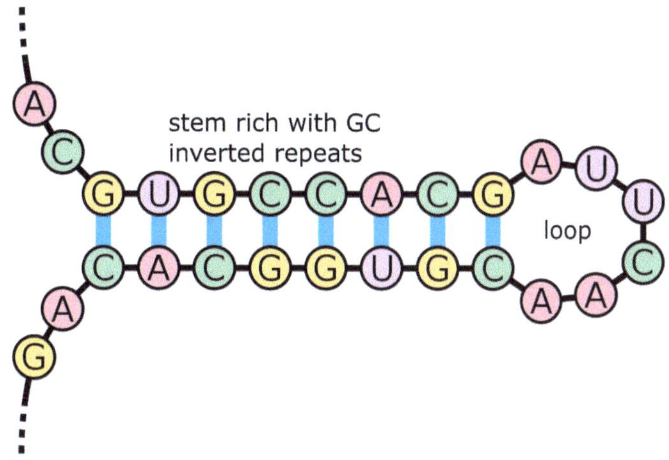

Fig. 1 Structural features of a hairpin. Loop and stem are labeled

1.2 Overview of Preexisting Prediction Methods

Even though the use of experimental assays to study the base pairing in RNA structures is the most accurate way, it is not feasible to study the ever-growing set of RNA secondary structures by in vivo and in vitro methods. Several computer algorithms and servers have been developed in the last 50 years, for studying the folding of RNA and predicting secondary as well as tertiary folding, and hence the intrinsic terminators in bacteria. Most programs look for an inverted repeat sequence, which is GC-rich and followed by a U-tail, which is the basic structure of a hairpin. Homology modeling [8], energy minimization [9, 10], stochastic context-free grammar [11], profile matching [12, 13] by sequence and structure alignment, and statistical thresholds [14] are the major methods used by different programs, separately or in combination. In recent years, machine-learning-based programs like iTerm-pseKNC [15] have also been created, which train models on preexisting data to classify the presence or absence of the ITT feature for an input sequence.

The intrinsic transcription terminator prediction method is heavily dependent on the knowledge about RNA folding, availability of computational power, larger training sets, and development of more optimized parallel algorithms to predict the folding of RNA sequences into their secondary structure. The programs of the past few decades and at present can be divided into three main categories:

1.2.1 Sequence Based

Based on a set of experimentally determined intrinsic terminators, consensus features like minimum GC content, stem and loop length, gaps, and length of U-tail are determined. Then a sequence of inverted repeats is sought out, from which only those that cross the feature thresholds are taken as potential terminators. A scoring matrix is often created to differentiate the true terminators from another secondary folding. Some software that use this method are WebGeSTer [9], ARNold [16], BhairPred [17], and others [18, 19] (that are simply based on decision boundaries).

The disadvantage of this strategy is the eminent bias in selecting features from only a few bacteria that have been thoroughly annotated, including *B. subtilis, E. coli* and *Mycobacterium* species. The advantages are that modern machine learning techniques like support vector machines (SVM) and artificial neural networks (ANN) can easily be applied here [15, 20].

1.2.2 Covariation-Based Model

These models identify homologues by creating an alignment of previously known terminator sequences and structures. Comparison of probable terminators to a "seed" alignment of representative noncoding RNAs is done. Then a Lod-score is calculated to calculate the probability of the match being significant or a random match.

This method includes probabilistic model-based programs like RNIE [12], Erpin [13], and Transterm [8]. The advantage of using these tools is that they can be used without a need for running complex experiments to obtain data for creating profiles. With advancement in technology, tertiary structures can also be added to improve specificity. However, since the covariance models are built on consensus alignments and signify similarity in sequences, any terminators that have noncanonical features or organisms that are evolutionarily far from the ones used for building the input profiles have a high chance to be missed out. These methods cannot be used with large sequences, else they would be very time-consuming and require huge computational powers to be used feasibly [21].

Many algorithms use sequence-based approach for finding the initial set of RNA secondary structures and then create a statistical distribution to filter out required structures, e.g., CONTRAfold [22], and predictions by Carafa et al. [23].

1.2.3 Pair Potential Based

This is a method where base pair energies are calculated from experimentally validated hairpins and other RNA secondary folding. An overall ΔG is also calculated to act as a threshold for selecting strong or weak terminators. These models consider the thermodynamics of RNA folding and are hence physics-based models, requiring initial experimentation to obtain a database of energies/potentials. Software for secondary structure prediction that uses dynamic programming or energy minimization and other optimization techniques are often used, like GTfold [24]. Most accurate programs today use this technique, either alone or in combination with the above two, for the prediction of intrinsic terminators, e.g., MFold [10], RNAMotif [25], and INTERPIN [26, 27].

1.3 Need for a New Method

Although many facets of the ITT-driven by product mRNA are understood [1], the common determinants required for such mechanisms to operate across bacterial organisms remain unclear [28].

Some programs previously tried to capture these genomic features to detect ITT-capable hairpins. However, ITT sites are known only in about one-third of gene ends, despite transcription termination being a conserved process across prokaryotes as well as eukaryotes. Here, a new group of hairpins, *cluster* hairpins, are discussed. They have the potential to fill the missing knowledge about ITT sites. The presence of such terminators were hinted previously, where groups of two hairpins were found in *M. tuberculosis*, and labeled as U-shaped hairpins [29]. Furthermore, according to our analysis and others [29–31], it was observed by in silico as well as in vivo methods that poly U-tail is not necessary for ITT.

Our work, INTERPIN, provides a valuable tool for researchers to make ITT site predictions in bacteria, which may aid in further feature annotations as well as understanding of the underlying mechanism of transcription termination. It will help to bridge the gap between presently known Rho-dependent and intrinsic terminators and might serve as a means to find possible drug targets for future bacterial studies.

2 Predicting Intrinsic Transcription Termination Sites

2.1 Predicting Hairpin Structures

We initially constructed a dataset with 13 bacterial genomes and tested our algorithm to predict transcription termination sites on them [26]. Here, we present the algorithm and the results on a dataset of 12,745 bacterial chromosomes across 10 bacterial phyla including—Firmicutes, Chlamydiae, Actinobacteria, Spirochetes, Planctomycetes, Fusobacteria, Cyanobacteria, Thermodesulfobacteria, Acidobacteria, and Proteobacteria. Our dataset includes both GC and AT-rich chromosomes; 6,989 (54.8%) are GC-rich, i.e., GC content of organism's Fasta sequence is >50%. We also found termination sites in 27,938 bacterial plasmids.

First, hairpin structures were formed for each interoperonic region (IR) (IR region between 3 ′ end of coding region of first operon and 5 ′ end of coding region of next operon; only sequences − 20 to +270 bp from stop codon were taken for each IR) using MFold [10], which calculates free energy for predicted structures at 37 °C, by assuming a two-state model [32]. The operon boundaries were calculated using the FGENESB module of the MolQuest software (http://www.molquest.com) [33]. The hairpin stability is calculated by penalizing various factors like length and sequence of the loop, stacking of the first mismatch from the initiation free energy, $\Delta G_{initiation}$ [34].

Next, outlier hairpins based on 95 percentile of stem and loop length were removed, where the extremely deviant hairpins were stem lengths 20–28 bp and loop lengths 25–35 bases. After outlier removal, a total of 26,784,334 IR were taken where hairpin was found for bacterial chromosomes. Frequency of hairpin stem lengths were highest between 2 and 9 bp and loop lengths between 4 and 11 bases; however, ~ 1% of hairpins show stem length between 9 and 20 bp and/or loop length between 12 and 28 bases, especially in Firmicutes.

2.2 Defining Termination Units (TUs)

The hairpin structures belong to a cluster if the distance between the end of one hairpin and the start of the next is <15 nucleotides. All such clusters were labeled ***cluster***. The choice of <15 nucleotides as hairpin separating distance for identifying hairpin cluster is based on the typical length of the nascent RNA nucleotide strand that can be enclosed in the RNA exit channel (5 nts) and DNA: RNA hybrid (9–10 bp) [1], and unavailable for folding into a hairpin at any given time. It was ensured that a subset of hairpins from a higher-size *cluster* is not present as a separate entry in a lower-size *cluster*. Aside from the *cluster*, those hairpins without any neighbor at <15 nucleotides were labeled as ***single***.

2.3 Classifying Identified *TUs*

The *cluster* or *single* hairpin nearest to the stop codon was termed as the *identified TU*, since this is expected to provide the most efficient ITT. It would also kinetically be the first hairpin to form and disrupt RNAP movement. Out of all operons, the algorithm was able to predict hairpins in 92.6% of the cases with 54% of the operons terminated by *cluster* hairpin and 38.6% by *single* hairpins. Intrinsic terminators were not found only in ~7.4% of the total operons.

Fusobacteria had the highest percentage of predicted *single* hairpin (55% of IR), while Actinobacteria had the highest percentage of predicted *cluster* hairpin (65% of IR). -proteobacteria and Fusobacteria had the highest percentage of operons where a hairpin is not identified (11.7%, 11.2% of respective IRs). Interestingly, both these phyla have low average GC% (34%, 29.7%, respectively, with ~ 1% chromosomes in each of these phyla GC-rich).

For bacterial plasmids, a total of 2,252,089 IRs were analyzed out of which hairpins were predicted in ~ 91% (2,045,200) IRs. *Cluster* TUs formed 63.5% of total predicted TUs and the rest (36.5%) were *single*. This distribution shows the prevalence and importance of predicting *cluster* hairpins in bacteria. Out of all analyzed plasmids, 43% have a GC-rich sequence.

2.4 Features of Identified *TUs*

2.4.1 Distance of Hairpins from the Stop Codon

The hairpins were found close to the stop codon with ~ 75% of hairpins lying within 25 bases from the stop codon. Greater than 60% of the respective *cluster* hairpins and >50% of the respective *single* hairpins lie within 25 bases from the stop codon, the maximum being for β-proteobacteria, i.e., 86.8% of *cluster* and 80.5% of

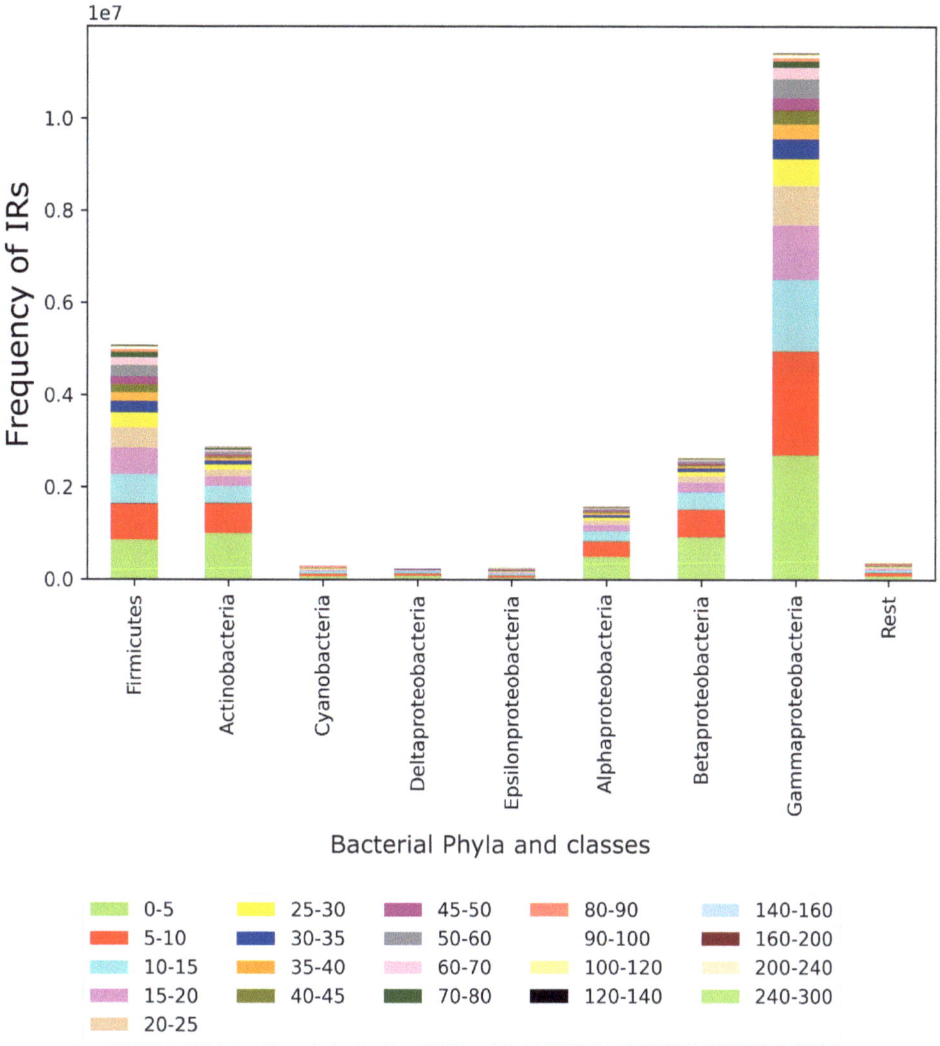

Fig. 2 The distance of hairpins from 3′ end of the coding region in each operon phylum- and classwise. Each color in a bar represents a distance range as shown in the legend, and the y-axis shows the frequency of hairpins for that range. Phylum and proteobacteria classes with IRs <150,000 have been clubbed together and shown in x-axis under bar "Rest," to improve the visibility of trends in the plot. These include Chlamydiae, Spirochetes, Planctomycetes, Fusobacteria, Thermodesulfobacteria, Acidobacteria, and other Proteobacteria

single hairpins. The phyla and classes also have the high average GC % of ~ 64%. Figure 2 shows phylum- and classwise changes in distance pattern from the stop codon.

If we look at the distance of plasmid hairpins from the stop codon, 83% *cluster* and 75% *single* hairpins are present within 25 bases from the stop codon. The farthest distance from the stop codon to a *cluster* hairpin is 258 bases and to a *single* hairpin is 260 bases.

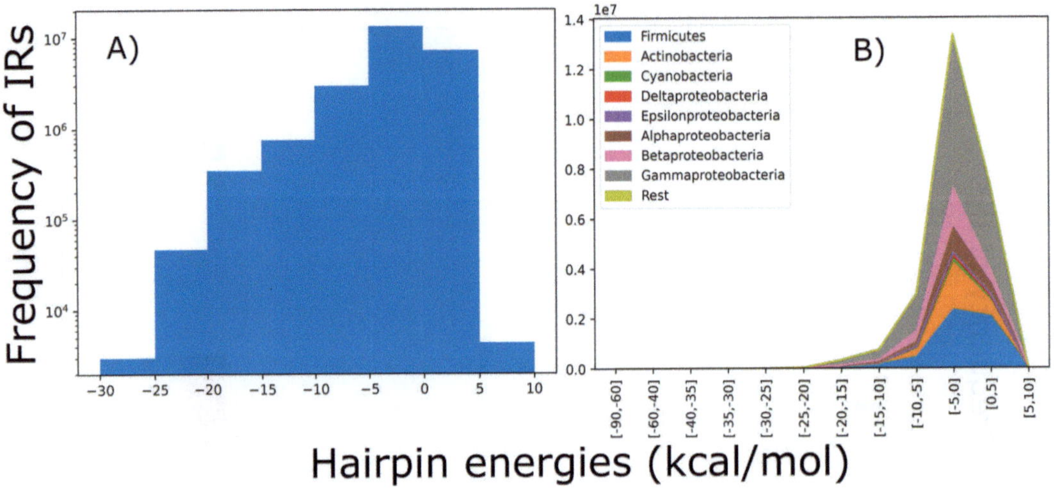

Fig. 3 Trends for hairpin energy distribution. (**a**) Figure shows the hairpin energies for all bacterial chromosomes combined. (**b**) Figure shows hairpin energy distribution phylum- and classwise with the help of a stack plot where distinct colors show bacterial phyla and classes. We can see that most hairpins in all phyla and classes have energies between − 10 and 10 kcal/mol

2.4.2 Energy Score of Identified *Hairpins*

The most favored energy values were between − 5 and 5 kcal/mol, with 83% of total hairpin energy scores in this range. The next major range was − 10 to − 5 kcal/mol, where the frequency falls to less than 12% (*see* Fig. 3a). Here, the energy score for *cluster* hairpin was calculated by taking the average of energy scores of all constituent hairpins. For all phyla and classes, >80% of hairpins energy scores were in the range − 5 to 5 kcal/mol (*see* Fig. 3b).

Similar to hairpins predicted in bacterial chromosomes, a majority (89%) of hairpins have energies between − 5 and 5 kcal/mol, followed by 9% of hairpins with energies in the range − 10 to − 5 kcal/mol.

2.4.3 Composition of Cluster *Hairpin*

A total of 14,459,285 *cluster* hairpin was found in all bacterial chromosomes combined. While 93% of these had between 2 and 5 hairpins in each *cluster*, a majority of them had 2 or 3 hairpins (i.e., 40% and 29%, respectively). All phyla and classes had an approximate average *cluster* hairpin density of 3 hairpins per *cluster*.

The maximum number of hairpins in any *cluster* was 18, in *Xanthomonas oryzae strain BB151-3*, a γ -proteobacteria with NCBI ID: NZ_CP036255. This organism is a rice plant pathogen. The operon ([2170021, 2170983, 2171413] where 2170021-2171413 is CDS and rest is IR region), corresponding to which this large *cluster* hairpin was found, is a IS110 family transposase. These transposases are needed for the efficient transposition of the insertion sequence or transposon DNA (https://www.ebi.ac.uk/interpro/entry/InterPro/IPR003346/). The distribution for size of the *cluster* hairpin is shown in Fig. 4.

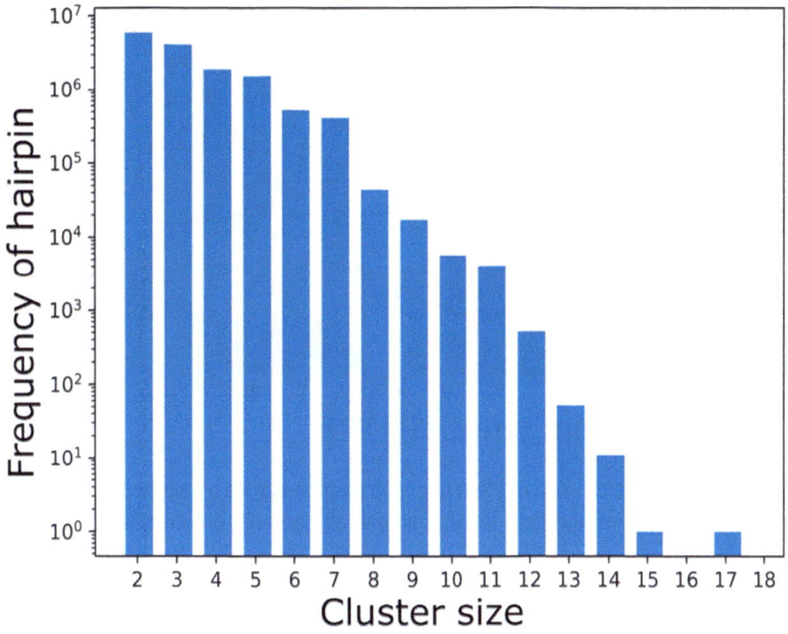

Fig. 4 The distribution of *cluster* hairpin size for all bacterial chromosomes

2.4.4 Effect of Stem and Loop Lengths on Hairpin Energy Score

Approximately, 73% of *cluster* hairpins had stem lengths between 2 and 5 bp and loop lengths between 4 and 11 bases, while these features account for 67% of *single* hairpins. On an average 86% of *cluster* hairpins and 84% of *single* hairpins had stem lengths between 2 and 9 bp and loop lengths between 4 and 11 bases. Figure 5 shows the overall stem and loop lengths distribution for these values in all bacteria combined.

Fusobacteria had the highest percentage of hairpins with large loop lengths (>20 bases), while Firmicutes had the highest percentage of hairpins with the largest stem length (>12 bp). This trend was followed for both *cluster* and *single* hairpins.

Hairpins predicted in plasmids have longer stems and loops as compared to those in chromosomes. The largest stem length is 47 bp and loop length is 37 bases; however, the percentage of such hairpins is very small. The majority of sizes are formed by hairpins with stem lengths between 2 and 9 bp and loop lengths of 4 and 11 bases. This group contains 87% of all *clusters* and 86% of all *single* hairpins.

Next, the relationship between hairpin energy scores and stem or loop lengths was investigated. The smallest hairpin with a stem length of 4 bases and energy scores between −4 and 6 kcal/mol constituted 24% of the total *cluster* hairpins and 20% of the total *single* hairpins. For *cluster* and *single* hairpins, the energy scores and stem length seemed to be inversely related. This can be seen in Fig. 6, where heatmaps clearly show an inverse linear relationship between energy scores and the stem lengths in all bacteria (Fig. 6).

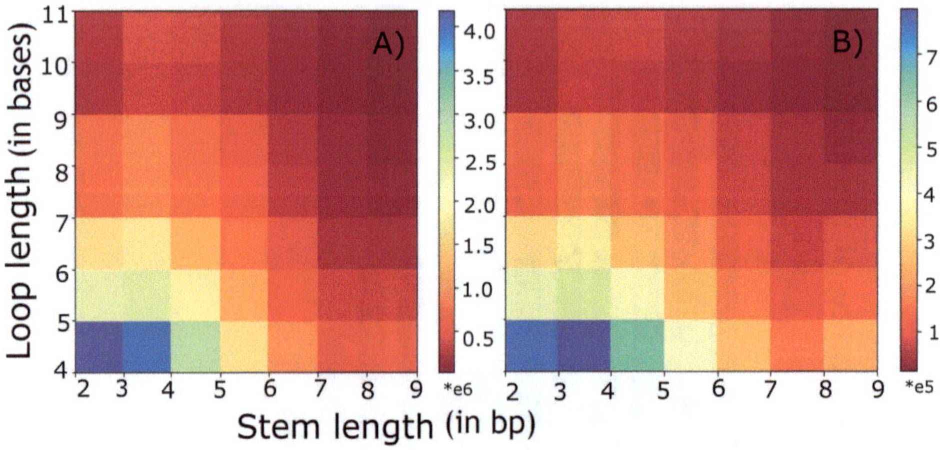

Fig. 5 Heatmap showing the distribution of stem and loop lengths across all bacterial chromosomes. Heatmap showing the distribution of stem and loop lengths across all bacterial chromosomes for (**a**) *cluster* and (**b**) *single* hairpins

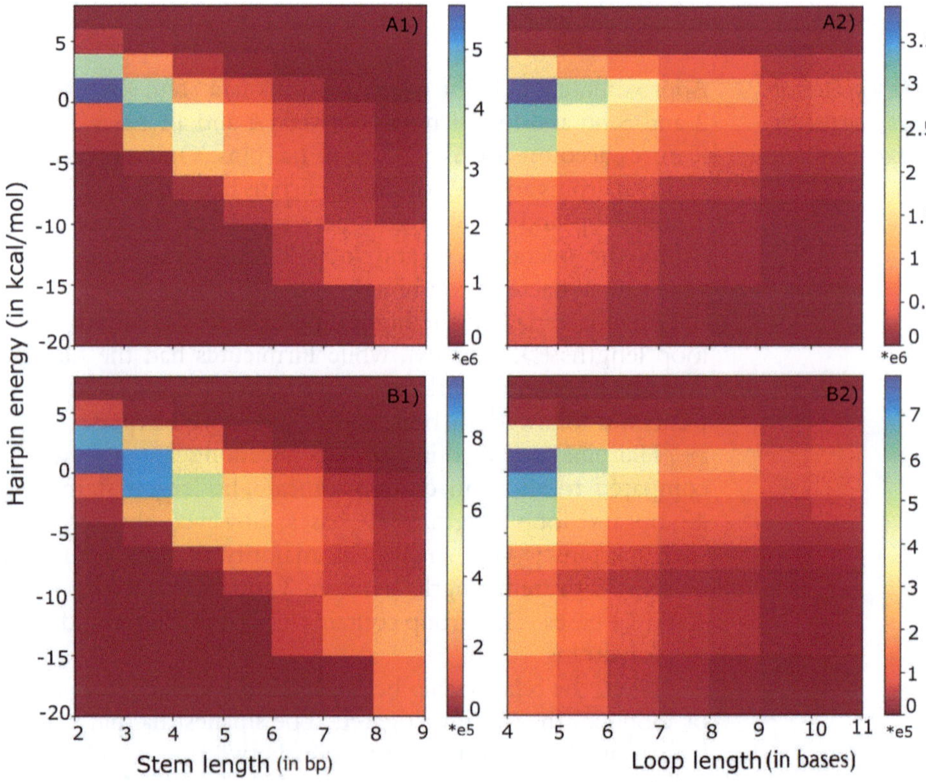

Fig. 6 Hairpin energy score distribution of stem and loop lengths from all chromosomes. Heatmap shows the hairpin energy versus stem (on left) and loop (on right) lengths distribution for all bacteria combined with *cluster* hairpins in A1, A2 and *single* hairpins in B1, B2

Similar relation was also seen for all phyla/classes taken. The relationship between energy scores and loop lengths is not similar, with no observed inverse or linear relationship. We suggest that hairpin stem lengths are regulated and related to energy, while hairpin loop lengths may not be regulated the same way. Very similar trends were also seen for plasmids.

2.5 Finding Poly U / Poly A Tails

Two possibilities for location of occurrence of hairpin with respect to the RNAP exit channel were noted from literature. One view says that the hairpin is formed just outside the RNAP exit channel [35, 36]. This would mean a minimum gap of 7 nt should exist between the hairpin and the RNA: DNA hybrid (5 nt in exit channel and 2 spacer nt), where the poly U pattern would be present. The other view says the hairpin is partially (5 bp stem) formed inside the exit channel [37], and the poly U is immediately downstream with possible 2–4 spacers. We calculated the occurrence of poly U/A pattern for both these distances.

DNA sequences, where Ts were replaced with Us, were taken. The algorithm then checked for As and Us to take care of genes both on the sense (+) and antisense (−) strands. The poly U or poly A pattern associated with the intrinsic terminators is supposed to be a continuous stretch that is not often interspersed with other bases.

2.5.1 Pattern Searched Immediately Downstream of Hairpin

For pattern search, a minimum criterion that at least three consecutive A or U bases must be present to label it as a true site in the IR was used. Only 10% of operons have a pattern within 4 nt from the hairpin end. We see that even upto 10 nt, only 29% of total operons have an accompanying pattern downstream of *identified* hairpins.

The pattern may not be perfect and can have mismatches with the presence of other bases in between. To account for this, patterns with ≤ 1 and ≥ 2 mismatches were found as well. With atmost 1 allowed mismatch, 16% of operons have a pattern within 4 nt from the hairpin end and 42% of operons have a pattern within 10 nt. With atmost 2 allowed mismatches, these numbers increase to 19% and 49% for distances 4 and 10 nt, respectively, from the hairpin end (*see* Table 1).

2.5.2 Patterns at Least 7 nt Downstream from Hairpin

Now the case where a hairpin is formed just outside of the exit channel, leaving a gap of minimum 7 nt (5 nt RNA exit channel and 2–3 nt spacer) between the poly U pattern and the hairpin 3′ end was considered. For a perfect matching pattern, 11% of operons lie within 10 nt from the hairpin end, while 26% are within 15 nt.

We get 18% and 43% of operons within 10 and 15 nt from the hairpin end, when at most 1 mismatch is allowed. With atmost 2 mismatches in pattern, 21% and 49% of operons are obtained within 10 and 15 nt from the hairpin end. These percentages are still low (*see* Table 2).

Table 1
The cumulative percentages for operons at various distances from the hairpin end, when finding a pattern immediately downstream of the hairpin end

Matching criterion	[0–2)	[2–4)	[4–6)	[6–8)	[8–10)	[10–15)	[15–20)	[20–30)	[30–50)	[50–70)	[70–100)
Perfect match	3.38	10.31	17.12	23.43	29.03	39.96	47.20	57.27	68.37	74.30	78.97
<=1 mismatch	5.29	15.63	25.46	34.43	41.98	55.16	62.89	72.21	80.88	84.86	87.81
<=2 mismatch	6.65	19.04	30.17	40.34	48.95	63.11	70.78	79.06	85.81	88.57	90.44

Table 2
The cumulative percentages for operons at various distances from the hairpin end, when finding a pattern at least 7 bases from the hairpin end

Matching criterion	[7–10)	[10–15)	[15–20)	[20–30)	[30–50)	[50–70)	[70–100)
Perfect match	10.95	26.47	36.86	50.98	65.49	72.65	78.04
<=1 mismatch	18.31	42.84	56.59	70.28	80.60	84.80	87.74
<=2 mismatch	21.22	48.96	64.17	77.16	85.34	88.27	90.13

3 Using RNA-seq Data for Identifying Termination Region

Public RNA-seq data was taken from the NCBI-SRA database http://www.ncbi.nlm.nih.gov/sra, ensuring to include only those datasets that had ≥70% alignment with reference chromosome, as of March 2022. The dataset had 380 nonredundant bacterial genomes. Each SRA file was first aligned with corresponding reference genomes. Properly paired reads with map quality ≥30 were filtered out. Next, for each IR, read depth was calculated.

From this data, 18 datasets did not have properly paired data or showed <5% alignment and were eliminated from the study.

3.1 Calculating Positive Slope Regions

For finding slopes, the cumulative frequency of reads for an operon was calculated. Slopes-sl_1 (that would represent the average read increase per window) for the whole region were found. RNA-seq data shows that even when the read coverage falls drastically in the IR, implying a termination has occurred, the reads are not completely absent in the region beyond the putative termination site. Therefore, to identify the termination site, all the regions where the count of reads increased after the stop codon were found. So, if the adjacent differences of sl_1 slope values are taken, the second derivative of the cumulative frequency curves is being calculated, defined by sl_2.

Calculating Slope sl_1

Let a and b be two points on the cumulative frequency plot at positions (i) and (i+w), where w is varying window size. Let read frequency at a be af and at b be bf.

Here, slope would be $\frac{(bf - af)}{w}$.

Slope values for different *sliding* windows "w" ranging from 5, 10, 15, and so on upto 30 were calculated. An example case is shown in Fig. 7.

Calculating Slope sl_2

Let $A = a_1, a_2, a_3, a_4, \ldots\ldots\ldots a_n$ be a set of sl_1 slopes for the IR region of an operon.

Then $sl_2 = a_2 - a_1, a_3 - a_2, a_4 - a_3, \ldots\ldots\ldots\ldots, a_n - a_{n-1}$.

sl_2 has positive and negative values as well as saddle points (points of inflection). In the IR, the first negative sl_2 slope after the stop codon was found, which is followed by a positive sl_2 slope. Such saddle points were taken, and the region extended till another drop in slope was seen. The changes in slope were prominently visible early in the IR using a sliding window size of 5, 10, and sometimes 15, but many such inflection points were lost when higher lengths of the window, i.e., 20–30, were taken. So, we took 10 as the sliding window size for our analysis (Fig. 7).

Now, we rationalize that termination must have taken place at the start of such region. This is because the drastic decrease in read frequency points to termination, and in order for an efficient termination, the cell is likely to pick an early termination site, and hence, the first such region was chosen for the presence of hairpin TU. Taking w = 10, if the average of all slope values is calculated in the coding region, then it is the coding (operon) region slope, and if it is calculated in IR, it is the IR slope. IR slope values range from 0 to 400 in our data, while coding region slope values range from 10^4 to 10^6. By taking the ratio of IR slope values: coding region slope values, i.e., 0–400: 10^4–10^6, would be in the range of ~ 0–0.004, showing that even though the slope values are positive in IR, the marginal increase of reads is negligible.

Hairpins closest to or present in this region were determined and found in 84% of these IRs. These were called the *RNA-seq derived* TUs. A match between the *RNA-seq derived* TUs and the *identified* TUs location was found. This offered an alternate approach to validate prediction against experimentally derived termination sites.

On an average, 66% of predicted TUs partially or fully overlapped with the *RNA-seq derived* TUs. The TUs matching or partially overlapping with *RNA-seq derived* ITT sites were *clusters* in 65.6% of the cases and *single* in the rest. These are the *matched* TUs.

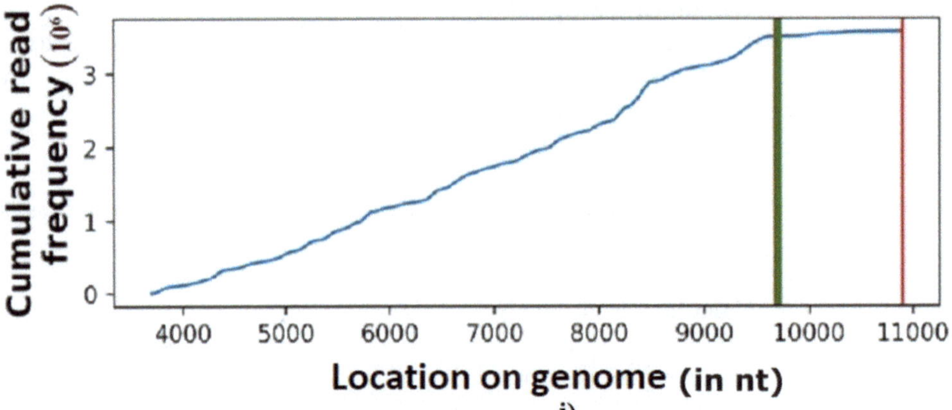

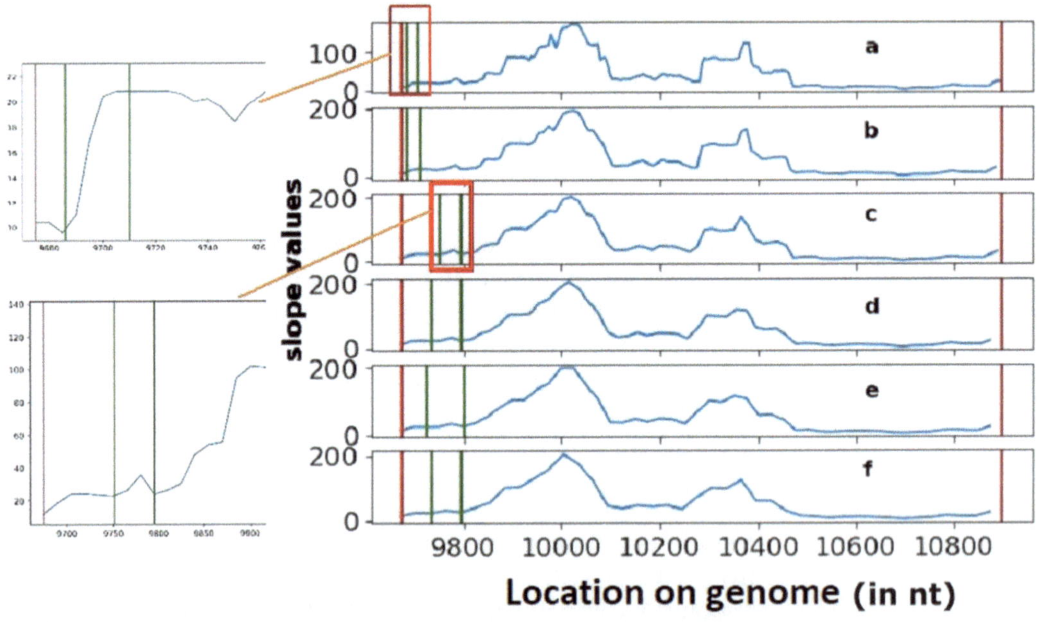

Fig. 7 Cumulative read frequency distribution. Plots show a cumulative read frequency distribution for an operon in *S. aureus* in (i). Their corresponding slope values in (ii) are calculated from the cumulative frequency distribution by taking "w" locations (**a–f** show $w = 5, 10, 15$, and so on up to 30) in a sliding window sequentially from the operon shown in (i). The x-axis shows the location on the chromosome, and the y-axis shows cumulative read frequency on that location in (i) and slope values in (ii). The red lines mark the start and end of the IR region. The green lines mark the start and end of the first positive slope region (pos^{st} and pos^{end}). As shown in the inset of plot (ii) slope values calculated from smaller size windows like 5 and 10 successfully capture all early inflection points in the curve (e.g., points P1 and P2), whereas increasing sizes further to >15 leads to a loss of such points. So, we take a window of size 10 for all further analysis. (This figure is adapted from [26])

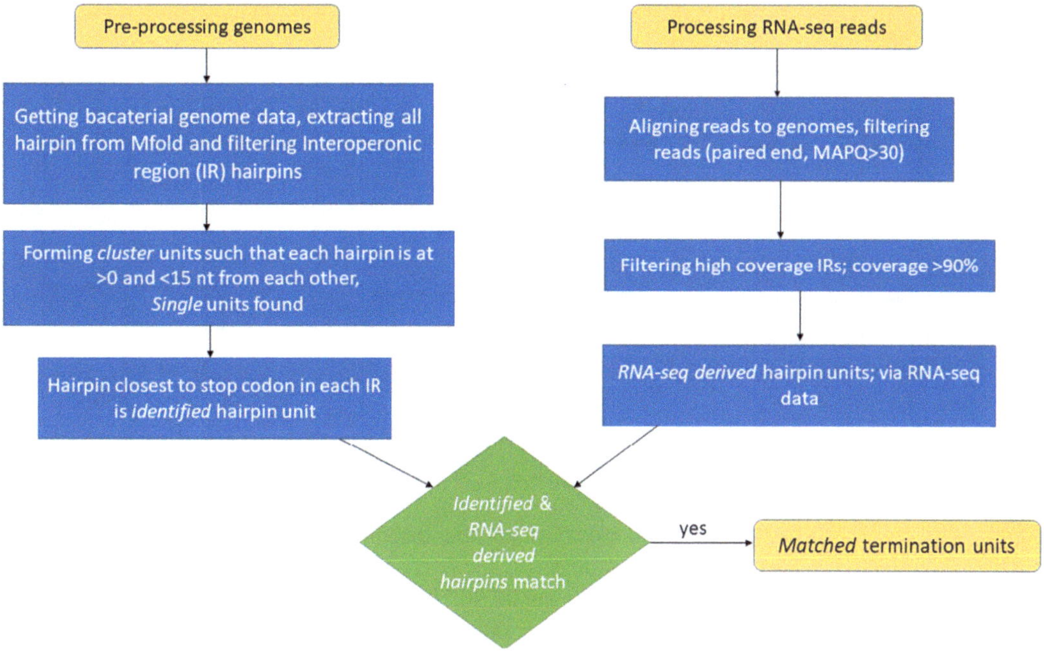

Fig. 8 Workflow of our analysis. A flowchart depicting the steps used to arrive at *matched* TUs and how *identified* and *RNA-seq derived* TUs are found. (Figure adapted from [26])

The transcription termination process in the cell is well-regulated. There might be more than one termination site to prevent read-through; hence, we find alternate termination sites, which may be both *cluster* or *single* hairpins. On including the alternate ITT sites predicted for each genome and matching their location to *RNA-seq derived* TUs, an additional 7.5% of IRs were seen to have an exact match, while 24% of IRs had an overlap between an alternate ITT site and *RNA-seq derived* TU, taking the cumulative overlap to 97.5% cases.

This shows that the predictions made by our method show good overlap and correspondence with experimentally determined termination sites as well, showing its ability to capture biological structures. Figure 8 shows the whole pipeline of the algorithm in the form of a flowchart.

4 Comparison of Predicted Hairpins with in vivo Validated Hairpins

A total of 1400 terminator sequences for *B. subtilis* (NC_000964) were taken from [38, 39], and 599 for *E. coli* K-12 MG1655 (NC_000913/U000960) were taken from [40, 41]. They were labeled as "experimental hairpins." Figure 9 summarizes the analyses. The statistics showed that the hairpins partially or fully overlap in 61% of the operons from *B. subtilis*, and 42% in *E. coli*.

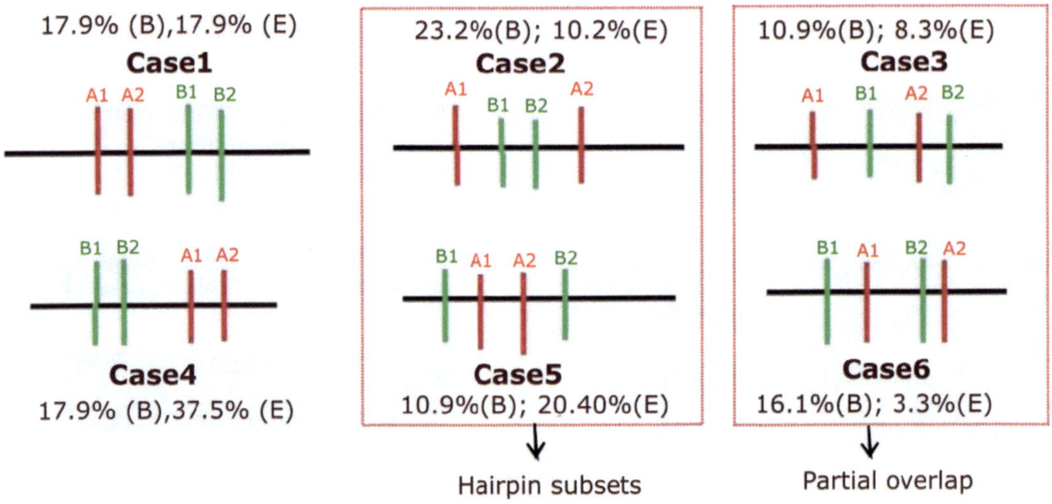

Fig. 9 Comparison of in vivo *validated* and *identified* hairpins. Diagram showing the comparison of experimental hairpins and *identified* hairpins for *B. subtilis* (B) and *E. coli* (E). The percentages are calculated with respect to the total experimental hairpins found in the IR: 967 for *B. subtilis*, and 520 for *E. coli*. Out of this, 30 and 13 cases in *B. subtilis* and *E. coli*, respectively, did not have a prediction

5 Tools and the INTERPIN Database

The algorithm was extended to cover ~25 million hairpins from 12,745 bacterial genomes, spanning 10 phyla. We have created a database, which is the largest collection of inferred ITT sites, and can be accessed through a public repository called INTERPIN (INtrinsic transcription TERmination hairPIN; http://pallab.cds.iisc.ac.in/INTERPIN/; Fig. 10) [27].

In the database, the chromosomes are organized into respective phyla/class, enabling search by phylum, organism name as well as NCBI ID. The web interface allows the results to be easily retrieved interactively. In this database, 58% of all the inferred hairpins are *cluster* hairpins.

Features and tools provided by the database are listed below:

- Prediction of ITT sites in bacteria. Results are presented in a summarized tabulated form as well as raw prediction data is provided.
- The genomic view of the operon, gene annotation, and TU predictions is provided by the integration of the IGV viewer into the database. The users can easily view predictions for selected region on the genome, w.r.t. the desired gene or location.

Fig. 10 The INTERPIN database logo

- The secondary RNA folded structure for the predicted hairpins is provided. The option to generate a 3D model for the same is also available.
- A detailed help page is provided to navigate through the database and infer the results tables.

More information can be found from the website http://pallab.cds.iisc.ac.in/INTERPIN/.

6 Discussion

In bacteria, the process of transcription and translation is concurrent [42, 43]. Operating both processes in concert requires precise kinetic control of the mRNA produced and the amino acid being incorporated by translation. It has been seen that terminator sequences present in the coding region are not able to terminate transcripts. This is because the closely trailing ribosome does not allow the folding of sequences to a hairpin structure [42]. It is only after the occurrence of a stop codon that decoupling takes place and hairpin sequences can fold into their secondary structure to disrupt the RNAP motion.

The lack of solid evidence on the wide prevalence of ITT units across diverse organisms has, therefore, been intriguing. For a basic ITT process to occur, the *single* hairpins are said to form at the rear end of the RNA exit channel [1] of RNAP. For a hairpin to form, the RNA needs to fold, while the transcription is occurring. Although previous studies proposed limits on hairpin lengths that can be accommodated near the RNA exit channel [1] and are required for the formation of a functional hairpin [23, 44], these have been observed only in select organisms such as *M. tuberculosis* [45] and not across most bacteria. Our observation suggests that a stretch of contiguous hairpins (hairpin *cluster*) is as effective as a single strong hairpin to cause ITT.

6.1 Why Formation of Cluster Hairpin Is Favorable?

The formation of *cluster* hairpin is more favorable for ITT than a *single* hairpin, as seen with the help of thermodynamic and kinetic factors. These are listed as below:

1. A more number of hairpins provide more RNA–protein interactions between the RNAP and the hairpins, which can help with increase in pause duration by stabilizing the RNAP complex.

2. Folding experiments show a preference for alternating structures (11 out of 14 to 40-length sequences) where there is interconversion between *single* and two hairpin states, the latter being the major one. So, more the number of nucleation sites, the more favorable for folding [46].
3. When long mRNA sequences fold, it was observed that short-range interactions have higher stability over long-range reactions and they are thus retained. In such cases, *cluster* hairpins would be preferred.

6.2 Presence of Poly U Pattern at Hairpin End

In 1999 [47] and again in 2007 [48], it was shown that oligonucleotides that mimicked hairpin stem could cause transcription termination in already stalled RNAP complexes. These oligonucleotides were able to cause termination without the aid of a poly U/A tail. This study also strengthens our observation that poly U-tail is unnecessary for ITT.

The length of poly U pattern is seen to vary across different genes and bacteria. Some papers also show that termination is possible in the absence of the poly U patterns, thus putting forward the idea that the consensus regarding the presence of a poly U pattern downstream of the hairpin 3′ end may not always hold, as suggested earlier. It should be noted that most experimental studies have been done with *E. coli* and *B. subtilis* only, and therefore, it is quite possible that these do not hold for other bacteria. We have seen the same in our results and shown how we were unable to find poly U patterns for the majority of the operons, even within 10 bp from 3′ end of the hairpin. We thus strongly suggest that earlier observation may have been a trait in *E. coli* and related organisms only.

7 Conclusion

ITT is a major regulatory process in bacteria. The incorrect folding or presence of nonefficient terminators will not be recognized as a stop signal by the RNAP and can cause a transcriptional read-through [49]. Failure to terminate can lead to the formation of unwanted proteins with adverse biological functions, due to the added cistrons in the newly formed mRNA. On convergent genes, transcriptional termination failure may also lead to a clash of the RNAP machinery transcribing adjoining convergent genes or the formation of transcripts that overlap [50]. The efficiency of termination can also affect the regulation of gene expression. Given the importance of transcription termination and the vast amount of emerging bacterial genome data, it becomes necessary to create models and algorithms that can predict ITT sites as well as contribute to a better understanding of the termination mechanism.

Through our work, and this chapter, we have tried to contribute to a better understanding and improved prediction of transcription terminations sites in bacterial genomes. We have highlighted the presence of *cluster* TU as a major group of termination sites that were not studied before us. The method is generic and can thus be applied to bacteria across all phyla/classes and for GC as well as AT-rich genomes.

References

1. Ray-Soni A, Bellecourt MJ, Landick R (2016) Mechanisms of bacterial transcription termination: all good things must end. Ann Rev Biochem 85(1):319–347. https://doi.org/10.1146/annurev-biochem-060815-014844
2. Kriner MA, Sevostyanova A, Groisman EA (2016) Learning from the leaders: gene regulation by the transcription termination factor rho. Trends Biochem Sci 41(8):690–699. https://doi.org/10.1016/j.tibs.2016.05.012
3. Ciampi MS (2006) Rho-dependent terminators and transcription termination. Microbiology 152(9):2515–2528. https://doi.org/10.1099/mic.0.28982-0
4. Lau LF, Roberts JW (1985) Rho-dependent transcription termination at lambda r1 requires upstream sequences. J Biol Chem 260(1):574–584
5. Dutta D, Chalissery J, Sen R (2008) Transcription termination factor rho prefers catalytically active elongation complexes for releasing RNA*. J Biol Chem 283(29):20243–20251. https://doi.org/10.1074/jbc.M801926200
6. Murakami KS (2015) Structural biology of bacterial RNA polymerase. Biomolecules 5(2):848–864. https://doi.org/10.3390/biom5020848
7. Sutherland C, Murakami KS (2018) An introduction to the structure and function of the catalytic core enzyme of Escherichia coli RNA polymerase. EcoSal Plus 8(1). https://doi.org/10.1128/ecosalplus.ESP-0004-2018
8. Ermolaeva MD, Khalak HG, White O, Smith HO, Salzberg SL (2000) Prediction of transcription terminators in bacterial genomes 1 1edited by F. E. Cohen. J Mol Biol 301(1):27–33. https://doi.org/10.1006/jmbi.2000.3836
9. Mitra A, Kesarwani AK, Pal D, Nagaraja V (2011) WebGeSTer DB—a transcription terminator database. Nucleic Acids Res 39:D129–D135. https://doi.org/10.1093/nar/gkq971
10. Zuker M (2003) Mfold web server for nucleic acid folding and hybridization prediction. Nucleic Acids Res 31(13):3406–3415. https://doi.org/10.1093/nar/gkg595
11. Knudsen B, Hein J (2003) Pfold: RNA secondary structure prediction using stochastic context-free grammars. Nucleic Acids Res 31(13):3423–3428. https://doi.org/10.1093/nar/gkg614
12. Gardner PP, Barquist L, Bateman A, Nawrocki EP, Weinberg Z (2011) RNIE: genome-wide prediction of bacterial intrinsic terminators. Nucleic Acids Res 39(14):5845–5852. https://doi.org/10.1093/nar/gkr168
13. Lambert A, Fontaine JF, Legendre M, Leclerc F, Permal E, Major F, Putzer H, Delfour O, Michot B, Gautheret D (2004) The ERPIN server: an interface to profile-based RNA motif identification. Nucleic Acids Res 32:W160–W165 (2004). https://doi.org/10.1093/nar/gkh418
14. Hofacker IL, Fontana W, Stadler PF, Bonhoeffer LS, Tacker M, Schuster P (1994) Fast folding and comparison of RNA secondary structures. Monatsh Chem 125(2):167–188. https://doi.org/10.1007/BF00818163
15. Feng CQ, Zhang ZY, Zhu XJ, Lin Y, Chen W, Tang H, Lin H (2019) iTerm-PseKNC: a sequence-based tool for predicting bacterial transcriptional terminators. Bioinformatics 35(9):1469–1477. https://doi.org/10.1093/bioinformatics/bty827
16. Naville M, Ghuillot-Gaudeffroy A, Marchais A, Gautheret D (2011) ARNold: a web tool for the prediction of rho-independent transcription terminators. RNA Biol 8(1):11–13 (2011). https://doi.org/10.4161/rna.8.1.13346
17. Kumar M, Bhasin M, Natt NK, Raghava GPS (2005) BhairPred: prediction of β-hairpins in a protein from multiple alignment information using ANN and SVM techniques. Nucleic Acids Res 33:W154–W159. https://doi.org/10.1093/nar/gki588
18. Hamada M, Sato K, Kiryu H, Mituyama T, Asai K (2009) Predictions of RNA secondary structure by combining homologous sequence

information. Bioinformatics 25(12):i330–i338. https://doi.org/10.1093/bioinformatics/btp228

19. Hoon MJLD, Makita Y, Nakai K, Miyano S (2005) Prediction of transcriptional terminators in Bacillus subtilis and related species. PLOS Comput Biol 1(3):e25. https://doi.org/10.1371/journal.pcbi.0010025
20. Nair TM, Tambe SS, Kulkarni BD (1994) Application of artificial neural networks for prokaryotic transcription terminator prediction. FEBS Lett 346(2):273–277. https://doi.org/10.1016/0014-5793(94)00489-7
21. Eddy SR, Durbin R (1994) RNA sequence analysis using covariance models. Nucleic Acids Res 22(11):2079–2088. https://www.ncbi.nlm.nih.gov/pmc/articles/PMC308124/
22. Do CB, Woods DA, Batzoglou S (2006) CONTRAfold: RNA secondary structure prediction without physics-based models. Bioinformatics 22(14):e90–e98. https://doi.org/10.1093/bioinformatics/btl246
23. Carafa YD, Brody E, Thermes C (1990) Prediction of rho-independent Escherichia coli transcription terminators: a statistical analysis of their RNA stem-loop structures. J Mol Biol 216(4):835–858. https://doi.org/10.1016/S0022-2836(99)80005-9
24. Swenson MS, Anderson J, Ash A, Gaurav P, Sükösd Z, Bader DA, Harvey SC, Heitsch CE (2012) GTfold: enabling parallel RNA secondary structure prediction on multi-core desktops. BMC Res Notes 5(1):341. https://doi.org/10.1186/1756-0500-5-341
25. Lesnik EA, Sampath R, Levene HB, Henderson TJ, McNeil JA, Ecker, DJ (2001) Prediction of rho-independent transcriptional terminators in Escherichia coli. Nucleic Acids Res 29(17):3583–3594. http://www.ncbi.nlm.nih.gov/pmc/articles/PMC55870/
26. Gupta S, Pal D (2021) Clusters of hairpins induce intrinsic transcription termination in bacteria. Sci Rep 11(1):16194. https://doi.org/10.1038/s41598-021-95435-3
27. Gupta S, Padmashali N, Pal D (2023) INTERPIN: A repository for intrinsic transcription termination hairpins in bacteria. Biochimie 214:228–236. https://doi.org/10.1016/j.biochi.2023.07.018
28. Gupta S, Pal D (2024) Detection of intrinsic transcription termination sites in bacteria: consensus from hairpin detection approaches. J Biomolecular Structure and Dynamics 1–11. https://doi.org/10.1080/07391102.2024.2325107
29. Unniraman S, Prakash R, Nagaraja V (2001) Alternate paradigm for intrinsic transcription termination in eubacteria. J Biol Chem 276(45):41850–41855. https://doi.org/10.1074/jbc.M106252200
30. Ingham CJ, Hunter IS, Smith MC (1995) Rho-independent terminators without 3′ poly-u tails from the early region of actinophage øc31. Nucleic Acids Res 23(3):370–376. https://www.ncbi.nlm.nih.gov/pmc/articles/PMC306685/
31. Mitra A, Angamuthu K, Jayashree HV, Nagaraja V (2009) Occurrence, divergence and evolution of intrinsic terminators across eubacteria. Genomics 94(2):110–116. https://doi.org/10.1016/j.ygeno.2009.04.004
32. Mathews DH, Sabina J, Zuker M, Turner DH (1999) Expanded sequence dependence of thermodynamic parameters improves prediction of RNA secondary structure11edited by I. Tinoco. J Mol Biol 288(5):911–940. https://doi.org/10.1006/jmbi.1999.2700
33. MolQuest (2021). http://molquest.com/molquest.phtml?topic=reference_license
34. Serra MJ, Barnes TW, Betschart K, Gutierrez MJ, Sprouse KJ, Riley CK, Stewart L, Temel RE (1997) Improved parameters for the prediction of RNA hairpin stability†. Biochemistry 36(16):4844–4851. https://doi.org/10.1021/bi962608j
35. Santangelo TJ, Roberts JW (2004) Forward translocation is the natural pathway of RNA release at an intrinsic terminator. Mol Cell 14(1):117–126. https://doi.org/10.1016/S1097-2765(04)00154-6
36. Toulokhonov I, Artsimovitch I, Landick R (2001) Allosteric control of RNA polymerase by a site that contacts nascent RNA hairpins. Science 292(5517):730–733. https://doi.org/10.1126/science.1057738
37. Kang JY, Mishanina TV, Bellecourt MJ, Mooney RA, Darst SA, Landick R (2018) RNA polymerase accommodates a pause RNA hairpin by global conformational rearrangements that prolong pausing. Mol Cell 69(5):802–815.e1. https://doi.org/10.1016/j.molcel.2018.01.018
38. Mandell ZF, Oshiro RT, Yakhnin AV, Vishwakarma R, Kashlev M, Kearns DB, Babitzke P (2021) NusG is an intrinsic transcription termination factor that stimulates motility and coordinates gene expression with NusA. eLife 10. https://doi.org/10.7554/eLife.61880
39. Mondal S, Yakhnin AV, Sebastian A, Albert I, Babitzke P (2016) NusA-dependent

transcription termination prevents misregulation of global gene expression. Nat Microbiol 1(1):1–8. https://doi.org/10.1038/nmicrobiol.2015.7

40. Ju X, Li D, Liu S (2019) Full-length RNA profiling reveals pervasive bidirectional transcription terminators in bacteria. Nat Microbiol 4(11):1907–1918. https://doi.org/10.1038/s41564-019-0500-z

41. Lalanne JB, Taggart JC, Guo MS, Herzel L, Schieler A, Li GW (2018) Evolutionary convergence of pathway-specific enzyme expression stoichiometry. Cell 173(3):749–761.e38. https://doi.org/10.1016/j.cell.2018.03.007

42. Li R, Zhang Q, Li J, Shi H (2016) Effects of cooperation between translating ribosome and RNA polymerase on termination efficiency of the rho-independent terminator. Nucleic Acids Res 44(6):2554–2563. https://doi.org/10.1093/nar/gkv1285

43. Wang X, MPA N, Jeon HJ, Lee Y, He J, Adhya S, Lim HM (2019) Processing generates 3′ ends of RNA masking transcription termination events in prokaryotes. PNAS 116(10): 4440–4445. https://doi.org/10.1073/pnas.1813181116

44. Wilson KS, von Hippel PH (1995) Transcription termination at intrinsic terminators: the role of the RNA hairpin. Proc Natl Acad Sci 92(19):8793–8797. https://doi.org/10.1073/pnas.92.19.8793

45. Unniraman S, Prakash R, Nagaraja V (2002) Conserved economics of transcription termination in eubacteria. Nucleic Acids Res 30(3): 675–684. http://www.ncbi.nlm.nih.gov/pmc/articles/PMC100295/

46. Nagel JHA, Flamm C, Hofacker IL, Franke K, Smit DHM, Schuster P, Pleij CWA (2006) Structural parameters affecting the kinetics of RNA hairpin formation. Nucleic Acids Res 34(12):3568–3576. https://doi.org/10.1093/nar/gkl445

47. Yarnell WS, Roberts JW (1999) Mechanism of intrinsic transcription termination and antitermination. Science 284(5414):611–615. https://doi.org/10.1126/science.284.5414.611

48. Shankar S, Hatoum A, Roberts JW (2007) A transcription antiterminator constructs a NusA-dependent shield to the emerging transcript. Mol Cell 27(6):914–927. https://doi.org/10.1016/j.molcel.2007.07.025

49. Platt T (1986) transcription termination and the regulation of gene expression. Annu Rev Biochem 55(1):339–372. https://doi.org/10.1146/annurev.bi.55.070186.002011

50. Proudfoot NJ (2016) Transcriptional termination in mammals: stopping the RNA polymerase II juggernaut. Science 352(6291): aad9926. https://doi.org/10.1126/science.aad9926

Chapter 20

RNA-Seq Analysis of Mammalian Prion Disease

Ambarish Kumar, Jamie Dixson, and Rajeev K. Azad

Abstract

A protein, which can attain a prion state, differs from standard proteins in terms of structural conformation and aggregation propensity. High-throughput sequencing technology provides an opportunity to gain insight into the prion disease condition when coupled with single-cell RNA-Seq analysis to reveal transcriptional changes during prion-based pathogenicity. In this chapter, we present a protocol for RNA-Seq analysis of mammalian prion disease using a single-cell RNA sequencing dataset procured from the NCBI GEO database. This protocol is a tool that can assist researchers in characterizing mammalian prion disease in a reproducible and reusable manner. Further, the resulting output has the potential to provide transcript biomarkers for mammalian prion diseases, which can be employed for diagnostic and prognostic purposes.

Key words Prions, Epigenetics, scRNA-Seq, Prion diseases, Biomarkers

1 Introduction

Prions are broadly characterized as proteins that are infectious from cell-to-cell when in a non-wildtype alternate structural conformation that possess the ability to recruit other similar protein molecules to the alternate conformation [1]. The most well-characterized prions form amyloids when in the prion form. The known amyloid prions are generally enriched with asparagine (N) and/or glutamine (Q). While poly(Q/N) regions often occur within the prion domain of most amyloid-forming prion proteins, some are dispersed throughout the entire amino acid sequence [2]. In addition to poly(Q/N), the namesake mammalian prion protein (PrP) contains octapeptide repeats consisting of polymeric (PHGGGWGQ). The wildtype conformation prion protein (PrPC) exists as an alpha-helix rich molecule whereas conversion to the prion conformation (PrPSC) results in transformation of much of the alpha helical structure to β-sheet [3]. In a broad sense, the conversion of wildtype prion-forming proteins to the prion form has been proposed as a molecular switch that might function as a non-Mendelian, reversible, and heritable adaptive mechanism.

Rajeev K. Azad (ed.), *Transcriptome Data Analysis*, Methods in Molecular Biology, vol. 2812,
https://doi.org/10.1007/978-1-0716-3886-6_20,

Additionally, in yeast and invertebrates, prions as molecular switches are often implicated as expression factors either at the transcriptional or translational levels [4]. Further, a prion-based switching mechanism appears to act in the control of flowering time in *Arabidopsis* [5]. This brings to light the possibility that prion diseases in mammals may be diagnosed early by monitoring the expression profiles of multiple genes. RNA-Seq datasets provide one such type of data for the detection of varied expression profiles. Mammalian PrP is the only widely accepted prion in mammals due to its easily observable intraspecific transmission. Interspecific infectivity is generally more limited due to an apparent species barrier. However, there are examples of interspecific transmission [6], most notably the infection of humans with Bovine Spongiform Encephalitis causing variant Creutzfeldt-Jakob Disease [3, 7]. Additionally, over the past decade it has become increasingly acceptable to refer to varied proteins implicated in diverse human neurodegenerative diseases as prions, prion-like proteins, prionoids, or any similar vernacular though not without considerable debate [5]. These include but are not limited to amyloid-β early on in the progression of Alzheimer's Disease (AD) and later Tau, α-synuclein in Parkinson's Disease (PD), and several other proteins involved in the progression of amyotrophic lateral sclerosis (ALS) and frontotemporal dementia (FTD) [3, 5–11]. Next-generation high-throughput sequencing has made it possible to characterize the transcriptional landscape of individual cells using several interrelated techniques. Those techniques are referred to as single cell RNA sequencing (scRNA-Seq) and their genesis can be traced to experiments involving the comparison of microarrays to the expression profiles obtained by sequencing RNA from a single cell derived from a mouse embryo [12]. Today scRNA-Seq is often used to identify transcriptional biomarkers that are either up- or down-regulated during disease progression. This has in some cases allowed for the identification of previously unresolvable cellular subtypes and may also allow for novel therapeutic strategies [13]. Additional applications have revealed that scRNA-Seq can be used to identify unique immunologic cell types as well as prognostic and diagnostic biomarkers in genetic disorders and retroviral, viral, and parasitic infections. Despite being unique in that PrP does not illicit an immune response [14], we have applied scRNA-Seq analysis to the transcriptomic characterization of sporadic Creutzfeldt-Jakob Disease (sCJD) in humans. Herein, we describe our use of that method using the Seurat R package [12, 15–19].

2 Materials

2.1 Dataset

The dataset used in this study was procured from the NCBI Gene Expression Omnibus. The NCBI GEO accession number of the dataset is GSE214374. There are ten samples of confirmed post-mortem sCJD and three samples of biopsy sCJD along with controls. Along with the sequencing files, there is a samples info file containing metadata of the human scRNA-Seq datasets. One sample is selected for the analysis in the study [20].

2.2 Tools

The analysis described herein was performed using Seurat R package on the TACC lonestar6 high-performance server (https://www.tacc.utexas.edu/). If datasets other than the one used herein are to be analyzed, execution of the included scripts on less powerful servers is perhaps feasible depending on the size of the dataset chosen. Annotation of the disease genes is performed using the PANTHER database available at https://www.pantherdb.org/. A full account of the resources used in this protocol is provided in Table 1.

3 Methods

3.1 Import the Dataset and Generate a Count Matrix

This step entails importing the sequence datasets and generating a gene expression count matrix. This matrix is an unnormalized sparse matrix of gene expression data for each cell type.

3.1.1 Import the R Libraries for the Seurat Package for scRNA-Seq Analysis and dplyr for Data Manipulation

```
library(Seurat)
library(dplyr)
```

3.1.2 Initialize a Variable for the Directory Containing Sequence Data

```
currdirc<-"C:/Users/14699/Downloads/testGSE2143174/"
```

3.1.3 Read the Matrix File for the Gene Expression

```
matrixfile<-paste0(currdirc,"12499_DGE.mtx")
```

3.1.4 Read the Cell Metadata File

```
cellmetadata<-paste0(currdirc,"12499_cell_metadata.csv")
```

Table 1
Resources used in the study

Resources	Name	Description
Dataset	Human scRNAseq	NCBI GEO accession—GSE214374
Software	Seurat [15–19]	R package for the single cell sequencing analysis
Annotation database	PANTHER	Knowledgebase of functional annotation of genes and proteins (https://www.pantherdb.org/).
Computational platform	TACC Lonestar6 server	High performance computing server (https://www.tacc.utexas.edu/)

3.1.5 Read the Feature File

```
featuresfile<-paste0(currdirc,"12499_genes.csv")
```

3.1.6 Generate Sparse Matrix for the Gene Expression Counts

If the analysis is performed on entire samples, one should create a Seurat object for each count matrix corresponding to each sample and save those Seurat objects in a designated location. Once all of the Seurat objects have been saved, they are merged into a single Seurat object. Creation of the sparse matrix for each sample is accomplished using the script below and merging of the Seurat objects can be accomplished using the method described at https://satijalab.org/seurat/archive/v3.1/merge_vignette.html.

```
data.seurat <- ReadMtx(mtx = matrixfile, cells = cellmetadata,
features = featuresfile, cell.column = 1, feature.column =
1, cell.sep = ",", feature.sep = ",", mtx.transpose = T, skip.
cell = 1, skip.feature = 1)
```

3.2 Create Seurat Object of the Gene Expression Count Matrix

In addition to the count data, Seurat objects store additional information including cell level and feature level metadata. The data format is designed for count matrix and analyses such as normalization, PCA, and clustering results. The self-contained nature of the Seurat object and the data format makes it convenient for use throughout the analysis. The input of this step is the unnormalized sparse matrix generated in the previous step. Thresholds are set to include only those cells which have at least 200 features and only those features which are expressed in at least three cells. This is accomplished using the following script.

```
data.seurat <- CreateSeuratObject(counts = data.seurat, min.
cells = 3, min.features = 200)
```

3.3 Preprocessing

With regard to the transcription-level data, we keep only those cells with number of genes expressed above and below pre-set threshold limits. This step is needed here to overcome any sequencing artifact or bias. This will also improve the resolution of gene sets which are more apt to convey biological signals in downstream analyses. The gene expression count matrix generated in the previous step is used as input and "nFeature_RNA" is the parameter name for the pre-set threshold limit to filter out cells based on number of expressed genes. Cells with more than 200 and less than 2500 genes expressed are considered in the analysis and these thresholds are set as shown in the script below.

```
data.seurat <- subset(data.seurat, subset = nFeature_RNA >
200 & nFeature_RNA < 2500)
```

3.4 Normalization of the Data

Cell-level normalization should be performed on the raw count matrices. The basic steps involved in normalization are:

(a) Dividing each gene expression value by total gene expression value across the cell type.
(b) Scaling the values by multiplying by a scaling factor.
(c) Log-transformation of the values to obtain final log normalized gene expression values.

These steps are implemented with the NormalizeData( ) function, with essential arguments of the function as data object which is a subset of gene expression matrix produced in previous step, normalization method, and scale factor as shown below. The scale factor is used to adjust the range of gene expression values during the normalization step shown in the script below.

```
data.seurat <- NormalizeData(data.seurat, normalization.meth-
od = "LogNormalize", scale.factor = 10000)
```

A technological challenge associated with the analysis described herein is the requirement and allocation of virtual memory for processing of a high dimensional count matrix such as required during normalization of the gene expression count data. A laptop of ~13 GB RAM may fail to process a large count matrix. This problem can be overcome by using a high-performance computer with sufficient RAM and the ability to allocate the required virtual memory for data normalization.

3.5 Identification of Highly Variable Features (Feature Selection)

Only the features (genes) that exhibit high cell-to-cell variation should be retained for further analysis; all others should be discarded. These genes are termed highly variable features. For this protocol, we retained only the top 2000 highly variable features for analysis. The number of variable features to be retained is set using the parameter "nfeatures." Variable features across the single cells are selected and ranked based on statistical distribution of gene expression. Variance stabilizing transformation (vst) is used in the "FindVariableFeatures" function as a selection and ranking method for variable features as shown in the following script.

```
Data.seurat <- FindVariableFeatures(data.seurat, selection.
method = "vst", nfeatures = 2000)
```

3.6 Scaling the Data

A linear transformation is performed on the selected features by scaling and centering the feature counts. This results in a mean expression value of 0.0 and a variance of 1.0. The "ScaleData" function uses a Seurat object of variable features as an input. This data object has gene names of the variable features present as row names. Those row names are extracted and passed on to the scaling function as an argument as shown below.

```
all.genes <- rownames(data.seurat)
data.seurat <- ScaleData(data.seurat, features = all.genes)
```

3.7 Perform Linear Dimensional Reduction

The next step entails dimensional reduction for the feature data using principal component analysis (PCA). All of the variable features are passed on as feature arguments in the RunPCA function along with the scaled data as an input.

```
data.seurat <- RunPCA(data.seurat, features=VariableFeatures
(object=data.seurat))
```

3.8 Clustering of Cell Types

Clustering of cells by cell type based on the expression data is performed using a graph-based cell embedding technique. This is implemented in two steps. The first step is building a graph of cells based on nearest neighbors and the second step is to obtain clusters of cells from the nearest neighbor graph using a pre-defined resolution as shown below. Herein the analysis nearest neighbor graph is built using the first ten principal components obtained from the dimensionally reduced dataset. These principal components are passed on as an argument for the "dim" parameter of the FindNeighbors function. Subsequently cells are clustered using the generated nearest neighbor graph. The function "FindClusters" uses the "resolution" parameter to generate a clustered cell community. Smaller resolution (below 1.0) will generate a greater number of clusters with lesser data in each clustered cell community and vice versa. The script to obtain the nearest neighbors graph and the clusters respectively can be found below.

```
data.seurat <- FindNeighbors(data.seurat, dims = 1:10)
data.seurat <- FindClusters(data.seurat, resolution = 0.5)
```

The parameter settings are somewhat permissive allowing the researcher to adjust them without massive swings in analysis results. Annotation of the cell clusters with the cell types is an important task in single cell sequencing analysis. This can be accomplished by a label transfer approach using the reference dataset with known cell types or by using R function to assign the cell type to the obtained cell clusters.

3.9 Nonlinear Dimensional Reduction (UMAP/tSNE)

Visualization of a dataset characterized by nonlinearity is challenging and cannot be easily dealt with using linear dimensional techniques like PCA. To visualize a nonlinear high dimensional dataset, a nonlinear dimensional technique, namely, Uniform Manifold Approximation and Projection (UMAP), needs to be implemented. This generates a projection of the dataset from a high dimensional space to lower dimensional space that is more easily visualized. The "RunUMAP" function uses clustered cell types as input and the "dim" parameter for selecting first "n" numbers of the dimensionally reduced feature set. The following command is used to implement this.

```
data.seurat <- RunUMAP(data.seurat, dims = 1:10)
```

While the nonlinear dimensional reduction function has two available options as methods: tSNE and UMAP, UMAP may be preferable over tSNE due to lower processing time.

3.10 Visualize the Clustered Cell Types

The output of the previous step is a dimensionally reduced dataset containing cell types and cell clusters. This dataset is used in the "DimPlot" function as input and the name of the reduction technique is input as a parameter and as shown in the script below. The graph of clustered cell types is visualized in two-dimensional space using a 2D scatter plot as shown in Fig. 1.

```
data.seurat.plot<-DimPlot(data.seurat, reduction = "umap")
data.seurat.plot + DarkTheme()
```

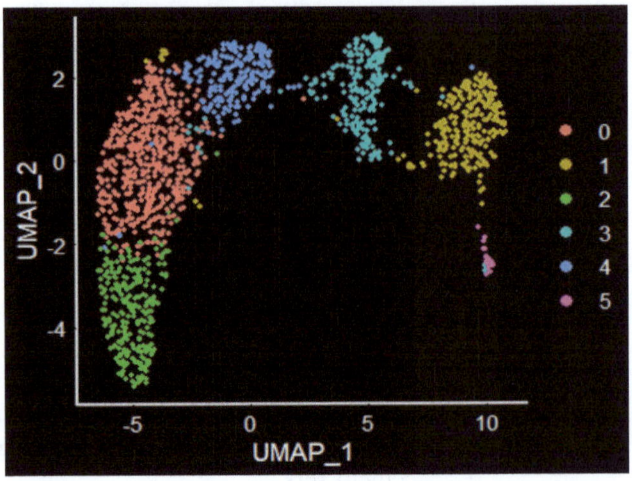

Fig. 1 Visualization of cell clusters: UMAP plot depicting different cell classes in different colors

3.11 Identification of Differentially Expressed Features

We now identify differentially expressed features across all clusters using quantitative measures and applying statistical testing of gene expression values. Dimensionally reduced data after clustering of the cell types are used in the function "`FindAllMarkers`" as input. Positive markers that are upregulated genes are obtained for the "TRUE" value of the parameter named "only.pos." One more filtering criterion is applied to consider only those genes which are not infrequently expressed in the cell population. A threshold value of 0.25 is passed as an argument to "min.pct" parameter to consider only those genes which are expressed by a minimum of 25% of the cells in different groups of cell types.

Further, the top two variable features across each cluster are obtained after cluster-wise grouping of the variable features. Obtained positive markers are arranged as per the average log-fold change value. This step is accomplished using the script below.

```
data.seurat <- FindAllMarkers(data.seurat, only.pos = TRUE,
min.pct = 0.25, logfc.threshold = 0.25)
data.seurat %>% group_by(cluster) %>% slice_max(n = 2, order_by
= avg_log2FC)
```

Output of the differential expressed genes is in r tibble data structure. For filtering or sorting the differentially expressed genes list, change the r tibble data structure to tabular format.

3.12 Functionally Characterize the Differentially Expressed Features Using the PANTHER Database

The steps to functionally characterize the differentially expressed genes using the PANTHER database are as follows. First, enter the Ensemble accession numbers of variable features into the "enter IDs" field of the PANTHER database. Next, select the list type—"-ID List", select organism—"Homo Sapiens" as applicable and then select analysis type—"Functional classification viewed in gene list." The results are served on a new page with various options. The user should select the option "PANTHER Families" under the toggle down menu—"Convert List to." To save the ontologies to a file on the local computer, the user should select the "File" option in the toggle down menu—"Send List to."

Ensemble accession numbers used in the dataset make it easier to annotate the genes list using the PANTHER database. Most human genes are not yet annotated in the database, and, therefore, there are many genes in the list that are unannotated and do not yield interpretable signals. The disease implications of the obtained gene ontologies can be found using a literature search of the differentially expressed genes and their functions.

3.13 Prepare Catalogue of Differentially Expressed Features

Catalogued differentially expressed genes in Table 2 are obtained after merging all 13 samples together. This merged data object generated 13 clusters. The top two differentially expressed genes from each cluster are shown in Table 2.

Table 2
List of disease genes

Gene	Gene name
ENSG00000176884	GLUTAMATE RECEPTOR IONOTROPIC, NMDA 1
ENSG00000116254	CHROMODOMAIN-HELICASE-DNA-BINDING PROTEIN 5
ENSG00000185736	DOUBLE-STRANDED RNA-SPECIFIC EDITASE B2
ENSG00000178568	RECEPTOR TYROSINE-PROTEIN KINASE ERBB-4
ENSG00000229618	NOVEL TRANSCRIPT
ENSG00000228566	UNCHARACTERIZED GENE
ENSG00000141668	CEREBELLIN-2
ENSG00000185666	SYNAPSIN-3
ENSG00000241956	LONG INTERGENIC NON-PROTEIN CODING RNA
ENSG00000272168	CANCER SUSCEPTIBILITY 15
ENSG00000110436	EXCITATORY AMINO ACID TRANSPORTER 2
ENSG00000251372	LONG INTERGENIC NON-PROTEIN CODING RNA 499
ENSG00000198886	NADH-UBIQUINONE OXIDOREDUCTASE CHAIN 5
ENSG00000198786	NADH-UBIQUINONE OXIDOREDUCTASE CHAIN 4
ENSG00000147488	HIGH MOBILITY GROUP PROTEIN 20A-RELATED
ENSG00000183230	CATENIN ALPHA-3
ENSG00000167123	INACTIVE GLYCOSYLTRANSFERASE 25 FAMILY MEMBER 3
ENSG00000107331	ATP-BINDING CASSETTE SUB-FAMILY A MEMBER 2
ENSG00000187416	LHFPL TETRASPAN SUBFAMILY MEMBER 3 PROTEIN
ENSG00000038427	VERSICAN CORE PROTEIN
ENSG00000182601	HEPARAN SULFATE GLUCOSAMINE 3-O-SULFOTRANSFERASE 4
ENSG00000120549	SICKLE TAIL PROTEIN HOMOLOG
ENSG00000206190	PHOSPHOLIPID-TRANSPORTING ATPASE VA
ENSG00000102755	VASCULAR ENDOTHELIAL GROWTH FACTOR RECEPTOR 1
ENSG00000125730	COMPLEMENT C3
ENSG00000148655	LEUCINE-RICH MELANOCYTE DIFFERENTIATION-ASSOCIATED PROTEIN

References

1. Dixson JD, Azad RK (2020) Prions: roles in development and adaptive evolution. J Mol Evol 88(5):427–434. https://doi.org/10.1007/s00239-020-09944-2
2. Ross ED et al (2005) Primary sequence independence for prion formation. Proc Natl Acad Sci U S A 102(36):12825–12830. https://doi.org/10.1073/pnas.0506136102

3. Walker LC, Jucker M (2015) Neurodegenerative diseases: expanding the prion concept. Annu Rev Neurosci 38:87–103. https://doi.org/10.1146/annurev-neuro-071714-033828
4. Guadagno AH, Medina SH (2023) The manifold role of octapeptide repeats in prion protein assembly. Pept Sci (Hoboken) 115(2):e24303. https://doi.org/10.1002/pep2.24303
5. Eraña H, Charco JM, Di Bari MA et al (2019) Development of a new largely scalable in vitro prion propagation method for the production of infectious recombinant prions for high resolution structural studies. PLoS Pathog 15(10): e1008117. https://doi.org/10.1371/journal.ppat.1008117. Published 23 Oct 2019
6. Jaunmuktane Z, Mead S, Ellis M et al (2015) Evidence for human transmission of amyloid-β pathology and cerebral amyloid angiopathy [published correction appears in Nature. 2015 Oct 22;526(7574):595]. Nature 525(7568): 247–250. https://doi.org/10.1038/nature15369
7. Collinge J (1999) Variant Creutzfeldt-Jakob disease. Lancet 354(9175):317–323. https://doi.org/10.1016/S0140-6736(99)05128-4
8. Goedert M (2015) Neurodegeneration. Alzheimer's and Parkinson's diseases: the prion concept in relation to assembled Aβ, tau, and α-synuclein. Science 349(6248):1255555. https://doi.org/10.1126/science.1255555
9. Sabate R, Rousseau F, Schymkowitz J, Ventura S (2015) What makes a protein sequence a prion? PLoS Comput Biol 11(1):e1004013. https://doi.org/10.1371/journal.pcbi.1004013. Published 8 Jan 2015
10. Soto C (2012) Transmissible proteins: expanding the prion heresy. Cell 149(5):968–977. https://doi.org/10.1016/j.cell.2012.05.007
11. Das AS, Zou WQ (2016) Prions: beyond a single protein. Clin Microbiol Rev 29(3): 633–658. https://doi.org/10.1128/CMR.00046-15
12. Tang F, Barbacioru C, Wang Y et al (2009) mRNA-Seq whole-transcriptome analysis of a single cell. Nat Methods 6(5):377–382. https://doi.org/10.1038/nmeth.1315
13. AlJanahi AA, Danielsen M, Dunbar CE (2018) An introduction to the analysis of single-cell RNA-sequencing data. Mol Ther Methods Clin Dev 10:189–196. https://doi.org/10.1016/j.omtm.2018.07.003. Published 2 Aug 2018
14. Sigurdsson EM, Frangione B, Wisniewski T (2002) Immunization for Alzheimer's disease. Drug Dev Res 56:135–142. https://doi.org/10.1002/ddr.10068
15. Butler A, Hoffman P, Smibert P, Papalexi E, Satija R (2018) Integrating single-cell transcriptomic data across different conditions, technologies, and species. Nat Biotechnol 36(5):411–420. https://doi.org/10.1038/nbt.4096
16. Stuart T, Butler A, Hoffman P et al (2019) Comprehensive integration of single-cell data. Cell 177(7):1888–1902.e21. https://doi.org/10.1016/j.cell.2019.05.031
17. Satija R, Farrell JA, Gennert D, Schier AF, Regev A (2015) Spatial reconstruction of single-cell gene expression data. Nat Biotechnol 33(5):495–502. https://doi.org/10.1038/nbt.3192
18. Hao Y, Stuart T, Kowalski MH et al (2023) Dictionary learning for integrative, multimodal and scalable single-cell analysis. Nat Biotechnol. https://doi.org/10.1038/s41587-023-01767-y. Published online 25 May 2023
19. Hao Y, Hao S, Andersen-Nissen E et al (2021) Integrated analysis of multimodal single-cell data. Cell 184(13):3573–3587.e29. https://doi.org/10.1016/j.cell.2021.04.048
20. Dimitriadis A, Zhang F, Murphy TE et al (2022) Single-nuclei transcriptomics of mammalian prion diseases identifies dynamic gene signatures shared between species. bioRxiv (Cold Spring Harbor Laboratory). https://doi.org/10.1101/2022.09.13.507650. Published online 15 Sept 2022

Chapter 21

In Silico Identification of tRNA Fragments, Novel Candidates for Cancer Biomarkers, and Therapeutic Targets

Ankita Singh, Shafaque Zahra, Simran Arora, Fiza Hamid, and Shailesh Kumar

Abstract

The identification of a wide variety of RNA molecules using high-throughput sequencing techniques in the transcriptome pool of living organisms has revealed hidden regulatory insights in the cell. The class of non-coding RNA fragments produced from transfer RNA, or tRFs, is one such example. They are heterogeneously sized molecules with lengths ranging between 15 and 50 nt. They have a history of being dysregulated in human malignancies and other illnesses. The detection of these molecules has been made easier by a variety of bioinformatics techniques. The various types of tRFs and how they relate to cancer are covered in this chapter. It also provides a summary of the biological significance of tRFs reported in human cancer. Additionally, it emphasizes the utilities of databases and computational tools that have been created by different research teams for the investigation of tRFs. This will further aid the exploration and analysis of tRFs in cancer research and will support future advancement and a better comprehension of these molecules.

Key words tRFs, tRNA fragments, Cancer, tncrNAs, Biomarker, tncRNA Toolkit

1 Introduction

Our understanding of cancer biology has suggested that genetics has an important role in not just etiopathogenesis and diagnosis but also treatment, development of resistance to treatment, as well as prognosis of various cancers. Identification of these genetic markers is possible only due to the application of bioinformatics and computational biology in eukaryotes. For elucidating the study of molecular mechanisms of novel non-coding RNAs, researchers have used a variety of bioinformatics approaches in addition to molecular biology methods leading to the breakthrough in cancer studies.

Rajeev K. Azad (ed.), *Transcriptome Data Analysis*, Methods in Molecular Biology, vol. 2812,
https://doi.org/10.1007/978-1-0716-3886-6_21,

1.1 Introduction to tRNA-Derived Small RNAs in Cancer

Of the entire human genome, only 1% is composed of protein-coding genes, and the rest are non-coding genes, our understanding of which is rather limited. Researchers have suggested infrastructural or regulatory functions of many of these non-coding RNAs (ncRNAs). The infrastructural ncRNAs include transfer RNAs (tRNAs) and ribosomal RNAs (rRNAs), while regulatory RNAs include small ncRNAs (sncRNAs), of size up to 200 nucleotides (nt) and long ncRNAs (lncRNAs), greater than 200 nucleotides in length [1]. Some of these ncRNAs can be used as biomarkers for the diagnosis, treatment, and prognosis of cancers.

The tRNA-derived fragments (tRFs) and tRNA halves (tRHs) are types of tRNA-derived small ncRNAs (tncRNAs/tDRs), which were previously thought to be products of random tRNA degradation. However, they are derived by precise processing and fragmentation of tRNAs. Although their function is not very well understood, recently accumulated evidence supports correlation with differentiation, development as well as metabolism in prokaryotes as well as eukaryotes [2]. They may play an important role in regulating protein translation. Thus, they can be a potential source of small regulatory RNAs. The presence of tncRNAs in several human cancer cell lines suggests their role in tumor development and progression [3, 4]. The human tRFs are generated from both pre- and mature tRNAs (both nuclear and mitochondrial tRNA) through Dicer-dependent or independent processes. However, tRNA halves are generated because of the cleavage of mature tRNAs at anticodon caused by endonuclease angiogenin (ANG), a member of the RNase A superfamily found in mammalian cells [5]. Angiogenin is typically inhibited and its release is induced by stress. The average lengths of tRFs and tRNA halves are ~20 nt and ~33 nt, respectively. Although their function, as well as mechanisms, are not exclusive, they may mediate the exchange of information between the nucleus and the mitochondria [6].

Four types of tRFs have been broadly identified and named as, tRF-1, tRF-5, tRF-3, and internal RFs (i-tRF) [7], and recently a 5′ leader-exon tRF was discovered in an animal model. Each type of tRFs is generated by specific ribonucleases and is regulated by specific pathways. The tRFs were assumed to silence gene expression through association with Argonaute (AGO) proteins [8]. They have also been shown to suppress human breast cancer expression probably by displacement of pro-oncogenic transcripts from RNA binding protein YBX1 [9]. There are two groups of tRNA halves (tRHs) discovered, namely, 5′ tRH and 3′ tRH [10]. The inhibition of translation through nicked tRNAs or association with AGO/Piwi, formation of a repression complex with other unknown proteins, and guiding mRNAs cleavage by interaction with RNase Z or RNase P are proposed possible functional models for tRNA halves. The natural, as well as synthetic 5′ tiRNAs (5′

tRHs) from alanine and cysteine inhibit translation by quantitative displacement of eukaryotic translation initiation factors like elF4G/A from uncapped mRNA transcripts [11].

These novel classes of tRNA-derived sncRNAs (tncRNAs) are expressed in various organisms from microbes to humans, thus suggesting an association with diverse cellular biological processes and diseases in humans. Research literature accumulated over time has demonstrated the role of tRFs as diagnostic markers and therapeutic targets in human cancers. The identification of these molecules through the development of deep sequencing technologies and computational tools and their importance in gene regulation with a role in tumor etiopathogenesis have raised the need for storage as well as management of massive data on tncRNAs.

1.2 Biological Relevance/Functions of tRFs in Various Cancer

The involvement of tRFs in cancer biology is suggested by the facts that (i) tRFs can potentially affect gene expression by interacting with AGO and Piwi proteins in a cell type-specific manner. (ii) They can interact with ribosomes and aminoacyl tRNA synthetase and modulate protein translation. (iii) Their interaction with RNA-binding proteins may be a possible mechanism for the development of cancers as well as metastasis. The class-wise description of various tRFs identified in relevance to several cancer types and reported in published literature has been enlisted in Table 1.

1.3 tRFs Identified as Highlighting the Prognostic Potential of tRFs or as Cancer Biomarkers

(i) tRNA-Thr (type: tRF-1) is a marker of progression in chronic lymphocytic leukemia (CLL) [12].

(ii) Elevated levels of tRNA-Thr-3 (tRF-3) in breast cancer patients are associated with an increase in overall survival and could probably suppress metastasis [16].

(iii) The percentage of tRFs produced from tRNA-Lys-CTT/Phe-GAA (tRF-5) correlated with progression-free survival in prostatic cancer [23].

(iv) tRNA-Val-TAC-3/Gly-TCC-5/Val-AAC-5/GluCTC-5 (tRF-5) can be used as a diagnostic biomarker for both hepatocellular cancers and breast cancers, as they are elevated in plasma-derived exosomes and in MCF7 extracellular vesicles in respective cancers [23].

(v) Sex hormone-dependent tRNA-derived RNAs (SHOT-RNAs), which are tRNA halves found in exosomes, can be used as biomarkers. As they accumulate specifically and abundantly in estrogen receptor-positive breast cancers androgen receptor-positive prostate cancer cell lines and luminal type breast cancer patient tissues [5, 28].

(vi) tRNA-ValAAC (tRH-5) is significantly downregulated in the clear cell subtype of renal cell carcinoma, and levels are inversely related to the stage of cancer and grade. Hence can be used as a diagnostic as well as a prognostic biomarker [30].

Table 1
Various classes of tRFs identified in various cancers and their biological relevance

tRF type	Cancer type	Biological relevance
tRF-1	B-cell chronic lymphocytic leukemia (CLL)	Downregulation of miR-3676 (tRF): which otherwise may inhibit expression of TCL1 by targeting 3′ UTR of TCL1[12]
	CLL and lung cancers	Downregulation and mutation of miR-4521/3676 [13]
	Hematopoietic malignancies and solid tumors	miR-4521/3676 is associated with Piwi-like protein 2 complexes to serve as oncogenes or tumor suppressors [13]
	Colon, breast, and ovarian cancers	Level of expression of tRFs changes with oncogene activation and progression of cancer. Overexpression of ts-101 and ts-46 can inhibit colony expression (by affecting chromatin structure, cell survival, cell proliferation, and apoptosis) [14]
		tRF-1001 depletion impairs cell proliferation: decreased DNA synthesis and increased G2-phase cells [15]
tRF-3	Breast cancer	Elevated 3′ tRNA-Thr levels are associated with an increase in overall survival [16]
		3′ tRNA-Thr downregulates E-cadherin and upregulates N-cadherin and MMP-2 by targeting TWIST-1 and could suppress epithelial-mesenchymal transition and metastasis. This downregulation is negatively associated with VEGF and positively with HER2 expression [12–14, 17]
	Lymphoma	CU1276 is capable of binding the endogenous 3′ UTRs of RPA1 and tRFs in cancer-suppressing mRNA transcription thus, inhibit cell proliferation and regulate molecular response to DNA damage [18]
	Hepatocellular carcinoma	Repression of LeuCAG 3′ tsRNA induces apoptosis through binding to RPS28 and RPS15 to promote protein synthesis. Thus Anti-Leu 3′ tsRNA locked nucleic acids (LNAs) can reduce tumor growth and the final volume of the tumor [19]
	Colorectal cancer	tRF/miR-1280 is associated with cancer growth and metastasis. Its in vitro upregulation represses the proliferation of cells and colony formation by probably targeting JAG2 and repressing the expression of Gata1/3 and miR-200b [20]
	T-cell leukemia/lymphoma (ATLL)	tRF-3019 is found in abundance in the HTLV-1 virus and can increase its reverse transcription, thus inhibiting the replication of reverse transcription [21]

(continued)

Table 1 (continued)

tRF type	Cancer type	Biological relevance
tRF-5	Breast cancer	tRFs affected cell signaling, energy metabolism, and immune response in a cohort of triple-negative carcinoma breast [22]
		Elevated in MCF7 extracellular vesicles
	Prostate cancer	Significant downregulation of tRFs in different stages
		Progression-free survival is associated with tRFs generated from tRNA-LysCTT and tRNA-PheGAA [23]
	Testicular germ cell tumor	tRFs generated from the 5′ end of mature RNAs and tRNA-GluGAG and tRNA-AspGAY are highly expressed [24]
	Liver cancer	High levels of tRF-5s (tRNA-ValTAC-3, tRNA-ValAAC-5, and tRNA-GlyTCC-5) and tRH-5s found in plasma-derived exosomes [25]
i-tRF	Breast cancer	i-tRFs generated from tRNA-Glu, tRNA-Asp, tRNA-Gly, and tRNA-Tyr act as tumor suppressors by competitively binding with YBX1-binding motifs via 3′ UTRs [9, 26]
	Prostate cancer	i-tRFs are the most abundant tRFs (66%) in prostate cancers
	Ovarian cancer	Increased i-tRFs with no statistical difference among benign, borderline, and cancer cases [27]
tRNA halves	Estrogen receptor positive and luminal type breast cancers, androgen receptor-positive prostate cancer	Sex hormone-dependent tRNA-derived RNAs (SHOT-RNAs): Augmented signaling pathways and production of tRNA halves promote cell proliferation. Their expression increases after the addition of sex hormones and their receptors [5, 28]
	Non-small cell lung cancer	tRNA-LeuCAG generating 5′tRH is found elevated [29]
	Clear cell renal cell carcinoma	5′tRH generated from tRNA-VaIAAC is downregulated [30]
	Head and neck Squamous cell carcinoma	5′tRH from isoacceptors of tRNA-Ala, Cys, and Tyr is increased, and from tRNA-Arg, Glu, Gly, Lys, Trp, and Val is decreased [31]

(vii) LeuCAG 3′ tsRNA is a potential diagnostic marker and a novel target for cancer treatment. Because tRH-5 generated is elevated in non-small cell lung cancer tissues and cell lines, and its levels are positively correlated with tumor progression. Anti-Leu 3′ tsRNA locked nucleic acids treatment for 4 weeks can reduce both the growth as well as the volume of tumor in a xenotransplanted mouse model of patient-derived orthotopic hepatocellular cancer [19].

1.4 Identification of tRFs Using Small RNA Sequencing (sRNA-Seq) Datasets

The tRFs exist not just in cell lines but they are also detected in various biofluids as well as cell culture supernatants. The sRNA deep sequencing data analysis helps to uncover the source types of RNAs as well as quantify their expression.

The tRFs/tRHs can be identified by (i) General tools, not specific like SPORTS [32]; (ii) Databases; (iii) Computational pipelines; and (iv) Integrated web servers, e.g., tRF2Cancer [33] with user-friendly interface.

Small RNA deep-sequencing data are now generated in enormous quantities due to improvements in next-generation sequencing (NGS) methods. These data are now exploited for screening novel RNAs including tRFs. Databases to obtain or manage tRFs data as well as tools to recognize tRFs have been developed to research tRFs from these data. We compiled the tools and databases for tRFs identification and collection in this review to assist researchers in selecting the most effective tools for their analyses and stimulating the creation or advancement of tools in the field. Table 2 enlists databases/pipelines/tools available till now for tRFs detection.

2 Materials

2.1 Hardware Requirements

- Operating System: 32-bit/64-bit Linux.
- Available RAM: Minimum of 4 GB.
- Free disk space varies with the number of samples and sequencing depth.

2.2 Software Requirements

Software Packages	tncRNA Toolkit, FastQC, fastp
Dependencies	tRNAscan-SE (v2.0.6), samtools (v1.10), bedtools (v2.29.2), Bowtiel (v1.3.0), HAMR (v1.2)
Python Dependencies	Python3, pandas (v1.1.0), biopython (v1.77)

Table 2
List of tools and databases available for tRFs exploration

Computational tool/pipelines for tRFs detection			
Tool	**Description**	**URL**	**References**
tDRmapper	Designed specifically for trimmed human small RNA-seq data (single-end, 50×) generated on the Illumina sequencing platform using cDNA libraries that were prepared using the Illumina TruSeq protocol	https://github.com/sararselitsky/tDRmapper	[34]
tncRNA Toolkit	A pipeline for the identification of tRNA-derived small ncRNAs (tncRNAs) from high-throughput sequencing data	http://www.nipgr.ac.in/tncRNA	[35]
tRF2Cancer	A web server to detect tRNA-derived small RNA fragments (tRFs) and their expression in multiple cancers	http://rna.sysu.edu.cn/tRFfinder/	[33]
tsRFun	A platform for decoding tsRNA functions by high-throughput small RNA-seq and CLIP-Seq data	https://rna.sysu.edu.cn/tsRFun/	[36]
tRFTars	Predict targets of tRFs in humans with a user-friendly interface	http://www.rnanut.net/tRFTar/	[37]
tRFs databases/web servers			
Database	**Description**	**URL**	**References**
tRFdb	A relational database of transfer RNA-related fragments	http://genome.bioch.virginia.edu/trfdb/	[38]
PtncRNAdb	Plant transfer RNA-derived non-coding RNAs (tncRNAs) database	https://nipgr.ac.in/PtncRNAdb	[39]
PtRFdb	Plant transfer RNA-derived fragments database	http://www.nipgr.ac.in/PtRFdb/	[40]
OncotRF	An online resource for the exploration of tRNA-derived fragments in human cancers	http://bioinformatics.zju.edu.cn/OncotRF/	[41]
MINTbase	A database for the interactive exploration of mitochondrial and nuclear tRNA fragments (tRFs)	https://cm.jefferson.edu/MINTbase/	[42]
tsRBase	A comprehensive database for the expression and function of tsRNAs in multiple species	http://www.tsrbase.org/	[43]
Olvedy et al. repertoire	A comprehensive database on tRFs in different stages of prostatic cancers	Website unavailable	[23]
tRFdb	A relational database of transfer RNA-related fragments	http://genome.bioch.virginia.edu/trfdb/	[38]

2.3 Installation Documentation

- Download the tncRNA Toolkit package from below link:
  http://172.16.28.201/tncRNA/tncRNA-Toolkit.tar.gz
- Extract the compressed tar file using the command:
  tar -xf tncRNA-Toolkit.tar.gz
  cd tncRNA-Toolkit/
- Installing bowtie1:
  cd util/
  unzip bowtie-1.3.0-src.zip
  cd bowtie-1.3.0-src/
  make
  cd ../../

This distribution includes the python3 script *tncRNAs.py* and other scripts in the "util" folder which are needed to run the main script.

3 Methods

The objective of this analysis is to identify tRNA fragments (tRFs) from small RNA-seq data (Fig. 1). The analysis begins specifically targeting the removal of 3′ adaptor sequences from small RNA-seq data. Subsequently, a filtering step is implemented to discard any sequences deemed low-quality, contaminated, or extraneous. The dataset under examination in this instance is sourced from the NCBI Sequence Read Archive (SRA) and originates from humans.

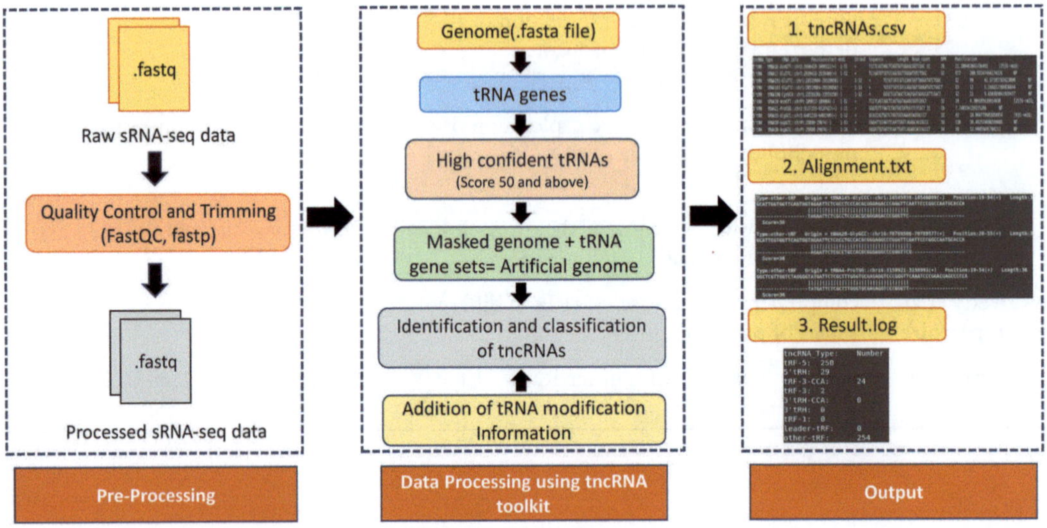

Fig. 1 Representation of pipeline followed by tncRNA Toolkit, for the identification of tncRNAs in small RNA-seq datasets

3.1 Obtaining sRNA-Seq Data and Reference Genome from NCBI

- The example data can be obtained from the Sequence Read Archive (SRA) under the accession number SRR029127. To directly download it, type

  ./fastq-dump --origfmt SRR029127

- Get the reference human genome by using NCBI Reference Sequence: GRCh38.p14

  wget_https://ftp.ncbi.nlm.nih.gov/genomes/all/GCA/000/001/405/GCA_000001405.29_GRCh38.p14/GCA_000001405.29_GRCh38.p14_genomic.fna.gz

3.2 Quality Assessment and Preprocessing of sRNA-Seq Data

FastQC (https://www.bioinformatics.babraham.ac.uk/projects/fastqc/) is the tool to check the quality of high-throughput sequencing datasets. It provides a comprehensive set of analyses and visualizations to assess the quality of sequence data. Moreover, FastQC creates an HTML report, including different metrics and charts useful for pinpointing problems in your data.

To use FastQC, user need to download and install it on their system. Once installed, it can run on the data files.

Download and Install FastQC

You can download FastQC from the official website: https://www.bioinformatics.babraham.ac.uk/projects/fastqc/

After downloading, follow the installation instructions provided on the website.

FastQC usage

./fastqc --outdir=/path/to/output_directory --threads=4 SRR029127.fastq

The FastQC generates a list of output files, including an HTML report, which can be opened in a web browser to visualize the quality metrics and assessments.

Consult the FastQC documentation for more details on how to interpret the results: https://www.bioinformatics.babraham.ac.uk/projects/fastqc/Help/

If the data is found to have bad quality reads or adapter presence, then use fastp. Fastp (https://github.com/OpenGene/fastp) serves as a comprehensive tool for quality control and preprocessing in high-throughput sequencing data analysis. It excels in tasks like adapter trimming, quality filtering, and other essential preprocessing steps. Utilize the subsequent command to filter your data, employing the length parameter to exclude reads shorter than 14 base pairs through fastp:

./fastp -i SRR029127.fastq -o SRR029127_trim.fastq –length required 14 --html= SRR029172.html --json= SRR029127.json

3.3 tRF Detection Pipeline

In this section, we employ the tncRNA Toolkit to identify small non-coding RNAs (tncRNAs) derived from transfer RNAs (tRNAs) within high-throughput sequencing data. Developed in Python3 and complemented by shell/bash programming, the toolkit excels in detecting tncRNAs originating from mature tRNAs, as well as their leader or trailer sequences (up to 50 base pairs). The toolkit categorizes these identified fragments into various types, including tRF-5, tRF-3(CCA), tRF-1, leader tRFs, 5′ tRH, 3′ tRH(CCA), and other-tRFs, contingent upon their cleavage site within the parental tRNA and their respective lengths (Fig. 1).

The initial step involves constructing the Bowtie (https://bowtie-bio.sourceforge.net/) index using the genome of interest, provided in a FASTA file, as demonstrated by the following command:

python3 tncRNAs.py -g GCA_000001405.29_GRCh38.p14_genomic.fna -s Homo_sapiens

This command will generate the bowtie index and needed files in "lib/indexes/Homo_sapiens" for *Homo sapiens*.

Note: The genome FASTA header should follow the format ">chr[Num]," where [Num] represents the chromosome number. Headers for mitochondrial and plastid sequences should begin with ">chrMt" and ">chrPt," respectively. This consistent format facilitates script automation and aids in distinguishing nuclear and organellar regions. Once the index construction is complete, users can proceed to identify tncRNA in single-end small RNA-seq data for the specified species using the subsequent command:

python3 tncRNAs.py -s Homo_sapiens -i SRR029127_trimmed.fq -o output_dir

Note: The species name provided in the above command should be identical to the name provided for building the bowtie index.

Upon successful execution of this command, three output files—namely, tncRNAs.csv, Alignment.txt, and Result.log—will be generated in the designated output directory. These files contain information pertinent to tncRNA identification and alignment details for further analysis.

3.4 tncRNA Toolkit Output Files

The tncRNA Toolkit generates three distinct output files when executed with the "-o" option in the specified directory, as outlined earlier:

1. **tncRNAs.csv**: The "tncRNAs.csv" file contains comprehensive information about each identified tncRNA. This includes details such as tncRNA type, parental tRNA locus information, tncRNA position on the tRNA (start-end), strand, sequence,

length, read count, RPM (reads per million), nature, and the position of modified nucleosides (Fig. 1).

2. **Alignment.txt**: The "Alignment.txt" file illustrates the alignment of each tncRNA sequence with its parental tRNA sequence. This alignment encompasses the mature tRNA, 5′ leader, and 3′ trailer precursor tRNA sequences. The file provides alignment scores and includes fundamental metadata derived from the "tncRNAs.csv" file (Fig. 1).
3. **Result.log**: The "Result.log" file serves as a log file that tracks statistical data related to various sub-types of tncRNAs identified within the sample. This log file offers insights into the analytical results obtained during the tncRNA Toolkit execution (Fig. 1).

4 Conclusion

tRFs are known to control a variety of cellular biological processes and are not only the result of random degradation with no purpose. Oncogenic potential and tumor suppression by tRFs are now well known, thus indicating their potential for use as diagnostic and therapeutic markers. Identification from random degradation will still be difficult, though. Both probability-based evaluations of reads enrichment on the genome and tRNAs and machine learning techniques have advantages and disadvantages. The occurrence, management, and prognosis of human cancer have long been the subject of molecular research. As our current understanding of tRFs is probably just the tip of the iceberg, the introduction of sRNA therapeutics, including research on targeting specific tRFs for therapy, as well as diagnosis and prognostic markers in various cancers, has a long way to go with a promising future.

5 Challenges and Limitations

It is extremely difficult to separate certain tRFs from a pool of sRNA degradation fragments that is essentially random. Additionally, when mapping RNA sequences on just the tRNA loci or the entire genome, there is a chance that the results will be unclear. This can be due to tRNA lookalikes causing false positives. Additionally, using improved sequencing techniques to handle tRNA nucleoside changes can produce tRFs that are more precise. It is essential for the scientific research of tRFs and for improving communication in the literature that tRFs have a standard name. The tRFs nomenclatures, however, differ between different programs and databases.

In addition to increased demands from new fields of study, there are still several challenges with the recent development of pipelines and databases. To close the gaps in the tRFs study, a thorough investigation is also required. The main focuses of modern computational techniques are identification and tRFs databases. New computational and statistical methods are needed to precisely identify tRFs from small RNA sequencing data. New specialized databases are also projected to accommodate the users' diverse needs. Additionally, new fields of tRFs study are growing, but they lack the requisite tools and techniques. Therefore, combining new computational and experimental methodologies will be a challenging yet fruitful endeavor for tRFs research.

References

1. Hüttenhofer A, Schattner P, Polacek N (2005) Non-coding RNAs: hope or hype? Trends Genet 21:289–297
2. Keam SP, Hutvagner G (2015) tRNA-derived fragments (tRFs): emerging new roles for an ancient RNA in the regulation of gene expression. Life (Basel, Switzerland) 5:1638–1651
3. Venkatesh T, Suresh PS, Tsutsumi R (2016) tRFs: miRNAs in disguise. Gene 579: 133–138
4. Zhu L, Liu X, Pu W, Peng Y (2018) tRNA-derived small non-coding RNAs in human disease. Cancer Lett 419:1–7
5. Honda S, Loher P, Shigematsu M, Palazzo JP, Suzuki R, Imoto I et al (2015) Sex hormone-dependent tRNA halves enhance cell proliferation in breast and prostate cancers. Proc Natl Acad Sci USA 112:E3816–E3825
6. Xu W-L, Yang Y, Wang Y-D, Qu L-H, Zheng L-L (2017) Computational approaches to tRNA-derived small RNAs. Non-Coding RNA 3:2
7. Yu X, Xie Y, Zhang S, Song X, Xiao B, Yan Z (2020) tRNA-derived fragments: mechanisms underlying their regulation of gene expression and potential applications as therapeutic targets in cancers and virus infections. *Theranostics* 11: 461–469. Ivyspring International Publisher
8. Kumar P, Anaya J, Mudunuri SB, Dutta A (2014) Meta-analysis of tRNA derived RNA fragments reveals that they are evolutionarily conserved and associate with AGO proteins to recognize specific RNA targets. BMC Biol 12: 78
9. Goodarzi H, Liu X, Nguyen HCB, Zhang S, Fish L, Tavazoie SF (2015) Endogenous tRNA-derived fragments suppress breast cancer progression via YBX1 displacement. Cell 161:790–802
10. Shen Y, Yu X, Zhu L, Li T, Yan Z, Guo J (2018) Transfer RNA-derived fragments and tRNA halves: biogenesis, biological functions and their roles in diseases. J Mol Med 96: 1167–1176
11. Zhu L, Ge J, Li T, Shen Y, Guo J (2019a) tRNA-derived fragments and tRNA halves: the new players in cancers. Cancer Lett 452: 31–37
12. Balatti V, Rizzotto L, Miller C, Palamarchuk A, Fadda P, Pandolfo R et al (2015) TCL1 targeting miR-3676 is codeleted with tumor protein p53 in chronic lymphocytic leukemia. Proc Natl Acad Sci USA 112:2169–2174
13. Pekarsky Y, Balatti V, Palamarchuk A, Rizzotto L, Veneziano D, Nigita G et al (2016) Dysregulation of a family of short non-coding RNAs, tsRNAs, in human cancer. Proc Natl Acad Sci USA 113:5071–5076
14. Balatti V, Nigita G, Veneziano D, Drusco A, Stein GS, Messier TL et al (2017) tsRNA signatures in cancer. Proc Natl Acad Sci USA 114: 8071–8076
15. Lee YS, Shibata Y, Malhotra A, Dutta A (2009) A novel class of small RNAs: tRNA-derived RNA fragments (tRFs). Genes Dev 23:2639–2649
16. Li LZ, Zhang CZ, Liu LL, Yi C, Lu SX, Zhou X et al (2014) miR-720 inhibits tumor invasion and migration in breast cancer by targeting TWIST1. Carcinogenesis 35:469–478
17. Niu R, Zhang L, Xi G, Wei X, Yang Y, Shi Y et al (2007) Up-regulation of Twist induces angiogenesis and correlates with metastasis in hepatocellular carcinoma. Cancer Res 67:5364
18. Maute RL, Schneider C, Sumazin P, Holmes A, Califano A, Basso K, Dalla-Favera R (2013) tRNA-derived microRNA modulates proliferation and the DNA damage response and is

down-regulated in B cell lymphoma. Proc Natl Acad Sci 110:1404–1409

19. Kim HK, Fuchs G, Wang S, Wei W, Zhang Y, Park H et al (2017) A transfer-RNA-derived small RNA regulates ribosome biogenesis. Nature 552:57–62
20. Huang B, Yang H, Cheng X, Wang D, Fu S, Shen W et al (2017) TRF/miR-1280 suppresses stem cell-like cells and metastasis in colorectal cancer. Cancer Res 77:3194–3206
21. Ruggero K, Guffanti A, Corradin A, Sharma VK, De Bellis G, Corti G et al (2014) Small noncoding RNAs in cells transformed by human T-cell leukemia virus type 1: a role for a tRNA fragment as a primer for reverse transcriptase. J Virol 88:3612–3622
22. Telonis AG, Rigoutsos I (2018) Race disparities in the contribution of miRNA isoforms and tRNA-derived fragments to Triple-negative breast cancer. Cancer Res 78:1140–1154
23. Olvedy M, Scaravilli M, Hoogstrate Y, Visakorpi T, Jenster G, Martens-Uzunova ES (2016) A comprehensive repertoire of tRNA-derived fragments in prostate cancer. Oncotarget 7:24766–24777
24. Rounge TB, Furu K, Skotheim RI, Haugen TB, Grotmol T, Enerly E (2015) Profiling of the small RNA populations in human testicular germ cell tumors shows global loss of piRNAs. Mol Cancer 14:1–13
25. Zhu L, Li J, Gong Y, Wu Q, Tan S, Sun D et al (2019b) Exosomal tRNA-derived small RNA as a promising biomarker for cancer diagnosis. Mol Cancer 18:1–5
26. Schaffer AE, Eggens VRC, Caglayan AO, Reuter MS, Scott E, Coufal NG et al (2014) CLP1 founder mutation links tRNA splicing and maturation to cerebellar development and neurodegeneration. Cell 157:651–663
27. Peng EY, Shu Y, Wu Y, Zeng F, Tan S, Deng Y et al (2018) Presence and diagnostic value of circulating tsncRNA for ovarian tumor. Mol Cancer 17:15–18
28. Valadi H, Ekström K, Bossios A, Sjöstrand M, Lee JJ, Lötvall JO (2007) Exosome-mediated transfer of mRNAs and microRNAs is a novel mechanism of genetic exchange between cells. Nat Cell Biol 9:654–659
29. Shao Y, Sun Q, Liu X, Wang P, Wu R, Ma Z (2017) tRF-Leu-CAG promotes cell proliferation and cell cycle in non-small cell lung cancer. Chem Biol Drug Des 90:730–738
30. Nientiedt M, Deng M, Schmidt D, Perner S, Müller SC, Ellinger J (2016) Identification of aberrant tRNA-halves expression patterns in clear cell renal cell carcinoma. Sci Rep 6:1–9
31. Martinez BV, Dhahbi JM, Nunez Lopez YO, Lamperska K, Golusinski P, Luczewski L et al (2015) Circulating small non coding RNA signature in head and neck squamous cell carcinoma. Oncotarget 6:19246–19263
32. Shi J, Ko EA, Sanders KM, Chen Q, Zhou T (2018) SPORTS1.0: a tool for annotating and profiling non-coding RNAs optimized for rRNA- and tRNA-derived small RNAs. Genomics Proteomics Bioinformatics 16:144–151
33. Zheng L-L, Xu W-L, Liu S, Sun W-J, Li J-H, Wu J et al (2016) tRF2Cancer: a web server to detect tRNA-derived small RNA fragments (tRFs) and their expression in multiple cancers. Nucleic Acids Res 44:W185–W193
34. Selitsky SR, Sethupathy P (2015) tDRmapper: challenges and solutions to mapping, naming, and quantifying tRNA-derived RNAs from human small RNA-sequencing data. BMC Bioinformatics 16:354
35. Zahra S, Singh A, Poddar N, Kumar S (2021) Transfer RNA-derived non-coding RNAs (tncRNAs): hidden regulation of plants' transcriptional regulatory circuits. Comput Struct Biotechnol J 19:5278–5291
36. Wang JH, Chen WX, Mei SQ, Yang YD, Yang JH, Qu LH, Zheng LL (2022) TsRFun: a comprehensive platform for decoding human tsRNA expression, functions and prognostic value by high-throughput small RNA-Seq and CLIP-Seq data. Nucleic Acids Res 50:D421–D431
37. Xiao Q, Gao P, Huang X, Chen X, Chen Q, Lv X et al (2021) tRFTars: predicting the targets of tRNA-derived fragments. J Transl Med 19: 1–15
38. Kumar P, Mudunuri SB, Anaya J, Dutta A (2015) tRFdb: a database for transfer RNA fragments. Nucleic Acids Res 43:D141–D145
39. Zahra S, Bhardwaj R, Sharma S, Singh A, Kumar S (2022) PtncRNAdb: plant transfer RNA-derived non-coding RNAs (tncRNAs) database. 3 Biotech 12:1–7
40. Gupta N, Singh A, Zahra S, Kumar S (2018) PtRFdb: a database for plant transfer RNA-derived fragments. Database 2018. https://doi.org/10.1093/database/bay063

41. Yao D, Sun X, Zhou L, Amanullah M, Pan X, Liu Y et al (2020) OncotRF: an online resource for exploration of tRNA-derived fragments in human cancers. RNA Biol 17:1081–1091
42. Pliatsika V, Loher P, Telonis AG, Rigoutsos I (2016) MINTbase: a framework for the interactive exploration of mitochondrial and nuclear tRNA fragments. Bioinformatics (Oxford, England) 32:2481–2489
43. Zuo Y, Zhu L, Guo Z, Liu W, Zhang J, Zeng Z et al (2021) TsRBase: a comprehensive database for expression and function of tsRNAs in multiple species. Nucleic Acids Res 49:D1038–D1045

INDEX

Rajeev K. Azad (ed.), *Transcriptome Data Analysis*, Methods in Molecular Biology, vol. 2812, https://doi.org/10.1007/978-1-0716-3886-6,

MIX
Papier aus verantwortungsvollen Quellen
Paper from responsible sources
FSC® C105338

If you have any concerns about our products,
you can contact us on
ProductSafety@springernature.com

In case Publisher is established outside the EU,
the EU authorized representative is:
Springer Nature Customer Service Center GmbH
Europaplatz 3, 69115 Heidelberg, Germany

Printed by Libri Plureos GmbH
in Hamburg, Germany